² £92.00

D1806628

Acknowledgement is made to G. van Oortmerssen et al, for the use of Figure 2, on p.9, which appears on the front cover of this book.

Computer Methods in Marine and Offshore Engineering

Proceedings of the Third International Conference on Computer Aided Design, Manufacture and Operation in the Marine and Offshore Industries, held at Key Biscayne, Florida, USA, during 15-17 January 1991.

Editor: T.K.S. Murthy

Computational Mechanics Publications
Southampton Boston

T.K.S. Murthy
Associate Director of External Programmes
Computational Mechanics Institute
Ashurst Lodge
Ashurst
Southampton SO4 2AA
U.K.

British Library Cataloguing in Publication Data
International Conference on Computer Aided Design,
    Manufacture and Operation in the Marine and Offshore
    Industries (3rd ; 1991 ; Key Biscayne, U. S. A.)
    Computer methods in marine and offshore engineering.
    1. Marine engineering. Applications of computer systems
    I. Title II. Murthy, T. K. S. (Thiruvalam K. S.) *1914-*
    623.870285

    ISBN 1-85312-122-3

ISBN 1-85312-122-3 Computational Mechanics Publications, Southampton
ISBN 1-56252-054-7 Computational Mechanics Publications, Boston, USA

Library of Congress Catalog Card Number 90-85217

Printed and bound by Billings and Sons Ltd, Worcester

PREFACE

This book contains the edited version of some of the papers presented at the Third International Conference on Computer Aided Design, Manufacture and Operation in the Marine and Offshore Industries (CADMO 91) held at Key Biscayne, Florida, USA in January 1991. The first of these important series of conferences took place in Washington DC, during September 1986 and the second in Southampton, UK in September 1988. The third conference discussed some of the most recent work in the field since 1988 and in particular the following topics:- Ship Design, Ship Hydrodynamics and Flow Simulation, Ship Motions and Seakeeping, Ship Structures, Ship Operation, Navigation and Propulsion, as well as Offshore Structures and Offshore Operations. The meeting demonstrated that considerable developments in the field had occurred in recent years and that these activities will continue to increase in the future. The book comprises selected papers written by international scientists.

The Conference was sponsored and organised by the Computational Mechanics Institute of Wessex Institute of Technology, Southampton, UK with the support of the members of the International Scientific Committee to whom the Editor is indebted.

T.K.S. Murthy
Southampton, January 1991

CONTENTS

SECTION 1: SHIP DESIGN

HOSDES and MARDES: Advanced Concepts for Integrated
CAD-systems
G. Glijnis, G. van Oortmerssen
   3

On the Optimum Structural Design of MARPOL Tankers
Considering Tank Arrangement
C.D. Jang, S.S. Na
   25

Statistical Analysis of Full Ships Resistance
*D.Kostov, S. Kyulevcheliev, S. Hongcui, J. Quinwei, G. Shenghan,
F. Youzhang*
   41

A Method to Assess the Seakeeping Behaviour of a Merchant
Ship in its Early Stage of Design
D. Boote, D. Bruzzone
   55

Shipbuilding Software Technology for the Nineties
G. Marshall, W. Horsham, D. Catley
   69

Computer Graphics in Warship Design
J.M. Duncan, P.H. Rutland, P.E. Gibbs
   85

SECTION 2: SHIP HYDRODYNAMICS AND FLOW SIMULATION

Computing Wave Resistance, Wave Profile, and Sinkage
and Trim of Transom Stern Ships
L. Cong, C.C. Hsiung
   99

A Computer Tool for Solving the Wave Resistance Problem for
Conventional and Unconventional Ships
J.J. Maisonneuve, G. Delhommeau
   113

2D-respective 3D-Flow Field Measurements in Open Channel
Flow compared with Numerical Models
W. Bechteler, H. Sattel, K. Schätz, A. Tasdemir
   127

Numerical Computation of the Flow Around a Diving Plane in
Water Waves
S. Fontaine, S. Huberson, J.L.Montagné
   141

First and Second Order Wave Forces on Floating Bodies From 153
Far-Field Relations
A.C. Fernandes, L.A.P. Levy

Calculation of Potential Flow Around a Given Profile by the 163
Method of Surface Vorticity
A. Kukner

SECTION 3: SHIP MOTIONS AND SEAKEEPING

The Use of High Definition Graphics in Seakeeping and 183
Ship Manoeuvring
P.A. Wilson

Modelling of the Combined Yaw and Sway for Ship 195
Manœuvring Models
M.M.A. Pourzanjani

SECTION 4: SHIP STRUCTURES

An Optimum Structural Design System Using Reanalysis Method 211
S.W. Park, J.K. Paik, I.S. Nho, H.S. Lee

An F.E. Approach to Yacht Structural Analysis 225
R.A. Shenoi

SECTION 5: SHIP OPERATIONS AND NAVIGATION

A Review of Mathematical Models used in Ship Manoeuvres 243
J. Chudley, M.J. Dove, N.J. Tapp

Development of Marine Autopilots 259
M.J. Dove, C.B. Wright

The Role of a Mathematical Model for Improvement of 273
Marine Navigation
M.J. Dove, K.M. Miller, C.T. Stockel

Safety Assessment of High Speed Vessel Traffics Using Computer 287
Simulation
M. Numano, K. Okuzumi, J. Fukuto, Y. Murayama

SECTION 6: SHIP PROPULSION

Numerical Prediction of Effects of Hull Variation on Propeller 299
Performance
D. Hally, D.J. Noble

Model Propeller Measurement Using a 3D Laser Scanner 311
T. Randell

SECTION 7: OFFSHORE STRUCTURES

Numerical and Experimental Study On Fatigue Analysis of   331
Offshore Tubular Joints
K.N. Cho, Y.S. Jang, W.I. Ha, C.D. Jang, S.J. Kang, D.H. Nam

Evaluation of Fatigue Damage on Fixed Jacket Platforms   345
T.A.P. Lopes, L.C.M. Meniconi

Shape Optimization on the Basis of Biological Growth with   357
Special Regard to Slanted Joints
M. Beller, C. Mattheck, J. Schäfer

Spectral Integrations for Dynamic Responses of Offshore   367
Platforms to Random Waves
Y.H. Chen, D.H. Tsaur

SECTION 8: OFFSHORE OPERATIONS

Computer Controlled Remote Testing of Marine Risers   387
in the Adriatic
A.K. Basu, A. Giuggioli

Exploitation of Compound Safety Knowledge for Computer-   407
Based Operational Monitoring and Control
G. Langli, B.A. Bremdal, D. Hirdes

SECTION 1: SHIP DESIGN

HOSDES and MARDES: Advanced Concepts for Integrated CAD-systems

G. Glijnis and G. van Oortmerssen
Maritime Research Institute Netherlands (MARIN), P.O.Box 28, 6700 AA Wageningen, The Netherlands

1 INTRODUCTION

Recent developments in hardware and software tools have stimulated the design of ships by means of computer programs. Especially during the past five years the number of such developments seem to grow at an increasing rate.

The developments of hardware concern the power of the CPU, (graphical) displays and peripherals as laser printers, plotters, etc.. With respect to software the increased use of general drawing packages, new programming languages, graphics standards and database systems can be mentioned.

Of the four types of database systems, the relational database type can be considered to be state-of-the-art. Most applications of this type are in the area of administrative automation. Some recent publications about the use of relational database systems are Byran [1] (describing some aspects of the ship data storage), Hills [2] (describing the use of a RDBMS for ship compartmentation), Allieri [3] (giving a general description of the use of a relational database management system (RDBMS) in a design system) and Schumann-Hindenberg [4] (describing the use of a RDBMS for ship compartmentation). Some other articles exist concerning Computer Aided Engineering or Computer Aided Manufacturing. None of these publications deals with the storage of software system data or the storage of numerical data of ship designs.

MARIN has developed two software systems for the conceptual and preliminary design of ships with the possibility for the customer to add new programs to these systems. The core of those systems consists of a relational database system for storage of information about the structure of the system and ship design information in the early stages of the design process.

Section 2 describes the user requirements for Computer Aided Design packages and the software concept implemented in MARIN's systems HOSDES and MARDES. Section 3 starts with a short theoretical description of relational database systems. The main part of this section will be a description of a relational database as used in HOSDES and MARDES. The limited length of this paper confines this description to non-geometrical data. In future papers this description will be completed.

2 THE CAD SYSTEMS HOSDES AND MARDES

The Maritime Research Institute Netherlands (MARIN) decided in 1983 to design and build an advanced Computer Aided Design system for the conceptual design of ships. In first instance this system (called HOSDES) has been realized for naval vessels of the frigate-destroyer type. In October 1988 the delivery of the software has been completed, although still new calculation programs are being developed and included. A derivative (MARDES) which is meant for merchant vessels has been developed recently.

Large parts of HOSDES and MARDES are identical, e.g. the definition of the hull form and internal arrangement and the calculation of hydrostatics, loading conditions, intact and damage stability. Only the calculation programs differ for different ship types. The following application programs are now available within the HOSDES/MARDES environment:

1. programs for estimation of the resistance and propulsion characteristics of different ship types with empirical formulae based on regression techniques.

2. programs for calculation of propeller particulars, based on systematical series.

3. programs for estimation of the seakeeping behaviour ships based on a strip theory for two or six degrees of freedom, or on empirical data.

4. a program for estimation of the manoeuvring behaviour of vessels, based on empirical data.

5. programs for estimation of the mass of different ship types, based on empirical formulae.

6. a program for selection of the propulsion system for frigates and destroyers.

7. program for the calculation of the range of frigates and destroyers with a specified mission profile.

8. programs for the transformation of the hull form with respect to the principal dimensions, the block coefficient and the longitudinal centre of buoyancy. Also a special method has been developed for vessels with deadrise and without a parallel midship. It is possible to interpolate new lines from existing ones.

9. programs for calculation and manipulation of the Sectional Area Curve.

10. a Concept Exploration Model for frigates and destroyers. This program calculates with a number of simple formulas a large number of ship designs covering almost all aspects of ship design. After calculation of these ships the rejection criteria are applied to obtain a small number of promising designs. See Colwell [8] for details of this program.

11. the Seasafe-package for calculation of hydrostatics, capacities, centres of gravity, stability cross curves, wind moment curves, grain heeling moments, critical KG's for intact and damage stability, loading conditions, floodable lengths curves, trim and heel, international regulations, time history of flooding, weight distribution, longitudinal strength (bending moments, shear forces, deflections), launching and tonnage.

12. the Eagle general drawing package, adapted to the specific requirements of the ship designer.

2.1 Specification Of User Requirements
Before starting the development of a Computer Aided Design system it is useful to make an overview of the wishes and requirements of the users of this system. They will act as guidelines for the software design. Some of these requirements are:

1. The CAD system has to contain application (calculation) programs for all aspects of interest for the ship design. Also the programs have to be suitable for different ship types. Preferably, several programs should exist, covering the same aspect, each suitable for a specific

area of application. Examples are separate resistance and propulsion programs for fishing
vessels, supply vessels, coastal trade vessels
and fast displacement ships.

2. There should be several calculation programs
covering the same aspect but with different
levels of detail and accuracy. The programs
vary from rules-of-thumb to for instance advanced flow calculations.

3. The designer should have complete freedom of
program execution. It is often not necessary to
perform each calculation for each ship type,
and the user must be free to choose the pre-
ferred execution sequence. Our philosophy is
that the designer really should be designer.

4. The system should be user friendly. This means
with a fast response, sufficient help information, self explaining, with a consistent program control. In practice, this means the introduction of one User Interface Management
System that takes care of the communication between the user and the system in a uniform way.

5. The data storage should be consistent and uniform with facilities for back-up, recovery, security and data protection.

6. The geometrical definition should be powerful
and easy to use.

7. The system should have powerful graphical facilities for both input and output functions. A
good use of graphics improves the user friend-
liness of the system.

8. The system should have facilities to take into
account the experience of the user in the form
of coefficients, relations or programs. Facilities should be available for easy inclusion of
such data and knowledge.

9. Another aspect of importance is a commercial
one. A CAD system should support the commercial
activities of the yard with 2D and 3D drawings,
inclusion of calculation results in reports,
etc..

10. If possible the system should support the designer in his task with advice on which programs to use, how parameters are calculated,

etc.. Such facilities can be called a Designer Guidance System.

11. It should be easy to incorporate new application programs that have become available for
new design aspects or for new ship types, etc..
Generally spoken, the system should easily allow extensions, preferably by the users themselves. It should not be necessary to return to
the software developer to incorporate user made
calculation programs.

In the next section a short description of the
HOSDES/MARDES system concept will be given.

2.2 The HOSDES/MARDES System Concept
The realisation of the system design objectives is reflected in fig. 1.

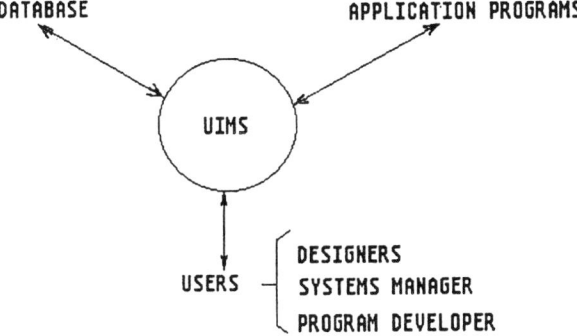

Fig. 1 Hosdes structure

In this figure the users, the data structure and the
application programs have been uncoupled. The
considerations behind this architecture are as
follows:

- if multiple data structures are to be avoided, a
single database has to be used. In this central
database each data item is stored once,

- if new programs are to be added, it is necessary
to execute each program separately, and not as a
subroutine of a bigger part. If calculation programs were incorporated as subroutines, it would
be necessary to modify the source code when adding new programs or deleting existing ones,
- if multiple user interfaces are to be avoided,
one program has to be defined that performs this
interface task, the so-called User Interface Management System (UIMS). This program forms the

link between the users, the database and the application programs.

The concept of separate executables makes addition of new programs by the user himself very easy. Also it is possible to incorporate several programs covering the same aspect with different accuracies for several ship types. Because a set of standardized subroutines is used for the User Interface Management System and the data handling a consistent data storage can be attained and a uniform user interface can be built. The User Interface Management System forms the heart of the system. It is a dedicated process through which all communication between the user, the application program and the datastructure is directed. Its main properties are:

- it is screen oriented which means that execution control is done by means of filling in predefined screens.

- all screens look uniform and are controlled in a uniform way. This is one of the main objectives of the system design.

- it has multi-tasking facilities that make simultaneous execution of several (time consuming) programs possible.

- one page of help information is on-line available at each field of a screen.

- it is built with a package of standardized input/output (I/O) routines.

The datastructure has been built with the relational database system Mimer. Use of a database system ensures the integrity and the consistency of the data. Furthermore such a professional database system has facilities with respect to back-up, recovery and security of data. Also the interface to the database has been built with a package of standardized I/O-routines. Because it is possible to add new programs or delete existing ones, it is necessary to store information about the programs, menus, parameters, units, help texts, etc. in the database. This information is stored in the system structure relations. User and project data have been defined because the system is multi-user and multi-tasking. Data that do not belong to a specific ship design (such as propulsion engines, propellers, equipment) are stored in ship independent information. Finally, the data concerning actual designs are stored in some ship dependent relations.

3 RELATIONAL DATABASE MANAGEMENT SYSTEMS

Fig. 2 is an illustration of a design of a RoRo-vessel
made with the MARDES system.

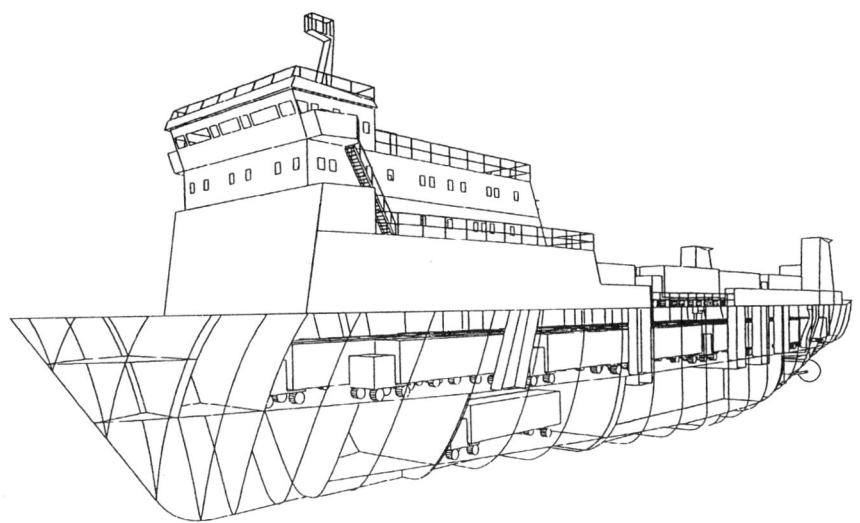

Fig. 2 Design of a RoRo-vessel

The design of such a ship is a complex process with
many aspects designers and data, several iterations
and alternatives and a often complex three-dimensional
geometry. The handling of such a large amount of
information requires special tools. The heart of a
ship design system consists of software for storage
and retrieval of information. Such a package (or
database) offers facilities for systematic, consistent
and uniform storage of data. Also for other aspects
are tools available, e.g. backup of data, recovery
after hardware or software errors, security of
information, simultaneous use of data, etc.. At
present four types of database management systems are
known: the hierarchical, the network, the relational
and the semantic database. Especially the relational
database is used very frequently for the automation of
administrations. Also for technical applications the
use of this type is growing. The advantages of
relational database systems are:

- a uniform structure and description can be defined,
- data redundancy can be avoided,
- inconsistensies can be avoided,
- data integrity is possible,
- common use of data is possible,
- a better availability of data can be attained,

- program development becomes easier,
- program maintenance will decrease,
- programs and data become independent.

Disadvantages however are:

- response times increase,
- mass data storage is not optimal,
- concurrent use is difficult,
- changes in an existing database design are costly.

In the next sections a very brief introduction will be given of the relational database theory and a description of the use within HOSDES and MARDES.

3.1 General Theory

Before giving a formal definition of a database relation the following table is an illustration of such a relation. A relation can be represented conveniently as a two-dimensional table with rows and columns. As an example the relation 'ship' in which some general data about ships are stored:

Relation SHIP:

SHPNAM	SHPTYP	SHPDAT	SHPTIM	MA	LPP	B
TEST	SUPPLY	900830	10:00	2500	60	7
RORO	RORO	900724	15:10	9500	116.5	20.4

where:
* SHPNAM: ship name (character).
  SHPTYP: ship type (character).
  SHPDAT: creation date (character).
  SHPTIM: creation time (character).
  MA   : displacement (real).
  LPP   : length between perpendiculars (real).
  B   : moulded breadth (real).

An objecttype is a general description of a real item or a concept of the real world. This description is given by means of a number of properties. Such a property is called an attribute. In the example the objecttype 'ship' (real thing or concept) can be described by a number of properties, e.g. name, ship type, creation date, creation time, displacement, length between perpendiculars, breadth, draught, depth, etc.. It is obvious that each description only forms a part of the total reality and cannot describe this total reality completely.

In a relational database relations are defined between several objecttypes and between an objecttype

and its attributes. According to the definition a database relation (shortly: relation) on the attributes
A1, A2 ... An is a subset of the powerset of the Cartesian product A1 x A2 x ... x An. The Cartesian product is the set of all possible combinations of the
attributes. The relation is a subset of these combinations. A relation is identified by a name and has one
or several attributes. An attribute or a combination
of attributes of a relation is a (primary) key if this
combination identifies one occurrence (called a tuple)
unambiguously and if a subset of this combination does
not. An attribute or a combination of attributes of a
relation R is called a foreign key if this combination
is not the primary key of R but the key of an other
relation. So the foreign key defines relations between
objecttypes. An attribute is represented by a column.
Each row is a tuple (one occurrence of the objecttype). The keys are identified with the asterisks
(_*). In the example the attribute 'SHPNAM' forms the
key of the relation SHIP. An extensive description of
the relational database theory can be found in Date
[5]. For selection, insertion, deletion or modification of data within a relational database system the
standard SQL (Structured Query Language) can be used.
An example of a selection clause is:

SELECT SHPID FROM SHIP WHERE SHPNAM='RORO'

Important advantages of relational database systems
are the compact data storage and the fact that it is
not necessary to know the internal structure of the
database. However, the consistent definition of the
database, the data integrity and the normalization
process are left to the responsibility of the database
designer.

3.2 Application Within HOSDES And MARDES
In this section a description will be given of the application of a relational database system and the definition of the relations used. The total number of
relations can be divided into relations concerning the
HOSDES/MARDES system structure, user and project data,
general ship independent data and ship dependent data.
It should be noted that only the most important aspects of the tables are given. With each relation one
or more tuples are given as an example.

3.2.1 System Data   In section 2.2 it has been stated
that all programs constituting HOSDES and MARDES are
separate executables. At present there exist ab. 120
programs of which are 30 shipbuilding applications.
All programs are put in a menu structure. The
organization of menus and programs is done by means of
the following relations:

Relation METHOD:

MTHID	LEVNO	MTDPSW	MTHSCR
DESP	1	SECRET	MARAP

where:
   * MTHID : method identification (character). This
     is the name of the program.
   * LEVNO : level number (integer). Four levels
     exist   within   the   system, reflecting   the
     increasing accuracy in the ship design process.
     This is one of the software system design
     objectives. The levels are:
     0 – system management level
     1 – geometry of the ship design not yet known,
        only in parametric form
     2 – external and internal geometry of the ship
        design is defined
     3 – geometry of appendages is defined MTDPSW:
        method   password   (character). Prevents
        unauthorized use. MTHSCR: name of screen
        file (character). Name of the file where the
        predefined screens of this program can be
        found.

In this relation all methods (programs) together with
the level number, passwords if applicable and the name
of the file where the pre–defined screens can be found
are stored. When a program is called via a choice at a
menu, the User Interface Management System (UIMS)
checks in the database whether a password exists and
looks for the pre–defined screens in the file in the
column MTHSCR.

Relation MENU:

MENID	LEVNO	SEQNO	MENTXT	MENTYP	MTHID
MAIN	1	3	Powering programs	M	APPOW
APPOW	1	0	Return to main menu	M	MAIN
APPOW	1	1	DESP – R & D Holtrop	P	DESP

where:
* MENID : menu identification (character).
* LEVNO : level number. See relation METHOD.
* SEQNO : sequence number (integer). Sequence number
          of the menu item in the menu.
  MENTXT: menu text (character). Text at the sequence
          number in the menu. This text gives which
          menu is the previous or the next or gives a

very short explanation of the program.
MENTYP: menu type (character). Can be 'M' (for menu
   - MTHID is the name of the previous or the
   next menu) or 'P' (for program)
MTHID : method identification. See relation METHOD.

With this relation the complete menustructure can be
built. Also it is easy to modify the layout of the menus and to add new programs to the system.

Relation CONNECT:

MTHID	PRGPAR	KYWPAR	FRMNAM	FLDNAM	DPPVAL
DESP	L	LPP	DESPF1	LPP	
L	LBP	LPP	REPOWF1	LBP	

where
* MTHID : method identification. See relation METHOD.
* PRGPAR: program variable name (character). Name of
   the variable as used within the program.
   KYWPAR: system name of this parameter (character).
   For system names for parameters only
   ISO/ITTC symbols are used.
   FRMNAM: name of the pre-defined screen where this
   parameter has to be filled in (character).
   FLDNAM: name of the field on the screen FRMNAM where
   this parameter has to be filled in
   (character).
   DPPVAL: default value for this variable (character).

The meaning of the connection file is as follows.
A number of existing calculation programs had to be
included in HOSDES and MARDES. Each program used specific, different names for variables with the same
meaning. To avoid reprogramming of each source code, a
kind of translation has been defined. The co-called
keyword KYWPAR is a unique parameter name with a
unique meaning within the database. It is possible
that several program variables refer to this system
parameter. For instance, the variables L of DESP and
LBP of REPOW refer to the same database parameter LPP,
meaning the length between perpendiculars. If a program needs the value of a variable, it looks into the
connection file to find the equivalent system parameter name. The value of this system parameter will be
fetched from the database and returned to the program.
If the value cannot be found in the database, the UIMS
looks which screen has to be shown to the user and
from which field the value has to be read. This value
will be returned to the program. When adding a new
program the programmer has to define the translation
program variable/system parameter/screen name/field
name by filling this relation.

Relation MESSAGE:

MSGID	MSGDSC
SHNOFO	Ship % not found

where:
MSGID : message identification (character). Mnemonic to be used in an application program.
MSGDSC: message to be displayed on the screen (character). A variable can be filled at the place of the % sign.

    Use of this relation means that the programmer of a new application program is free to define new messages to the user. These new messages also can be used by other programs. When a message has to be displayed, the program sends the message mnemonic and some parameter values to the UIMS. The UIMS looks into the database to obtain the corresponding text, fills in the parameter values and displays the complete message.

Relation SYSPAR:

PARID	PARCAT	PARDSC	PARSI	PARTYP
LPP	GE	Length between perp.	m	R

where:
* PARID : parameter identification (character). The unique name of a parameter within the system.
   PARCAT: category of the parameter (character). Each parameter belongs to one of the categories Geometry (GE), Powering (PO), Engine (EN), Mass (MA), Hydrostatics (HY), Seakeeping and manoeuvrability (SM), Strength (ST), Vibrations (VI) or Performance (PE).
   PARDSC: description of the parameter (character).
   PARSI : unit of the parameter (character). All values are stored within the database as SI units.
   PARTYP: Type of the parameter (character). Can be R (real), I (integer) or C (character).

    This relation contains all parameter names that are allowed within the system and the database. For system parameters the ISO/ITTC list of shipbuilding symbols for computer applications is used (ref. [6]). The programmer has to be aware of this list and he has to be very careful to use the correct variables and the correct meaning thereof. For instance, he should not define a system parameter LBP with the meaning

'length between perpendiculars' if a parameter with
this meaning already exists as LPP. This would be
disastrous for the consistency of the database. The
list of parameters can be expanded when adding a new
program.

Relation MTHPAR:

MTHID	PARID	PARDIR
DESP	LPP	I

where:
* MTHID : method identification. See relation METHOD.
* PARID : parameter identification. See relation
          SYSPAR.
  PARDIR: determines whether a parameter is input (I),
          output (O) for a program or both (B).
     This relation contains all parameters that are
used by a program as input or output parameter or
both. In this way it is possible to establish chains
of programs and to determine which influence a change
of one parameter will have on others. The consistency
between data can be established and maintained. This
facility will be used by the Design Guidance Subsystem
programs.

Relation VALIDAT:

MTHID	PARID	MINVAL	MAXVAL	VALTYP
DESP	LPP	0.0	500.0	S

where:
* MTHID : method identification. See relation METHOD.
* PARID : parameter identification. See relation
          SYSPAR.
  MINVAL: minimum value of the parameter (real).
  MAXVAL: maximum value of the parameter (real).
  VALTYP: type of the validation (character). A vali-
          dation can be H (hard), meaning that the gi-
          ven value has to satisfy this validation or
          S (soft), meaning that the program has not
          been designed for the given value with res-
          pect to applicability and accuracy.

     In the relation VALIDAT boundaries for program
variables can be defined. It is advantageous to store
these boundaries in the database instead of programming them in the source. Modification of the boundaries is easy, e.g. when it has become clear that a
program is valid for a larger area of the parameter.
When the program receives a value from the UIMS, it
looks into the database to find the validation bound-

aries. If they are not satisfied, the value is asked
again by the UIMS. Values violating the hard boundaries will never be accepted because the program will
end abnormally with e.g. a division by zero, data
overflow or an illegal array index.

3.2.2 User and Project Data   It is envisaged that ship
design are made within the scope of projects. A project can be e.g. a design for a certain customer. For
a project new parameters, design objectives, rejection
criteria and relations can be defined. Design objectives are the goals of the design. They are independent variables for the design process, e.g. a design
speed of 20 knots. If design objectives are not satisfied, this can not lead to rejection of a specific design. Rejection criteria are parameters that have to
be satisfied in the design. They are dependent variables of the design process, e.g. a maximum breadth of
32.24 m. If one (or more) rejection criterion is not
satisfied, a flag will be set to indicate that this
design will be rejected. This design is not deleted
from the database because the rejection criteria might
be changed. Relations can be empirical formulae. For
design objectives, rejection criteria and relations
also database relations have been defined. One of the
software system design was to develop a multi-user
system. Therefore user names have to be known to the
system. It is possible to assign several user names to
a specific project. In this way several aspects of the
design can be handled by several users.

Relation USER:

USRID	USRPSW
GERARD	SECRET

where:
* USRID : user identification (character). Name of the
         user within HOSDES/MARDES.
  USRPSW: password (character).

     The relation USER contains all users that are au-
thorized to use the HOSDES/MARDES system. It is not
necessary that these names are equal to the names that
are known by the computer itself. Protection against
unauthorized use is obtained by means of a password.
Also the user has to be assigned to a project via the
relations PROJECT and USRPRJ.

Relation PROJECT:

PRJID	PRJPSW	PRJDSC	CPUSEC
DEMO	SECRET	Demonstration project	4300

where:
* PRJID : project identification (character).
  PRJPSW: password (character).
  PRJDSC: project description (character).
  CPUSEC: total number of CPU seconds used for a project.

This relation contains the names of the projects, together with a password for protection of the project data. Also the used computer seconds and some other values are stored for each project.

Relation USRPRJ:

PRJID	USRID	SHPNAM	SHPOPT
DEMO	GERARD	RORO	1

where:
* PRJID : project identification. See relation PROJECT.
* USRID : user identification. See relation USER.
  SHPNAM: name of the previous ship used (character). See relation SHIP.
  SHPOPT: option number of the previous ship used (integer). See relation SHIP.

The relation USRPRJ forms the connection between the relations USER and projects and contains all users that are assigned to projects. Also the latest ship that is used is stored. This ship will be displayed as a default for the ship name when a program is started.

3.2.3 General Ship Independent Data   Some data exist that do not belong to one specific ship design, e.g. propulsion engines, propellers or equipment parts. These items are stored independently from ship design in the so-called Miscellaneous Data Management Subsystem. At present only storage of propulsion engines has been defined.

Relation ENGINE:

ENGID	ENGTYP	ENGHGT	ENGLEN	ENGMAS	ENGMCN	ENGMCR
6SWD280	D	2.500	4.000	18.000	16.667	1700000

where:
* ENGID : identification of the engine (character).
* ENGTYP: engine type (character), e.g. D (diesel), G
         (gas turbine) or S (steam turbine).
  ENGHGT: engine height (real). The total height of
          the engine.
  ENGLEN: engine length (real). the total length of
          the engine.
  ENGMAS: engine mass (real). The total weight of the
          engine.
  ENGMCN: engine revolution rate at MCR (real).
  ENGMCR: Maximum Continuous Rating (MCR) of the
          engine (real). Also the Maximum Overload ra-
          ting and the corresponding revolution rate
          are stored.

        The relation ENGINE only gives the identification
of the engine and some general data. Because the delivered brake power and the specific fuel consumption
depend on the number of revolutions is the relation
between revolution rate, delivered power and fuel consumption is stored in the relation ENGCHR.

Relation ENGCHR:

ENGID	ENGN	ENGPB	ENGSFC
6SWD280	10	1020000	0.14
6SWD280	13.333	1360000	0.13
6SWD280	15	1530000	0.14
6SWD280	16.667	1700000	0.15

where:
* ENGID : engine identification (character). See
          relation ENGINE.
* ENGN : revolution rate (real).
  ENGPB : delivered power at the revolution rate ENGN
          (real).
  ENGSFC: specific fuel consumption at the revolution
          rate ENGN (real).

        In this relation the relation between engine re-
volution rate, delivered power and specific fuel consumption is stored. The number of points is unlimited.

3.2.4 Ship Dependent Data Most store data belong to
actual ship designs. These data can be divided into
numerical and geometrical data. In this paper only the
numerical data are described in some detail. The geometry definition is described in Glijnis [7]. The corresponding database structure will be published in a
future article. The relations cover general data of
the design, parameter values (so-called characteris-

tics), graphs of independent and dependent values and the history of characteristic data.

Relation SHIP:

PRJID	SHPNAM	SHPOPT	SHPVRS	USRID	SHPID	SHPBF	SHPCF
DEMO	RORO	1	2	GERARD	56	N	Y

where:
* PRJID : project identification (character). See relation PROJECT.
* SHPNAM: ship name (character).
* SHPOPT: ship option number (integer). This is a design variant.
* SHPVRS: ship version (integer). The version number accounts for the accuracy level and is equal to the level number where the user is working. For the level number see the relation METHOD.
* USRID : user identification (character). See relation USER.
  SHPID : ship identification (integer). This unique number is used within all other relations with ship dependent data.
  SHPBF : built flag (character). Determines whether the design has been built.
  SHPCF : consistency flag (character). Determines whether the design is consistent.

The relation SHIP forms the most important one of the database. Here the connection between name, option and version number and other relations with design dependent data is defined. In fact a number of additional values of lesser importance have been defined for the relation SHIP. Difference has to be made between the so-called Global and the Local Database. In the Global Database built vessels or all designs that are considered to be 'good' are stored. These designs can be used by all users of the system and can therefore act as a starting point for a new design. The ships in the Global Database are maintained by the HOSDES/ MARDES system manager. For the Global Database an analogue relation SHIPG has been defined. The ships in the Local Database are actual designs that belong to a specific user and a specific project. So the maintenance of these design is done by the users themselves. The relation SHIP and all following relations belong to the Local Database.

Relation CHARACTERISTIC:

SHPID	PARID	PARVAL	MTHD	MTHID	CONFID	CHARCF	CHARRF
56	RT	400000	C	DESP	100.0	Y	N

where:
* SHPID : ship identification (integer). See relation SHIP.
* PARID : parameter identification (character). See relation SYSPAR.
  PARVAL: parameter value (real). The actual value of the parameter.
  MTHD : way of input (character). Determines whether a value has been defined by means of user manual input (I), calculated with a method (M) or calculated with an expression (C).
  MTHID : method identification (character). See relation METHOD.
  CONFID: confidence level of the parameter value (real). This is a measure of the accuracy or the confidence in the parameter value and can be between 0 and 100 %.
  CHARCF: consistency flag (character). Indicates whether the value is consistent with the input data.
  CHARRF: rejection flag (character). Indicates whether the ship will be rejected on the basis of the parameter value.

In this relation all single non-geometrical parameter values are stored. All values are stored in SI units. If necessary the values can be changed into user friendly values by the application programs. The consistency flag indicates whether a value is consistent with its input values. In the relation RELATION (not being described here) the combination method/input parameter/output parameter is stored. In this way the dependency of the calculated (dependent) values on the input (independent) values can be established. If a value is calculated by a method (e.g. the total resistance RT by the resistance method of Holtrop and Mennen) and later the input values are changed without running this method again, the consistency flag for RT is set. This means that the calculated value does not belong to the actual input values.

Relation AXES:

SHPID	INPARI	PARID	MTHID	AXVRS	AXDSC	CONFID	MTHD
56	V	RT	DESP	1	Demo	100.0	C

where:
* SHPID : ship identification. See relation SHIP.
* INPARI: independent   parameter   identification
          (character). This is the parameter along the
          horizontal axis.
* PARID : dependent parameter identification (char-
          acter). This is the parameter along the
          vertical axis.
* MTHID : method identification (character). See rela-
          tion METHOD.
* AXVRS : graph version (integer). Up to 10 graph
          versions can exist for a given combination
          of independent parameter, dependent para-
          meter and method.
  AXDSC : graph description (character). Comment at
          the graph.
  CONFID: confidence level of the parameter value. See
          relation CHARACTERISTIC.
  MTHD : way of input. See relation CHARACTERISTIC.

The relation AXES stores some general data of
graphs such as both parameters, the used method for
calculation, a description where comment can be given,
and a confidence level (analoguous to the relation
CHARACTERISTIC). The actual values are kept in the
next relation.

Relation COORD:

SHPID	INPRI	PARID	MTHID	AXVRS	INDNR	INPARV	PARVAL
56	V	RT	DESP	1	1	10	400000
56	V	RT	DESP	1	2	15	500000

where:
* SHPID : ship identification. See relation SHIP.
* NPARI : independent parameter identification. See
          relation AXES.
* PARID : dependent   parameter identification.   See
          relation AXES.
* MTHID : method   identification   (character).   See
          relation METHOD.
* AXVRS : graph version. See relation AXES.
* INDNR : sequence number (integer). From 1 to n.
  INPARV: value of the independent parameter (real).
  PARVAL: value of the dependent parameter (real).

The relation COORD is the actual storage of the
points constituting the graph, consisting of a independent parameter (plotted along the horizontal axis)
and a dependent parameter (plotted along the vertical
axis). All values are stored in SI units. Next to the
relations AXES and COORD two other relations have been
defined for storage of multi-dimensional graphs.

Relation HISTORY:

SHPID	PARID	PARDT	PARTM	PARVAL	MTHD	MTHID
56	RT	900831	09:00	400000	C	DESP

where:
* SHPID : ship identification. See relation SHIP.
* PARID : parameter identification. See relation SYSPAR.
* PARDT : parameter date (character). Gives the date of creation of the parameter value.
* PARTM : parameter time (character). Gives the time of creation of the parameter value.
  PARVAL: parameter value (real).
  MTHD : way of input (character). See relation CHARACTERISTIC.
  MTHID : method identification (character). See relation METHOD.

In the relation HISTORY all historical values of the parameters of a ship design are stored, together with the date and time of creation and the used method for definition. With this relation the designer can obtain an overview which methods have been used and how the parameters have been varied. Also he will be able to recall the design sequence of a certain ship design. These programs belong to the Design Guidance Subsystem.

4 CONCLUSION

The software systems HOSDES and MARDES have been developed to support the designer of naval vessels and merchant vessels, respectively. The calculation programs of both systems are tailored for these ship types. Several programs with different levels of detail and accuracy are available covering different aspects of ship design. A relational database system has been applied for storage of ship design data and system information. Such a system offers facilities for uniform, consistent data storage without redundancy and for back-up, recovery and security of data. The following information is stored within the database: ships, ship parameter values, graphs, historical values of parameters, propulsion engines, users and projects, programs, menus, messages, system parameters, validation boundaries of parameters and the connection between program variables, system parameters. The definition of the geometry will be described in future publications. An important new aspect of the software concept is the possibility of extending the system with new (existing) programs by the user. This is achieved by an architecture of separate programs communicating in a uniform way with the user by means of a common User Interface Management program. Information about the system itself is stored in the database as well, which makes dynamic changes of the contents of the system possible.

5 REFERENCES

1. Byran, A. et al., A relational data base for naval architecture, pp. 8.1-8.17, Proceedings of 25th anniversary symposium of CETENA, Genoa, Italy, 1987, CETENA, Genoa, 1987.

2. Hills, W. et al., An efficient compartmentation method for use in preliminary ship design, in Computer Applications in the automation of shipyard operation and ship design VI (Ed. D. Lin, Z. Wang and C. Kuo), pp. 141-153, Proceedings of the ICCAS 1988 Conference, Shanghai, People's Republic of China, 1988, North-Holland, Amsterdam, New York, Oxford, Tokyo, 1989.

3. Allieri, E. and G.A. Sartori, An integrated system for storage, retrieval and manipulation of ship design information, in Computer Applications in the automation of shipyard operation and ship design VI (Ed. D. Lin, Z. Wang and C. Kuo), pp. 33-42, Proceedings of the ICCAS 1988 Conference, Shanghai, People's Republic of China, 1988, North-Holland, Amsterdam, New York, Oxford, Tokyo, 1989.

4. Schumann-Hindenberg, U., Interactive design of ship compartmentation, in Computer Applications in the automation of shipyard operation and ship design V (Ed. P. Banda and C. Kuo), pp. 343-352, Proceedings of the ICCAS 1985 Conference, Trieste, Italy 1988, North-Holland, Amsterdam, New York, Oxford 1985.

5. Date, C.J., An Introduction to Database Systems, Addison Wesley, Reading, Amsterdam, London, Manila, Singapore, Sydney and Tokyo, 1977.

6. International Organization for Standardization, SHIPBUILDING SYMBOLS FOR COMPUTER APPLICATIONS, draft proposal DP 7463 (1981-8-14).

7. Glijnis, G., HOSDES/MARDES: a new concept for CAD-systems, Schip en Werf, Vol. 56, pp. 196-202, 1989.

8. Colwell, J.L., Users manual for the SHOP5 system: A Concept Exploration Model for monohull frigates

and   destroyers,   Defence   Research   Establishment
Atlantic, Dartmouth, Canada, 1988.

On the Optimum Structural Design of MARPOL Tankers Considering Tank Arrangement

C.D. Jang(*) and S.S. Na(**)

(*) Department of Naval Architecture, Seoul National Univ., Seoul, Korea

(**) Daewoo Shipbuilding and Heavy Machinery Ltd., Seoul, Korea

Abstract

An efficient approach to the minimum structural weight design of MARPOL tankers considering tank arrangement is suggested and applied to an actual ship of 280K DWT VLCC. By this approach, it is possible to find optimum tank arrangement and optimum structural design of actual oil tankers. For the minimum structural weight design, it is most recommendable to increase the tank length within the limit of MARPOL convention.

1. Introduction

More attention have recently been paid to the optimum structural design to minimize structural weight and construction cost of ships since several oil crises.

On the minimum structural weight design of oil tankers many studies[1]-[6] have been performed. As for the longitudinal members of oil tankers Moe, Lund and Kitamura[1]-[3] performed fundamental work on the minimum weight design on the basis of classification society rules. For the transverse members some extensive efforts[4]-[6] on the minimum weight design based on the direct strength calculation were made. Jang and Na[9] also suggested an efficient method for the minimum transverse structural weight design of actual oil tankers of VLCC class.

However most of the previous studies were simply carried out under a given tank arrangement which may have had great effect on the optimum structural design of oil tankers. In spite of its importance there exists very few studies on the optimum structural design of oil tankers considering tank arrangement because enormous computer efforts and costs would be needed to find the optimum value. Therefore the Authors have developed an efficient approach to the minimum structural weight design of MARPOL tankers with optimum tank arrangement.

In order to find the optimum design in this work, the tank size is determined to satisfy MARPOL convention, the longitudinal members are directly designed by classification society rules. And the web frame space, position of longitudinal bulkhead and scantlings of web frame structures are so determined that they can minimize the hull weight of tankers.

For the transverse structural analysis, Generalized Slope Deflection Method is applied as an efficient method. Also Equivalent Curved Beam Theory is adopted to calculate the stresses in the corner parts of web frame[9].

Hooke and Jeeves direct search method is applied to easily find optimum discrete design values of actual scantlings of structural members available in shipyards.

The minimum structural weight design program is developed based on direct search method combining longitudinal and transverse structural design program.

Some optimum design examples of an actual ship of 280K DWT VLCC are given and total structural weight and design scantlings are compared with those of a existing ship.

2. Minimum Weight Design of Longitudinal Strength Members

The minimum weight design procedure of longitudinal members is as follows. All the scantlings of longitudinal members except deck part can be determined by minimum rule requirements and deck part members should be designed to minimize the midship area which satisfies longitudinal hull girder strength.

2.1 Object function

Object function is the midship area of the middle part of ship.

2.2 Design variables

Deck thicknesses are adopted as primary design variables since the bottom structures are generally heavier than deck structures in oil tankers.

2.3 Constraints

For constraints, it is considered that the deck thicknesses should not exceed the rule minimum required thicknesses and also the hull section modulus should not exceed the rule required hull section modulus.

2.4 Optimization technique

Optimization technique based on Hooke & Jeeves direct search method is applied to easily obtain discrete value.
This method is to search for the optimum point using local search and global pattern movement.
Because of little searching time it is widely used for the engineering design purpose to overcome local minimum.

3. Minimum Weight Design of Transverse Strength Members

The minimum weight design procedure of transverse members by structural analysis is as follows.
The middle part of ship structures for one web space is modelled as a web frame structure. Structural analysis is carried out for the model to obtain the member stresses by generalized slope deflection method.
The transverse members are designed to minimize the web frame area which satisfies allowable stresses adopting optimization technique.

3.1 Transverse structural analysis method

As transverse structural analysis methods generalized slope deflection method(GSDM) is applied for uniform parts of beam and equivalent curved beam theory considering the effect of external loads for the corner parts of beam.
As shown in Fig.1, when the bracketed beam is deformed under distributed load, end moments and forces, the generalized slope deflection equations can be derived as follows.

$$
\begin{aligned}
M_A &= k_0\{2F_A\theta_A + G_A\theta_B - (2F_A + G_A)\phi\} - m_A \\
M_B &= k_0\{2F_B\theta_B + G_B\theta_A - (2F_B + G_B)\phi\} + m_B \\
Q_A &= -\frac{M_A + M_B}{L} + \frac{W(L-a)}{L} \\
Q_B &= Q_A - W \\
P_A &= EA(u_B-u_A)/L_A \\
P_B &= P_A
\end{aligned}
\tag{1}
$$

$$
\text{where, } k_0 = \frac{2EI_0}{L} , \quad \phi = \frac{v_B-v_A}{L}
$$

where F_A, F_B, G_A, G_B are coefficients to express the bracket effect and m_A, m_B are fixed end moments[9].
Also as shown in Fig.2, when the corner part is subjected to end moments, forces and lateral loads, the stresses at bracket edge are expressed as follows.

$$\sigma_a = \frac{M_\theta}{\rho[A'-A]} \left[\frac{\rho}{a} - \frac{A'}{A} \right] + \frac{P_\theta}{A} \qquad (2)$$

where,

$$M_\theta = M_1 - P_1(L_1-y_{12}-\rho_1\cos\theta) + Q_1\rho_1 \sin\theta$$
$$+ (W_1 + W_{m1}) L_1 \tan\theta (\rho_1\sin\theta-\overline{y_1}) / 2$$
$$P_\theta = P1 \cos\theta + Q1 \sin\theta + (W_1+W_{m1}) L_1 \tan\theta \sin\theta / 2$$
$$A = Af1 + tw (b'-a) + Af2$$
$$A' = Af2 \, \rho / a + Af1 \, \rho / b' + tw \, \rho \, \log (b'/a)$$
$$Af1 = \text{inner flange section area}$$
$$Af2 = \text{outer flange section area}$$

3.2 Object function

Object function is the sum of volume of uniform beam and bracketed part except shell plate.

3.3 Design variables

Design variables are web height and thickness, flange breadth and thickness for each member. Also radii of corner parts are considered.

To reduce the design variables the plate flange thickness is assumed the same as that of shell plate. And plate flange breadth is taken by the effective breadth for one web space.

3.4 Constraints

As for constraints the maximum stress should not exceed the allowable equivalent stress and shear stress, and also minimum thicknesses to prevent web buckling are considered.

4. Minimum Weight Design of Cargo Hold Part Considering Tank Arrangement

Optimum design procedure of the cargo hold part considering tank arrangement is as follows.

From the given cargo hold length except fore and aftbody the cargo hold weight is obtained by unit longitudinal weight from chapter two and unit transverse weight from chapter three considering tank length, web space and the location of longitudinal bulkhead.

Also the hullweight can be obtained by the summation of the cargo hold weight, transverse bulkhead weight and fore/aftbody weight etc.

The transverse bulkhead weight and fore/aftbody weight can be estimated by the hullweight estimation formula.

4.1 Hullweight estimation

As shown in Fig.3, the cargo hold weight is obtained by the unit longitudinal weight and unit transverse weight by dividing the ship structure into cargo hold part, forebody and aftbody part.

Also the transverse bulkhead and fore/aftbody weight are estimated by the following weight estimation formula and the hullweight can be obtained by the summation of the weight mentioned above and cargo hold weight.

$$\text{Hull Weight(Ton)} = W_1 + W_2 + W_3 + W_4 + W_5$$
$$+ W_6 \times NB_1 + W_7 \times NB_2 + W_8 \times NB_3 + W_9 \times NB_4 \quad (3)$$

where,

$W_1 = (W_l \times L_h + W_{tc} \times N_{tc} + W_{tw} \times N_{tw}) \times \{0.4 + 0.6(3 + C_B)/4\}$

$W_2 = 160 \times (L_f \times B \times D \times C_B/1000)^{0.728}$

$W_3 = 530 \times (L_a \times B \times D \times C_B/1000)^{0.469}$

$W_4 = 75 \times (L_a \times B \times 5/1000)^{0.67}$

$W_5 = 15 \times (L_a \times B \times 5/1000)$

$W_6 = 9 \times (B \times D/100)^{1.53} \times (2B_1/B) \times \sqrt{k}$

$W_7 = 9 \times (B \times D/100)^{1.53} \times (2B_2/B) \times \sqrt{k}$

$W_8 = 0.4 \times W_6$

$W_9 = 0.4 \times W_7$

NB_1, NB_2 = Numbers of water tight bulkheads in center and wing tanks

NB_3, NB_4 = Numbers of swash bulkheads in center and wing tanks

N_{tc}, N_{tw} = Numbers of web frames in center and wing tanks

W_l = Weight of longitudinal members per unit length

W_{tc}, W_{tw} = Weights of web frames in center and wing tanks

k = material constant

4.2 Object function

As shown in Fig.4, the object function is defined as the cargo hold weight considering the numbers of transverse bulkhead, the numbers of web and the location of longitudinal bulkhead as follows.

$$F = W_l \times L_h + W_{tc} \times N_{tc} + W_{tw} \times N_{tw} \quad (4)$$

4.3 Design variables

Design variables are deck plate thicknesses for the longitudinal member and the shape of web section for the transverse member.

Furthermore additional design variables are adopted such as the number of transverse bulkheads, the number of webs and the location of longitudinal bulkhead as parametric design variables as shown in Fig.4.

4.4 Constraints

Constraints are the required thicknesses at deck, the required
hull section modulus at bottom and deck for the design of
longitudinal members and allowable stresses of each transverse
member for the design of transverse members.

Also MARPOL(SBT,PL etc.) convention is adopted as constraint
for the arrangement and is to be checked by the package program
(SIKOB).

5. Results and Discussions of Optimum Structural Design

5.1 Design ship

As shown in Fig.5, the design ship is selected as recently
built existing 280K VLCC(DnV classification).

As shown in Fig.7, the minimum weight designs are carried out
for several selected tank arrangements satisfying MARPOL
convention to find the optimum tank arrangement.

5.2 Design loads

As shown in Fig.6, the design loads are selected as full load
conditions, ballast load conditions and abreast load conditions.

5.3 Selection of tank arrangements

As shown in Fig.7, the tank arrangements are selected as five
cases that the breadths of center tank are varied 41~46% of ship
breadth and the numbers of transverse bulkhead are varied for
wide range.

As shown in Table 1, the principal particulars are determined
to satisfy MARPOL convention.

5.4 Results and Discussions

As shown in Table 2, the results obtained by the minimum weight
design of the longitudinal members are similar to those of
existing ship.

As shown in Table 3, transverse structural weight can be
reduced about 6~7% for that of existing ship by the minimum
weight design of the transverse members.

The design procedure for the minimum weight design of cargo
hold part considering tank arrangement is as shown in Fig.8.

As shown in Table 4, the weights obtained by the minimum weight
designs with the optimum tank arrangement are compared with those
of existing ship for the longitudinal weight, transverse weight,
cargo hold weight and hull weight respectively.

The hull weights of ALT I, III are lighter than those of ALT II, IV because the depth can be decreased to adjust cargo capacity. Also the weights of ALT I, III are lighter than that of existing ship(ORIGIN) due to reduced number of transverse bulkheads.

From those results it is known that the depth should be determined considering the breadth of center tank not only to ensure the required cargo capacity but also to give minimum structural weight simultaneously. It is desirable to design the tank length as large as possible like ALT I, III within the limit of MARPOL convention.

As shown in Fig.9, midship design is carried out for the existing tank arrangement(ORIGIN) and shows the similar scantlings compared with those of existing ship.

Finally, it is known that tank arrangements adopted here are generally acceptable, because the maximum hullweight difference is about 500 tons, it is important that more attention should be paid to determine optimum tank arrangements at initial design stage.

6. Conclusions

The main conclusions of this work can be summarized as follows:

(1) An efficient approach to the minimum structural weight design of MARPOL tankers considering tank arrangement is suggested and applied to actual ship of 280K DWT VLCC.

(2) Minimum weight design program for whole structure of MARPOL tanker is developed based on direct search optimization technique combined with longitudinal and transverse structural design program.

(3) By this program, it is possible to find optimum tank arrangement and optimum structural design of actual oil tankers.

(4) For optimum structural design, it is known that the tank depth should be determined not only to ensure the required cargo capacity but also to give minimum structural weight simultaneously.

(5) It is desirable to increase tank length within the limit of MARPOL convention for minimum structural weight design of some actual oil tankers.

Acknowledgements

The Authors are very grateful to Mr. Jae Seon Yum, Department of Naval Architecture, Seoul National University, for his assistance during preparation of this paper.
Part of this work reported was supported by Non Directed Research Fund, Korea Research Foundation, 1989.

References

[1] Moe, J. and Lund, S.   Cost and Weight Optimization of Structures with Special Emphasis on Longitudinal Strength Members of Tankers, Norwegian Institute of Technology Report SKB II/M7, Trondheim, 1967.
[2] Moe, J.   Integrated Design of Tanker Structures, European Shipbuilding, No. 3-4, 1972.
[3] Kitamura, K.   Studies on Optimization of Ship Structures (3rd report) - optimum design of longitudinal members of tanker, SNAJ., vol. 132, 1973.
[4] Moe, J.   Optimum Design of Statically Indeterminate Frames by means of Nonlinear Programming, Norwegian Institute of Technology Report SKB II/M12, Trondheim, 1968.
[5] Lund, S.   Tanker Frame Optimization by means of SUMT Transformation and Behavior Models, Norweigian Institute of Technology Report SKB II/M17, Trondheim, 1970.
[6] Kavlie, D. and Moe, J. Automated Design of Frame Structures, J. of the Structural Division, ASCE, vol. 97, No. ST1, 1971.
[7] Daewoo Shipbuilding, Hullweight Estimation Guide at Marketing and Initial Design Stage, 1988.
[8] Na, S. S.   Structural Analysis and Minimum Weight Design of Ship Structures by Generalized Slope Deflection Method, Seoul National University Ph.D. Thesis, 1988.
[9] Jang, C.D. and Na, S.S.   Computer Aided Optimum Structural Design of Actual Oil Tankers(VLCC), Computational Mechanics Publications , 1988.
[10] Hooke, R. and Jeeves, T.A. Direct Search Solution of Numerical and Statistical Problems, J. of the Assoc. for Computing Machinery, vol. 8, No. 4, 1961.

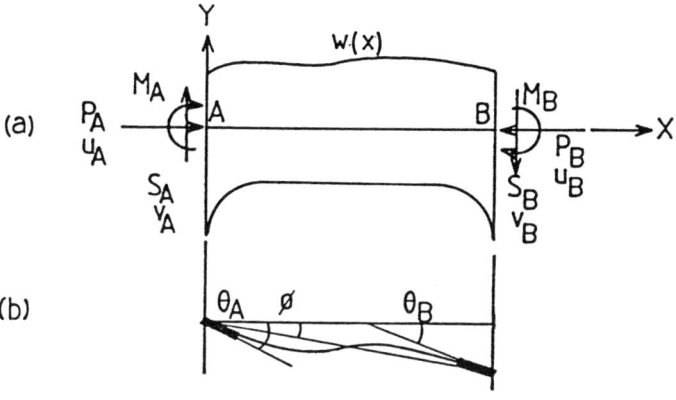

Fig. 1 Bracketed Beam for GSDM(2-D)

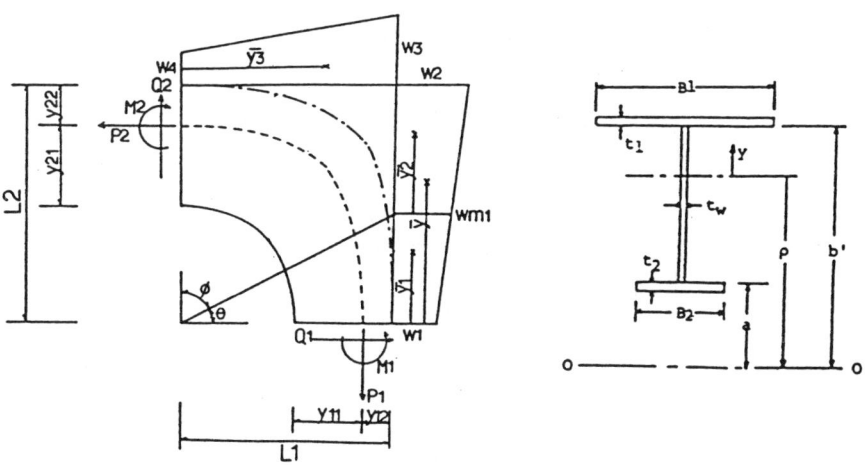

Fig. 2 Equivalent Curved Beam Model with External Loads

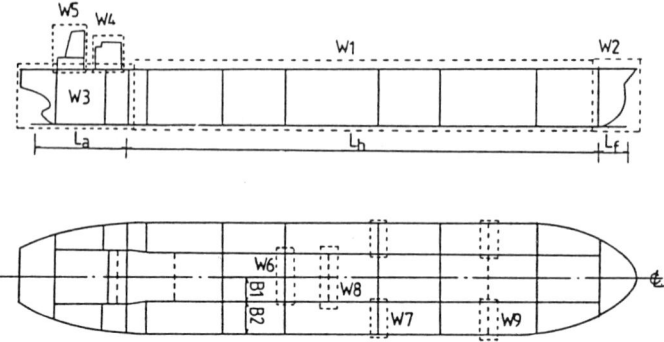

Fig.3 Division of Ship Structure for Hullweight Estimation

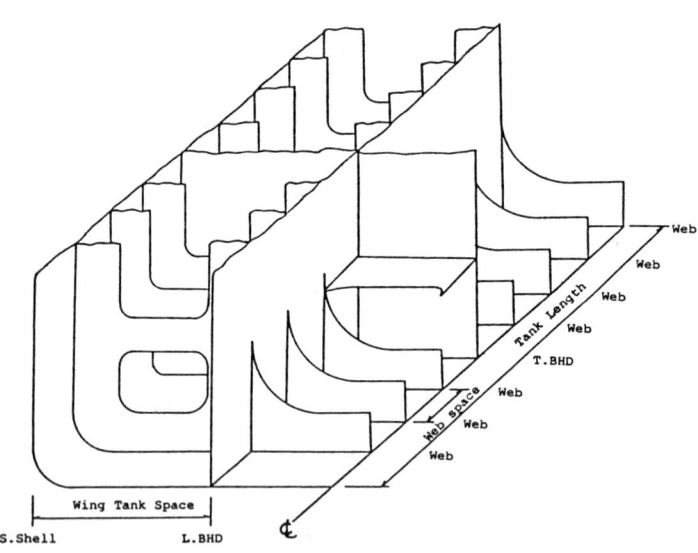

Fig.4 Parametric Design Variables

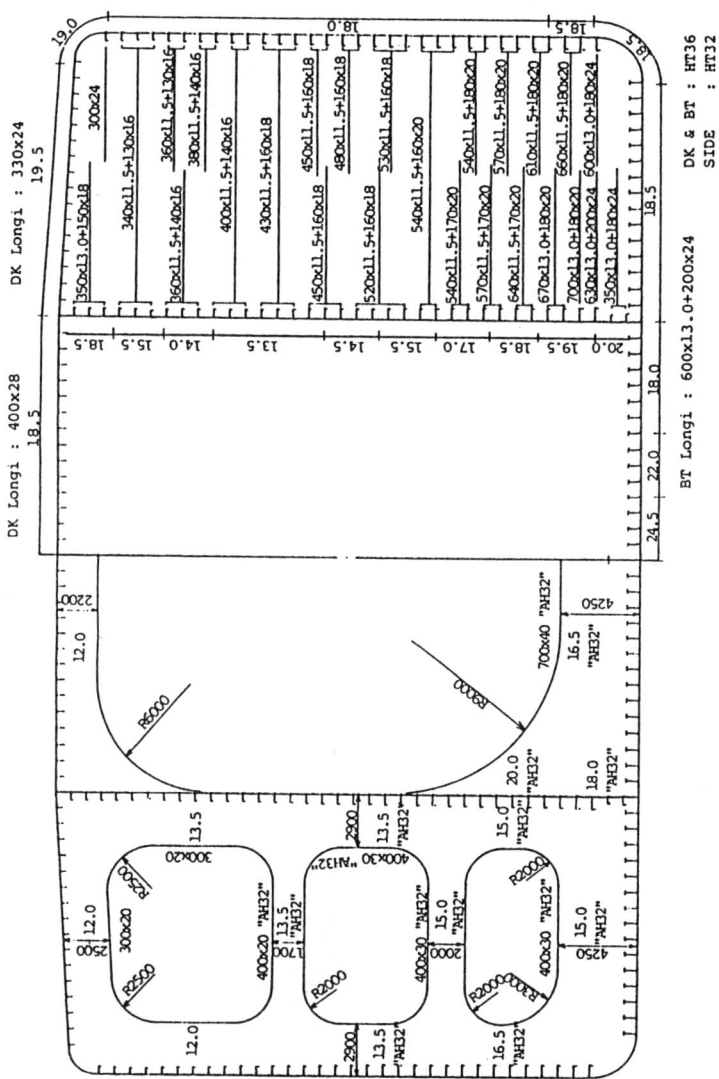

Fig. 5   Design Ship(280K VLCC:DnV Classification)

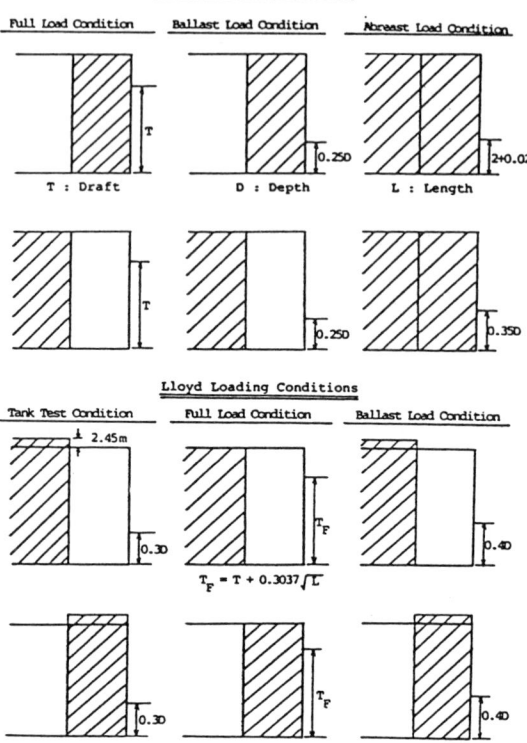

Fig. 6   Loading Conditions

Particulars			ORIGIN	ALT I	ALT II	ALT III	ALT IV
Length (m)			315.0				
Breadth (m)			57.2				
Depth (m)			30.4		30.9	30.3	31.2
Draft (m)			20.8				
Block Coeff.			0.828				
Hold Length(m)			250.7		251.04	250.8	
Forebody (m)			13.25		12.91	13.15	
Afterbody (m)			51.05				
No. of T.BHD	C.T.	O.BHD	6	5	5	6	6
		S.BHD	2	3	3	0	0
	W.T.	O.BHD	5	5	6	5	5
		S.BHD	2	0	0	0	0
Frame Space (mm)	Trans.		5450		5230	5700	
	Longi.		870		840	870	840

Table 1   Characteristics for Each Design Ship

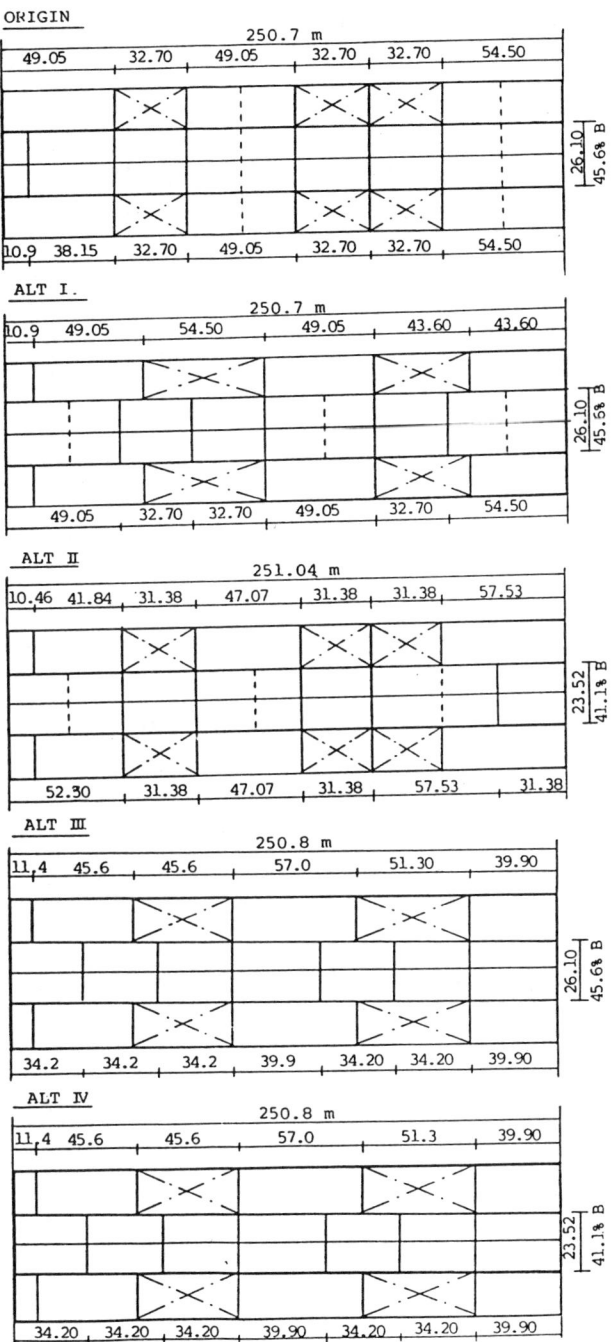

Fig. 7   Selection of Tank Arrangements

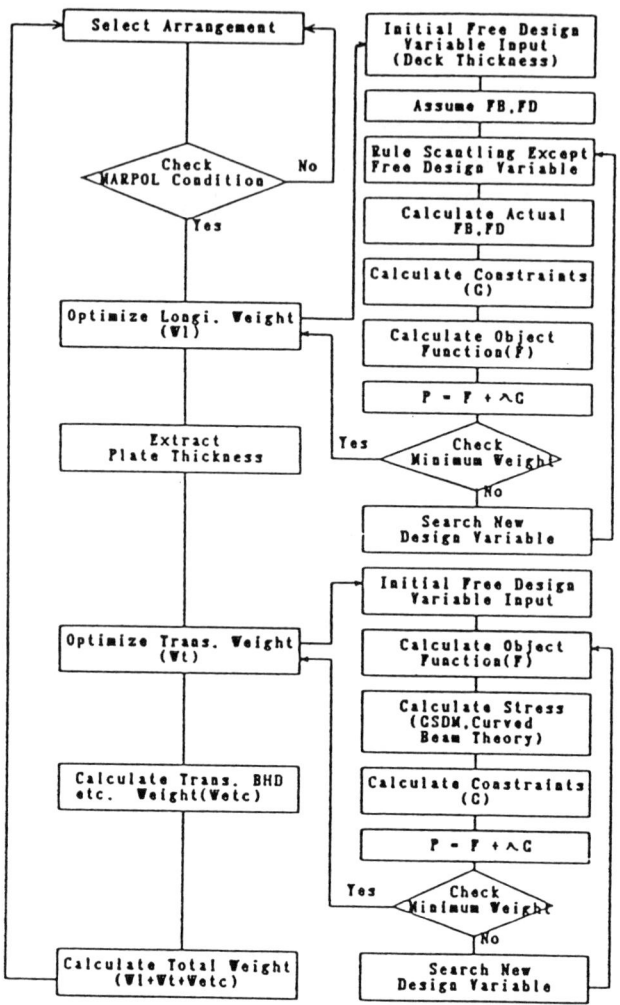

Fig. 8 Optimum Design Procedure for Tank Arrangement

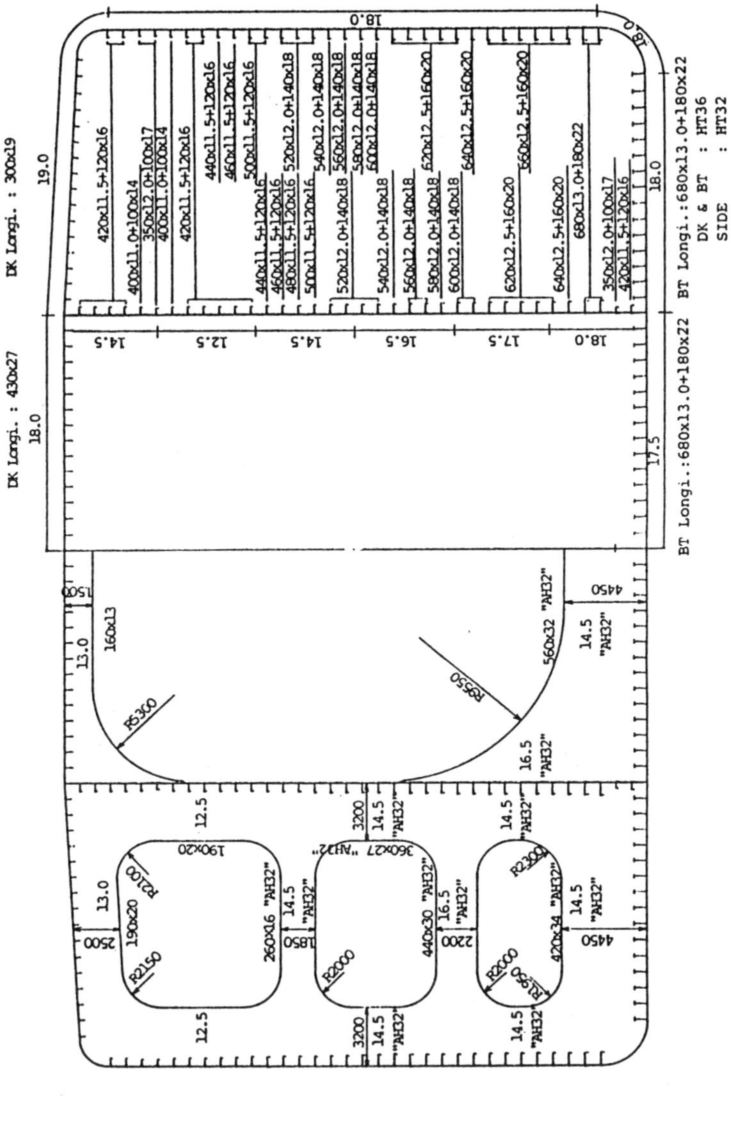

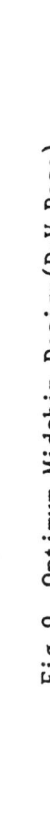

Fig. 9 Optimum Midship Design(DnV Base)

Table 2   Optimum Design Results for Longi. Member

INITIAL POINT	Fo (*E7mm3)	(Fi-Fo)/ Fi*100	ZB (%)	ZD (%)	ITER.
Xo	0.6790	-0.2	113.8	103.4	15
Xo-e	0.6786	-0.1	113.8	103.3	15
Xo-2e	0.6804	-0.4	113.9	104.0	15

where, e : convergence limit
       Xo: initial value
       Fi=0.6777E7 mm3

Table 3   Optimum Design Results for Trans. Member

STARTING POINT	DELTA	Fo (*1.E7 cm3)	(Fi-Fo)/Fi *100 (%)	ITER.	cpu (sec)
Xo	e	0.6386	-2.1	196	552
	2e	0.6162	1.5	209	584
	4e	0.6305	-0.8	161	453
Xo-2e	e	0.6107	2.4	324	904
	2e	0.6161	1.5	244	684
	4e	0.6390	-2.1	160	450
Xo-4e	e	0.6243	0.2	282	789
	2e	0.5838	6.7	285	799
	4e	0.5880	6.0	202	565

where, e : convergence limit
       Xo: initial value
       Fi=0.6256E7 cm3

Table 4   Comparison of Hullweight with Variation of Tank Arrangement

unit: Ton

	Existing Ship	ORIGIN	ALT I	ALT II	ALT III	ALT IV
Longi. Weight	13193	13211	13319	13032	13499	13577
Trans. Weight	5545	5480	5669	6033	5619	5943
T. BHD Weight	4018	3978	3527	3965	3452	3580
Cargo Hold Weight	22756	22669	22515	23030	22570	23100
Etc.	5812	5892	5892	5913	5875	5959
Hullweight	28568	28561	28407	28943	28445	29059

Statistical Analysis of Full Ships Resistance

D.Kostov, S. Kyulevcheliev(*) and S. Hongcui, J. Quinwei, G. Shenghan, F. Youzhang(**)

(*) Bulgarian Ship Hydrodynamics Centre, 9003 Varna, Bulgaria
(**) China Ship Scientific Research Centre, Wuxi, Jiangsu, China

ABSTRACT

Geometrical and experimental resistance data of 140 high-block single-screw merchant ship hull forms ($0.75 < C_B < 0.85$) tested at CSSRC and BSHC have been combined in a common data base.

Statistical analyses of these data have been performed independently at the two Centres attempting to find out 1) the hull form parameters (both integral and local) affecting most strongly the resistance and 2) the structure of this relationship for the investigated class of forms.

The paper describes the data sample and presents the two sets of statistical results and comparative analysis of them.

INTRODUCTION

Resistance prediction formulae derived by regression analysis of experimental data are widely used and are gaining further popularity as useful tools for early trade-off decisions in the generation and/or improvement of the hull form.

It is well known that the larger the amount of
data processed the wider the applicability range and
the higher the reliability of such regression
relations. That is why, in the framework of a joint
research project accomplished by the China Ship
Scientific Research Centre (CSSRC) and the Bulgarian
Ship Hydrodynamics Centre (BSHC) it has been decided
to combine the data bases of the two Centres and to
perform independent processing and analyses. Only
high-block (CB>0.75), single-screw ships at full
load draught have been subject of this project.
These are mainly bulk-carriers and tankers.
Geometrical and resistance data for 140 hull forms
have been collected, exchanged and processed.

NOMENCLATURE

A(I)   — non-dimensional prismatic curve ordinate
             at section I;
B/T   — beam/draught ratio;
CA   — correlation allowance;
CABT   — relative cross-section of bow bulb, %;
CB   — block coefficient;
CF   — frictional resistance coefficient;
CLPR   — relative length of bow bulb, %
CS   = $S/V^{2/3}$ wetted surface coefficient;
CT   — total resistance coefficient;
CVPR   = 2/3*CABT*CLPR/CB — relative volume of bulb
CW   — wave resistance coefficient;
DFBOW   = (A(18)-A(19.5))/L/B — entrance slope of
             prismatic curve ;
DFST   = (A(3)-A(1.5))/L/B — run slope of prismatic
             curve ;
DYWL   = YWL(19.5)/L/B — entrance angle of
             non-dimensional waterline;
Fn   — Froude number;
IE   — half angle of entrance at station 19.5;
IR   — half angle of run at station 0.5;
L/B   — length/beam ratio;
LCB   — longitudinal centre of buoyancy, %;
LPP   — length between perpendiculars;
LVOL   = $L/V^{1/3}$ — length/displacement ratio;
LWL   — length of waterline;
S   — wetted area, m² ;
V   — displacement volume, m³ ;
YWL(I)   — non-dimensional ordinate of main waterline
             at section I.
   (Stations are numbered from 0 at AP to 20 at FP.)

Subscripts:
M   — model
S   — ship

DESCRIPTION OF THE DATA SAMPLE

The ship resistance is represented by the model scale total resistance coefficient CT_M as a function of the Froude number Fn, i.e. the "raw material" that can be treated with different scaling procedures. The wetted area of the hull, represented by the non-dimensional coefficient CS, has been also subject of analysis, since it is equally important for accurate resistance prediction.

The geometrical parameters, agreed upon to be included in the common data base, can be classified in two groups :
- global (integral) parameters, such as principal dimensions (or their ratios), fullness coefficients, longitudinal centre of buoyancy.
- local (different) parameters, such as sectional area curve, design waterline curve, several characteristic body sections, bow bulb parameters.

Histograms of some of the main parameters are given in Fig. 1, showing their distributions and ranges of variations, and hence the applicability range of the regression formulae derived from these data.

Each of the two research teams, depending on its approach to the analysis of the data, has selected different set of parameters out of the information described above.

CSSRC ANALYSIS

Multivariate polynomial regression has been applied directly to the total resistance coefficient

For higher prediction accuracy, the experimental resistance values have been preliminary extrapolated to full-scale ship length LPP = 200 m. using the 3-D method, i.e. :

$$CT_S = (1+k)CF + CW + CA \quad , \quad\quad\quad (1)$$

where
$\quad$ CF $\quad$ — ITTC-57 friction line ;
$\quad$ 1 + k — form-factor obtained from model tests
$\quad\quad\quad\quad$ by Prohaska's method with CW assumed to
$\quad\quad\quad\quad$ be proportional to Fn_6 ;
$\quad$ CA $\quad$ — $(105(k_s/LWL)^{1/3}-0.64).10^{-3}$ $\quad\quad\quad$ (2)
$\quad\quad\quad\quad$ with $k_s = 150.10^{-6}$ — ITTC-78
$\quad\quad\quad\quad$ correlation allowance

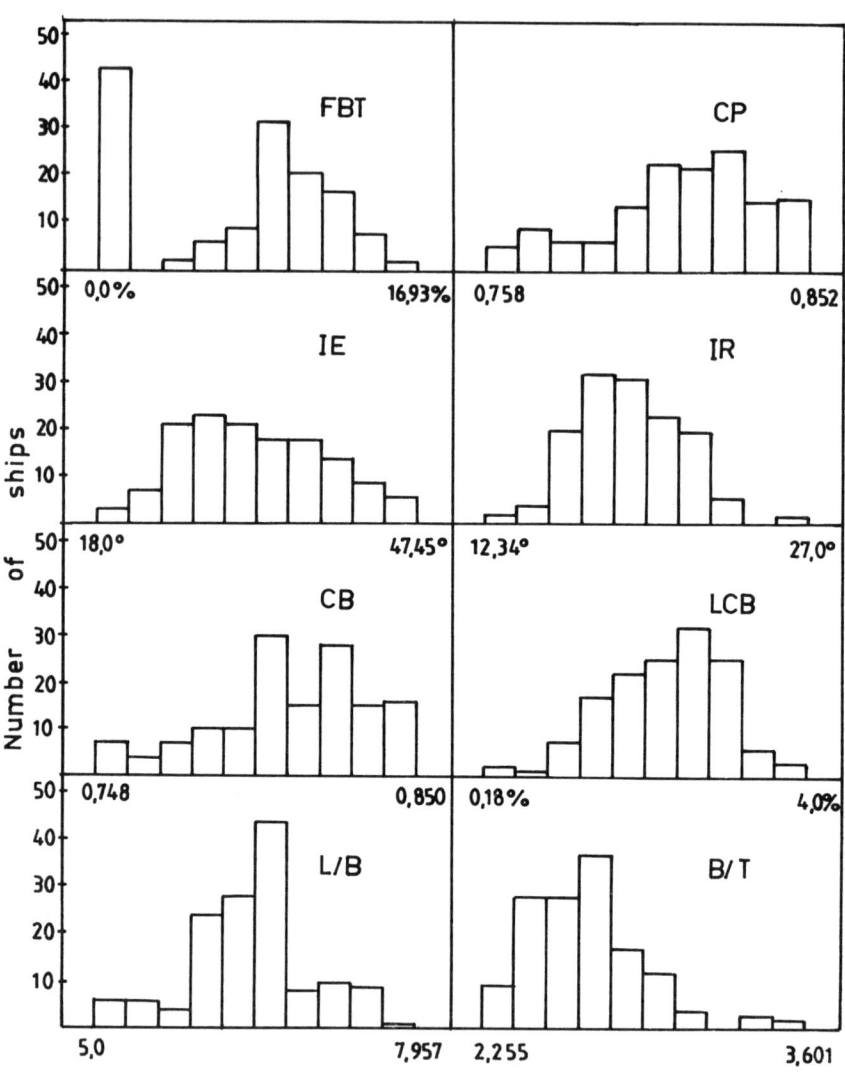

Fig. 1. Histograms of main particulars of
hulls included in the sample

The geometrical parameters assumed to have a significant influence on resistance, and included therefore in the analysis, have been selected on the basis of general physical considerations and common sense.

Thus, the regression equations are searched for in the following forms :

- total resistance coefficient of 200 m ship at one Fn

$$CT.10^3 = \sum_{0}^{N1} \sum_{0}^{N2} \sum_{0}^{N3} \sum_{0}^{N4} \sum_{0}^{N5} \sum_{0}^{N6} \sum_{0}^{N7} B_{ijklmnp} \cdot$$

$$(L/B)^i (B/T)^j CB^k LCB^l IE^m IR^n CABT^p \tag{3}$$

- form-factor

$$1 + k = \sum_{0}^{N1} \sum_{0}^{N2} \sum_{0}^{N3} \sum_{0}^{N4} \sum_{0}^{N5} \sum_{0}^{N6} \sum_{0}^{N7} C_{ijklmnp} \cdot$$

$$(L/B)^i (B/T)^j CB^k LCB^l IE^m IR^n CABT^p \tag{4}$$

- wetted surface coefficient

$$CS = \sum_{0}^{N1} \sum_{0}^{N2} \sum_{0}^{N3} \sum_{0}^{N4} D_{ijkl} \cdot (L/B)^i (B/T)^j CB^k CABT^l \tag{5}$$

In the above equations IE and IR are in radians and LCB is multiplied by 100.

The applied regression analysis algorithm can be described briefly as follows.

Firstly, the first m lowest-order terms of the corresponding equation are taken (m is the number of ships).

Secondly, the remaining column vectors are orthogonalized by means of Gram-Schmidt's method and only the first n with modulus larger than a preset control limit are left. In this way, with a control limit 0.01 less than 60 terms are usually left in the equations.

Finally, stepwise regression analysis is performed to obtain the regression coefficients. These coefficients are presented in Tables 1,2 and 3 for Equations (3), (4) and (5) respectively.

Table 1
CSSRC Regression Coefficients
of Wetted Surface Coefficient

ijkl	B_{ijkl}
0	-4.93216
1000	0.76721
100	2.82777
10	7.56256
1	70.25653
2000	0.03541
1001	-5.82957
110	-2.80653
101	-23.49146
2	37.50073
2101	0.15859
301	0.66549
22	-55.81682

Table 2
CSSRC Regression Coefficients
of Form Factor

ijklmnp	$B_{ijklmnp}$
0	1.33996
10000	-2.46858
1000	1.88809
1	-21.50229
1001000	-0.10458
1000001	1.33673
101000	-0.18206
10100	7.33308
2000	-0.19261
1100	-1.46432
1010	-0.92627
200	-9.20728
110	2.28074
101	32.42510
20	6.04899
11	22.04251
1002000	0.01559
101001	1.03658
1200	1.26057
1011	-7.26519
300	4.24423
201	-23.57922
2000101	-0.14784
1000030	-1.58231
300001	-0.12359
102010	0.11639
1120	-1.75782
3200000	0.00010

BSHC ANALYSIS

In this case the original model resistance data are treated directly. First, the equation

$$CT_M = (1+k).CF + pFn^6 \qquad (6)$$

is fitted to the whole Fn-range of each experimental curve $CT_M(Fn)$ by means of linear least-squares curve fitting.

Table 3. CSSRC Regression Coefficients of Total Resistance Coefficient

Fn = 0.10		Fn = 0.12		Fn = 0.14	
ijklmnp	$B_{ijklmnp}$	ijklmnp	$B_{ijklmnp}$	ijklmnp	$B_{ijklmnp}$
0	54.02743	0	17.83997	0	0.19929
1000000	-9.05609	10000	-42.79622	1000	2.44635
10000	-54.27757	1	11.58154	1	-59.68499
1000	-9.20978	1100000	0.24918	1001000	-0.23199
100	7.32541	1010000	-0.79862	1000100	0.70049
10	36.71400	1001000	0.05484	1000010	-0.63461
1010000	12.08646	1000100	0.48965	1000001	6.21845
1001000	0.40202	1000010	-1.00095	200000	-0.29798
1000100	-1.36629	200000	0.24211	110000	1.18803
1000010	-1.07958	110000	-4.24713	101000	-0.24081
1000001	-6.16141	100100	0.83752	100100	-1.41036
200000	2.88037	100001	-6.26620	100010	4.90382
110000	-15.05944	20000	38.25492	2000	-0.30723
101000	-0.28041	10100	-8.56024	1100	-1.62573
100100	1.92599	10010	17.19703	1001	7.75246
100010	-13.09738	2000	0.17642	101	53.23380
100001	5.94641	1100	0.85476	20	-12.03717
11000	8.77785	1010	-2.03270	11	59.62957
10001	20.18537	1001	3.15764	2	-83.44050
2000	0.28751	11	13.81975	1002000	-0.04796
1001	-4.79163	2	-160.47470	100002	59.85481
200	-3.82913	1002000	0.36100	1101	-7.30137
11	52.43489	100002	49.14479	1011	-12.07969
3000	-0.03438	1102	-2.64628	12	-231.48500
		3	-11.44897	2000101	-0.91258
		1000030	237.76040	300001	-0.43058
			-1.51608	101200	0.68159
				310	-5.36471

The assumption that Eq.(6) holds for the whole Fn-range is made since the high-block coefficient ships operate at low speeds, and Eq.(6) is exactly the present-day knowledge concept about the low-speed resistance. The validity of this assumption is proven by the high values of the coefficient of determination of the linear curve fit — it is higher than 0.90 for all experimental curves, represented with 12 – 20 points each.

The relationships of the coefficients (1+k) and p and the wetted surface coefficient CS with the geometrical parameters of the hull form have been established by regression analysis.

Table 3. (Continued)

Fn = 0.16		Fn = 0.18		Fn = 0.20	
ijklmnp	$B_{ijklmnp}$	ijklmnp	$B_{ijklmnp}$	ijklmnp	$B_{ijklmnp}$
0	0.98764	0	60.94193	0	167.45180
1	64.47050	1000000	-2.81700	10000	-421.91730
2000000	-0.04666	10000	-111.71870	1000	5.09489
1010000	0.69667	1000	4.47765	100	-16.90124
1000001	-7.81519	100	-37.51922	1	-63.19616
100100	0.36320	2000000	0.07016	1001000	-0.73546
100001	-6.10275	1010000	5.28806	1000100	0.82628
1001	-15.05418	1001000	-0.38544	1000001	10.61125
200	-3.61840	101000	-0.26507	200000	-0.35741
2	-251.29130	100010	2.37871	100010	8.00466
1002000	-0.02176	20000	35.18396	20000	265.37930
1001001	1.35696	10100	47.88946	10100	24.64495
100002	51.12622	10001	-10.64964	10001	-46.12627
3000	0.10069	2000	-0.33397	2000	-2.10080
2100	-0.48824	1010	-2.03064	1001	55.17357
2001	5.64022	20	-5.39999	110	-22.78806
1200	2.45001	11	41.15837	20	-17.64571
1101	-6.09424	101001	2.15617	11	48.75102
201	-18.48632	1101	-9.72582	2	195.01300
102	108.76430	102	111.16460	1002000	0.27997
3	708.27300	12	-195.24600	1001001	-5.89895
3000001	0.03680	300001	-0.36073	100002	-78.05501
103000	-0.01534	2003000	0.00142	3000	0.10775
3001	-0.56999			2001	-5.01570
2011	-1.96342				
1003	-202.22180				
211	31.98721				

The selection of the parameters, influencing most significantly the dependent variables 1+k, p and CS, has been performed by correlation analysis. The following parameters have been tested :
- global — CB, L/B, B/T, LVOL, LCB
- local — DYWL, DFST, DFBOW, CABT, CLPR, CVPR.

The correlation analysis has supplied a kind of statistical proof of the Taniguchi's "separability principle" :
- the form—factor 1+k, representing the viscous resistance of the hull, depends most strongly (besides on the main particulars) on the parameter of the stern DFST ;

   — the coefficient p connected with the wave
resistance of the hull depends most strongly on the
bow parameters DYWL and CVPR.

   It has been decided to develop two sets of
regression formulae :
   — "simple" formulae, including only principal
dimensions and LCB, i.e. parameters usually used at
the very initial feasibility studies and design
stages.
   — "complete" formulae, including also local
hull form parameters, used when the hull form is
being generated or is already defined.
   Stepwise, multiple regression has been
performed with critical F-value 2.5. Only terms
containing up to second degree of each independent
variables have been tested. This restriction has
been decided because of the rather limited number of
observations (140 hulls, further reduced by the
number of the rejected outliers in each specific
case), and it has been thought that a formula with
less terms would be more stable and more suitable
for prediction purposes (rather than curve fitting).

   The two sets of regression coefficients are
given in Tables 4 and 5.

Table 4. BSHC "Simple" Regression Formulae

SO		1 + k		p	
TERM	COEFFICIENT	TERM	COEFFICIENT	TERM	COEFFICIENT
1	3.22972	1	0.436512	1	8.412647
LCB	-0.484198	L/B	0.060210	CB$\ast$LCB	-28.592635
LVOL²	0.031737	B/T	0.146810	CB² $\ast$LCB	42.052192
LVOL$\ast$LCB	0.097837	LCB	0.477040	CB² $\ast$LCB²	- 1.497510
		L/B$\ast$LCB	-0.047266		
		B/T$\ast$LCB	-0.062142		

   In the case of the wetted surface coefficient
it has been tried to take account of the main
inhomogeneity of the sample — part of the hulls have
bow bulbs and the rest have not. From the bulbous

hulls wetted surface coefficient, the wetted surface coefficient of bulb has been subtracted, the latter being evaluated by an empirical formula derived earlier on other occasion:

$$CS = 0.021.[1+0.0452.(CABT)^{0.573}].$$

$$CLPR.\left(\frac{L/B}{CB^2.B/T}\right) \qquad\qquad (7)$$

SO is bulbless hull form wetted surface coefficient subjected to regression analysis.

Table 5. BSHC "Complete" Regression Formulae

SO		1 + k		p	
TERM	COEFFICIENT	TERM	COEFFICIENT	TERM	COEFFICIENT
1	-4.346125	1	-0.200673	1	905.7191
LVOL	3.135828	B/T	1.083259	CB	-2379.0477
DFST	32.246855	(L/B)²	0.008256	CVPR	2.87289
LVOL²	-0.220319	(B/T)²	-0.141865	CB²	1575.2390
LVOL*DYWL	0.216953	L/B * B/T	-0.050778	DYWL²	-2027.1760
LVOL*DFST	-5.791685	B/T*DFST	0.696456	CB*DYWL	180.6973
				CB*CVPR	-3.616093

VALIDATION ANALYSIS

Both the research teams have performed the usual and necessary analysis of variance and analysis of residuals. Due to space restrictions only a few results are presented here.

In both the analyses the residuals have been investigated statistically and proven to be normally distributed with zero mean.

The errors of the CSSRC regression model can be described by :

	Equation for		
	CS	1 + k	CT
Stand.deviation %	1.06	1.98	1.69-2.98
Relative % error at 95% level	< 2.1	< 3.96	< 5.96

The F-test ratio of all regression equations is much larger than the corresponding critical value. This is considered to be an indication for appropriate selection of the significant parameters.

In the BSHC approach the relationship resistance versus geometrical parameters is implicit. From practical point of view the accuracy of resistance prediction is important, and not simply the prediction of 1+k and p. That is why CT has been calculated in all input observation points (several points for every hull CT curve) and compared with the experimental values. Similar comparison has been made for the wetted surface coefficient :

	% of observations with error	
	$\mid CT \mid$ < 10%	$\mid CS \mid$ < 2%
Simple formulae	95.6	87.1
Complete formulae	97.0	87.8

Accuracy of 10% for CT and 2% for CS is considered to be practically acceptable.

Two hull forms, not part of the processed sample, have been tested as application examples. Their main particulars are given in Table 6. Only the so called "simple" formulae have been applied from the BSHC sets, because data necessary for the "complete" formulae have not been available.

Fig.2 show the results. In the two investigated cases the prediction accuracy of the two regression models is similar.

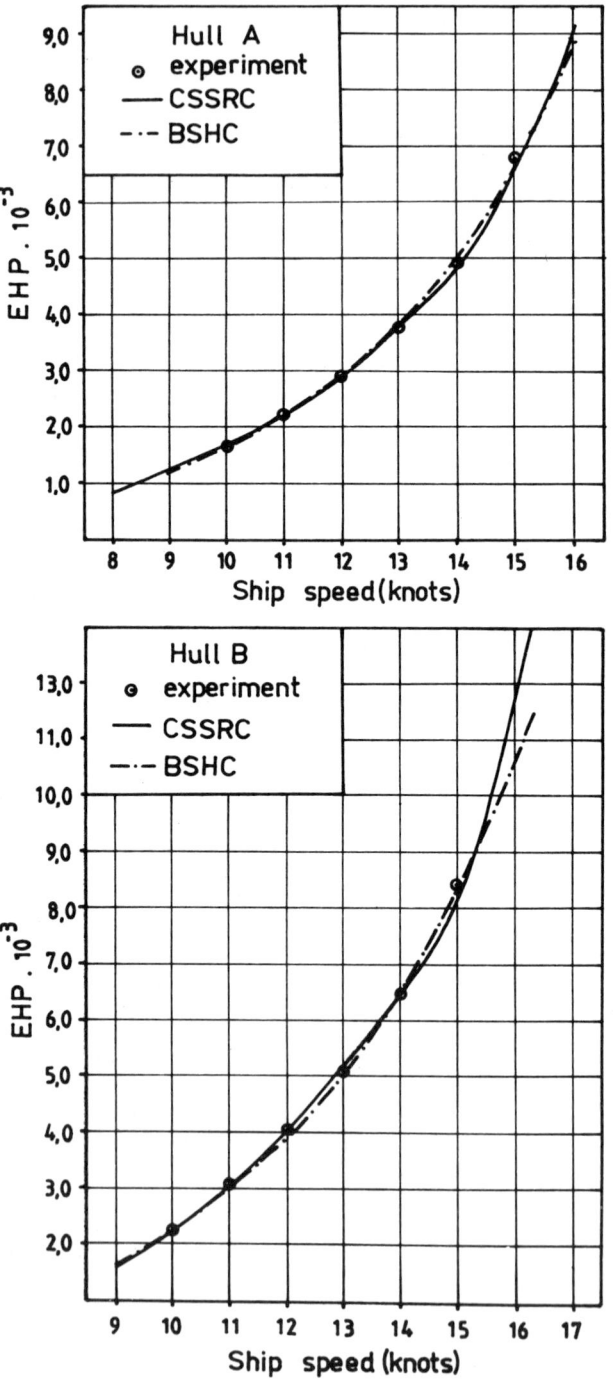

Fig. 2. Prediction performance validation

Table 6. Principal Dimensions of Validation Hull
        Forms

	Hull A	Hull B
V (m³)	40 896	72 899
LWL (m)	174.5	219.5
L / B	5.986	6.677
B / T	2.771	2.576
CB	0.825	0.842
LCB (%LPP)	2.885	2.343
IE (deg)	43.0	42.5
IR (deg)	23.0	17.5
CABT (%)	12.0	13.0

CONCLUSIONS

        Experimental resistance data for 140 high-block
hull   forms   (CB>0.75)   have   been processed
statistically.

        The CSSRC straightforward approach results in
regression   formulae   with   greater   number of
coefficients and better curve fit accuracy.

        The BSHC   model   is   characterized with a
physically justified structure of the resistance
formula   and   smaller   number   of regression
coefficients and may   be expected to have better
stability as a prediction tool.

        The derived empirical relations are applicable
for parametric studies and/or optimization at the
initial stages of ship design.

        The BSHC   "complete" formulae operate with
higher number of local form parameters and may be
therefore useful for generation and improvement of
the very form, and not only parametric studies of
principal dimensions.

ACKNOWLEDGEMENTS

        The joint research project, part of which is
reported here, has been funded by China National
Natural Science   Foundation and   the Bulgarian
Ministry of Science and Education.

A Method to Assess the Seakeeping Behaviour of a Merchant Ship in its Early Stage of Design

D. Boote and D. Bruzzone

Institute of Naval Architecture, University of Genova, Via Montallegro, 1 16145 Genova, Italy

ABSTRACT

In the early stages of ship design, the first choice of the main dimensions could be performed utilizing several methodologies available in literature. A further optimization process is then necessary and, in doing so, useful tools are furnished by standard series and data bases for the prediction of performances in calm sea. Such a philosophy may also be applied to seakeeping responses in order to predict the behaviour and the relative merit of a series of different design solutions all verifying the same contractual requirements and constraints. In this paper a procedure is presented to compare different ship designs in their early stage from the seakeeping point of view. The first phase of the proposed procedure consists in the computation of the seakeeping quantities of interest, utilizing a matrix of seakeeping data calculated for a standard series of merchant ships. In the second phase the responses are weighted on the base of the environment in which the ship will operate and a merit index, similar to that proposed by Bales, is determined. As an illustrative example, the procedure has been applied to four different classes of Ro-Ro ships.

INTRODUCTION

The design of ships is mainly based on contractual requirements. As it is well known, infinite solutions exist that satisfy a given number of starting contractual requirements. For each type of ship, the best solution may be found employing various kind of optimization criteria which are mainly based on some economical indexes. The final choice may be operated in a more or less sophisticated way, starting from the comparison of a limited number of design possibilities up to the utilization of computer codes which optimize an objective function. However the results become very difficult to satisfy the initial design requirements and, as well as, all the other imposed constraints. As a consequence, it may happen that the chosen solution is not the truly optimun one because, for instance, all the necessary parameters have not been taken into account, or because those which have been considered are not among the most significant ones. Furthermore, these kinds of procedures employ numerical coefficients, such as specific costs, which must be continuously updated and whose approximate level may often bring to little reliable results.

Usually, the seakeeping characteristics are not taken into account in the first phases of the design of a merchant ship, with the exception of the wave shear and bending moment generally estimated through empirical formulas proposed by Classification Societies. For the aforementioned reasons, a preliminary evaluation of seakeeping qualities is very difficult, but even more difficult is the task to determine its influence on the design itself. On the other hand, as the definition of the ship characteristics and the prediction of its performance in still water is tried through empirical procedures, so the same method may be attempted also for its motion performances. For instance, in the case of the preliminary evaluation of calm water resistance, use is made of empirical formulas, of statistical regressions based on a number of similar ships data or , finally, of systematic series.

In the last years, some very interesting papers have been published that deal with procedures to account for seakeeping characteristics in an early stage of ship design. Some of them are based on seakeeping calculations performed upon systematic series of hulls and results for the motions, wave bending moment and added resistance in waves are given as a function of some major ship dimensions and hull form parameters [1],[2],[3], other works are based on experiments as [4],[5],[6]. Then there are methods trying to optimize the hull form themselves, even reaching a first definition of the body plan [7]. Bales [8] proposed a procedure to define a merit index which is also followed by other authors [9],[10].[11].

The present paper may be situated in the first cited category as it is based on calculations executed for a standard series. It allows for results that, even if questionable in an absolute sense, however may be useful in comparing a number of early solutions for the main dimensional parameters. In fact some reservations can be made because the independent hull parameters considered are relatively few and the hull under consideration may have characteristics also remarkably different from those of the series.

The series which has been chosen is the BSRA [12] and the computed results may be accessible when some hull form parameters, Froude number and sea state are known. The sea state is here defined through a two parameters wave spectrum thus introducing two further independent variables: the significant wave height and the mean period. In such a way it is easy to determine the desired responses for different sea states making their long term evaluation possible. They may be weighted for each sea state using wave data relative to the considered zone.

As in the larger part of the previously cited papers, the results here reported are relative only to head seas, following the hypothesis that, in this case, one has generally the highest responses, at least among those here considered.

THE SEAKEEPING DATA SET

In this era the computers diffusion is more and more capillary and also the smallest design offices have their own personal computer and the software necessary to evaluate ship motions in a short time. In the early stage of the design it is often necessary to explore the effect of the variation of the major dimensions and hull forms parameters and this task cannot be performed systematically and in a short time using standard seakeeping computer codes. But a set of data to preliminary evaluate seakeeping performances, as those reported in the present or in other published works, is well suited to be stored in computer mass storage files and it may be quickly accessed and retrieved to have a rapid estimate of the quantity of interest for a large number of parameters or of sea state conditions even in a very simple computer environment.

As previously mentioned, the present method is based upon a lot of calculations performed for the BSRA series [13]; thus it is suitable for single screw merchant vessels with cruiser stern. The question if such a series can represent the present day hull forms, generally fitted with bulbous bows and transom sterns, may be asked. Nevertheless it must be pointed out that the aim of this work is that of giving a tool to be utilized in a preliminary stage, mainly for the comparison and the choice of the principal ship dimensions. On the other hand, similar procedures are also followed for the determination of the calm water resistance, when a preliminary evaluation of the required power is necessary. In that case too, systematic series, whose hull forms are not completely matching with those of the ship under consideration, are often used. The alternative, in some cases, consists in having recourse to simplified formulas. It is obvious however, that the class of hulls for which these results are applicable is quite well defined.

The choice of a systematic series has imposed some constraints on the number of parameters that could be considered. From the point of view of the hull geometry the choice has been restricted to the block coefficient C_B and to the L/B and B/T ratios. The ship speed was represented by its Froude number. Such a choice has been published at first by Loukakis and Chryssostomidis [2].

The sea state conditions have been represented by a two parameters wave spectrum defined by the significant wave height and the mean period. The seakeeping responses have been expressed by their root mean squares values for a ship having the standard length of 400 feet.

The calculated responses are heave, pitch, added resistance in waves, wave bending moment as well as absolute acceleration, relative motion and velocity at the forward and after perpendiculars. Synthesizing, each of the considered values has been expressed through a relationship of the type:

$$rms_k = rms_k(C_B, \frac{L}{B}, \frac{B}{T}, F_n, T_m, H_s) \tag{1}$$

where the index k refers to the k-th seakeeping response. Since C_B values considered by BSRA series vary approximately in the range 0.62 to 0.82, extrapolations have been performed to extend the hull geometry in a C_B range from 0.58 to 0.86 with a step of 0.07. The non-dimensional ratios L/B and B/T assumed the values 5.5 - 6.5 -7.5 and 2 - 3 - 4 respectively; 45 hulls have then been assessed. Froude number has been varied in the range between 0.1 and 0.3 with step 0.05.

Among the available two-parameters spectra it was assumed that expressed by the following equation:

$$S(\omega) = AB\omega^{-5} \exp\left(-B\omega\right)^{-4} \tag{2}$$

where:

$$A = 0.25 H_s^2 \qquad\qquad B = (0.817 \tfrac{2\pi}{T})^4$$

Owing to this wave spectrum formulation the responses calculations could be performed for a series of mean periods and for only one significant wave height referred in the following as "standard wave height" h_{st}. The root mean square relative to a different wave height can be easily found by multiplying the previously obtained result by the ratio between the actual and the standard significant wave height. As regards the added resistance in waves the ratio must be raised to the 2nd power.

Seakeeping calculations have been carried out by a computer code for ship motions analysis derived by the original SCORES program [14] which uses a standard strip theory and computes hydrodynamic coefficients on the basis of Lewis sections. In fig.1 are shown, as an example, the responses versus the mean period for $C_B = 0.65$, L/B = 5.5 and B/T = 3 for the various Froude numbers considered.

Since the results refer to a standard form 400 feet long, it is necessary to transfer them to actual ship length through the similitude ratio λ. Given a set of dimensions for which one of the seakeeping computed responses is desired, it is necessary, once the parameters C_B, L/B and B/T have been determined to transfer the mean wave period to the standard conditions, so it must be divided by the square root of the similitude ratio $\lambda = L/L_{400}$.

At this stage the 'standard value' of the root mean square may be searched for into the matrix of seakeeping data stored inside a direct access file; the 'standard value' so found has to be transferred to the design conditions multiplying it by the similitude ratio raised to an opportune exponent according to the considered response as shown in table I.

Table I - Similitude ratio exponents for seakeeping responses.

response	k	exponent	unit
heave	1	0	[m]
pitch	2	-1	[deg]
bend. mom.	3	3	[t*m]
add.res	4	1	[t]
accel.	5-6	-1	[m/s^2]
rel. vel.	7-8	-.5	[m/s]
rel. mot.	9-10	0	[m]

RANK INDEX ASSESSMENT

One among the aims of the present paper is to show the applicability of the data previously described through their utilization in determining a seakeeping ranking index for a given number of early defined sets of hull dimensions. As an example, this investigation has been carried out for a set of Ro-Ro ships; in fact, for this category, seakeeping responses have a remarkable influence on ship operativity for various reasons mainly relative to the high cruise speed and the consequent accelerations to which their particular type of cargo is subjected.

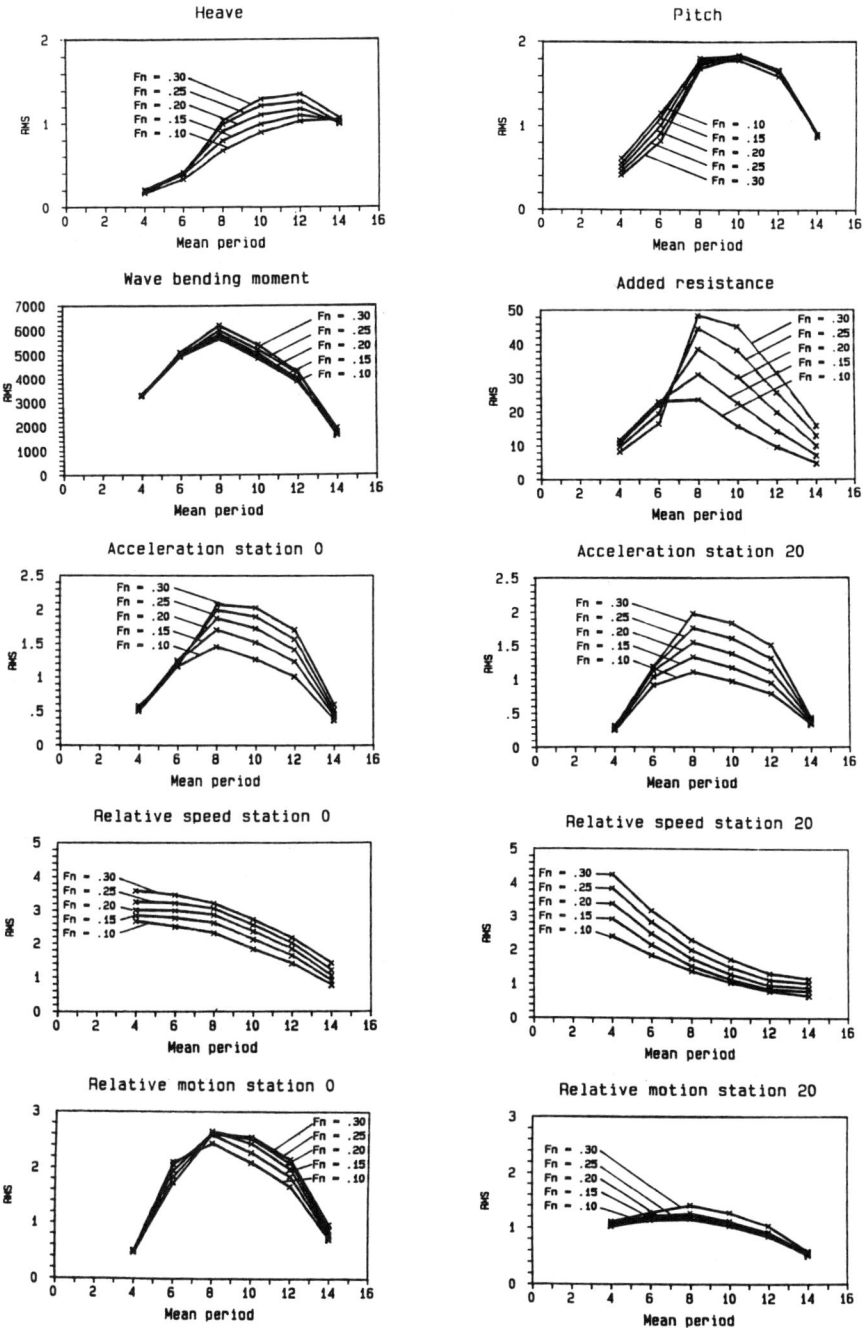

Fig. 1 - Example of responses from the seakeeping data file (C_B=0.65, L/B = 5.5, B/T = 3).

So the seakeeping behaviour directly influences the cruise speed keeping and, as a consequence, the economical efficiency and the security degree of the ship. Moreover, attention has been focused on Ro-Ro ships because, for their hull forms, they represent an extreme application for the seakeeping series we are dealing with.

Thus, the need to consider seakeeping performance among other classical design parameters, appears evident. To be able to compare different solutions, already in a first design phase, it is necessary to define a merit index, which allows for the overall dynamics characteristics in wave induced motions.

Whereas other published works [8],..,[11] consider ship categories for which the range of displacements is quite limited, in the case herein reported, this datum may assume even remarkably spread values. For this reason the investigation described in the following has been carried out for four values of the displacement which may be placed in the neighbouring of the extrema and intermediate with respect to the above mentioned range. These categories differ from each other not only for their dimensions but even for their operating field itself which, on the other hand, influences some other hull form parameters as well as the speed.

The starting data of a merchant ship design generally consist in deadweight, range and speed. Excellent procedures exist which allow to define a set of dimensions and hull parameters satisfying these requirements [15],[16]. Nevertheless the solution is not unique so one can have infinite sets of dimensions equally fulfilling the given prerequisites and constraints.

Owing to the fact that a rather strict link exists between deadweight and displacement, it was decided to perform the comparison keeping the displacement constant for each category. The displacements considered are 6000 t for small ro-ro ships, 35000 t for medium-large ships and, finally, 10000 t and 20000 t for intermediate sized ships.

For each of the four considered displacements, ten vessels with similar dimensional characteristics and a displacement nearby to that chosen for each group have been extracted from a data-base of actual ships built after 1980. Afterwards the dimensions of each ship have been normalized to the chosen displacement for the correspondent class multiplying them by the cube root of the ratio between the reference and the original displacement. In table II, III, IV and V the main normalized dimensions and parameters for the selected hulls are reported. Each group of ships does not claim to be exhaustive of the correspondent class, but they only wish to represent just a set of possible solutions for the proposed design.

At this stage, as one of the aims of the present study is the comparison among the various hulls proposed, an opportune seakeeping rank estimator must be defined. The responses which have been considered are the heave, the pitch, the added resistance in waves, the wave bending moment, the vertical absolute accelerations, the relative motions and velocities at the forward and after perpendiculars. Each of them have been evaluated by the seakeeping series described in the previous section, rather than utilizing a computer program. The employment of a computer code for ship motions calculation would have required the availability of a considerable mass of data regarding hull geometry

Table II - Main hull parameters for displacement $\Delta = 6000$ t

SHIP	Lbp	B	T	L/B	B/T	Cb
A1	101.95	18.49	5.27	5.51	3.51	0.589
A2	107.29	18.04	5.02	5.95	3.59	0.602
A3	100.54	18.23	5.25	5.51	3.47	0.608
A4	108.19	17.41	5.25	6.21	3.31	0.592
A5	116.01	16.95	5.09	6.85	3.33	0.585
A6	106.46	16.59	5.35	6.42	3.10	0.620
A7	115.91	18.51	4.68	6.26	3.96	0.583
A8	126.95	17.31	4.56	7.33	3.80	0.584
A9	107.63	18.22	5.13	5.91	3.55	0.582
A10	100.91	15.90	6.16	6.34	2.58	0.592

Table III - Main hull parameters for displacement $\Delta = 10000$ t

SHIP	Lbp	B	T	L/B	B/T	Cb
B1	127.05	21.28	6.15	5.97	3.46	0.586
B2	131.87	21.46	5.78	6.15	3.71	0.597
B3	132.49	19.68	6.43	6.73	3.06	0.582
B4	122.39	20.58	5.91	5.95	3.48	0.655
B5	137.95	19.08	5.62	7.23	3.40	0.660
B6	136.23	19.89	5.62	6.85	3.54	0.641
B7	129.36	18.78	6.05	6.89	3.10	0.664
B8	146.37	19.72	5.61	7.42	3.51	0.602
B9	137.38	21.56	5.45	6.37	3.95	0.604
B10	121.42	21.36	5.77	5.68	3.70	0.652

Table IV - Main hull parameters for displacement $\Delta = 20000$ t

SHIP	Lbp	B	T	L/B	B/T	Cb
C1	158.85	27.98	7.41	5.68	3.77	0.581
C2	138.63	21.99	9.01	6.30	2.44	0.696
C3	170.21	23.61	6.79	7.21	3.48	0.701
C4	172.33	26.04	7.35	6.62	3.54	0.580
C5	162.16	24.60	8.20	6.59	3.00	0.585
C6	158.36	23.82	7.41	6.65	3.21	0.684
C7	149.25	25.21	7.26	5.92	3.47	0.700
C8	168.32	26.27	7.32	6.41	3.59	0.591
C9	144.31	25.47	8.48	5.67	3.00	0.614
C10	164.91	25.77	7.36	6.40	3.50	0.612

Table V - Main hull parameters for displacement $\Delta = 35000$ t

SHIP	Lbp	B	T	L/B	B/T	Cb
D1	193.39	29.02	8.91	6.66	3.26	0.669
D2	175.42	28.95	10.19	6.06	2.84	0.647
D3	175.65	25.20	10.91	6.97	2.31	0.693
D4	194.58	29.92	9.87	6.50	3.03	0.583
D5	189.79	30.07	10.11	6.31	2.97	0.580
D6	156.98	27.12	11.24	5.79	2.41	0.700
D7	194.58	29.92	9.90	6.50	3.02	0.581
D8	175.75	31.53	8.90	5.57	3.54	0.678
D9	188.81	31.03	9.65	6.09	3.22	0.592
D10	170.36	27.36	10.42	6.23	2.63	0.689

and weight distribution; as a consequence a quite heavy labour would have been required. On the contrary, the seakeeping series allows a quick estimate of the response under consideration, having at one's disposal only the main geometric characteristics of the ship.

The environment has been taken into account by referring to a two-variate frequency distribution as those reported by wave atlas. The wave statistics may be handled at different levels of sophistication; for instance it would be useful to adopt, for each class of ships, a set of frequency histograms for various Marsden squares according to the main route the ship is designed to operate. In this work a two-dimensional histogram relative to a large area, as North Atlantic, has been assumed; this solution has been judged acceptable owing to the comparative nature of this design stage.

From the original two-variate tables, reported in [17], a simplified bi-dimensional histogram has been derived by reducing the variability range of the wave height and mean period, so disregarding the extreme values for each variable with low occurrence frequencies.

A matrix of w_{ij} values has been created by which each particular response relative to a given H_j, T_i couple has to be weighted, according to the relationship:

$$rms_k = \sum_{j=1}^{N_h} \sum_{i=1}^{N_T} w_{ij} rms_i \frac{H_j}{h_{st}} \tag{3}$$

where:
k is the index relative to k-th response;
N_h is the number of wave heights;
N_T is the number of wave mean periods;
h_{st} is the 'standard wave height.

In table VI the weighting factors w_{ij} considered in the numerical example are presented.

Table VI - Percentage of occurencies for each H_j,T_i

		Wave periods (s)							
		5.5	6.5	7.5	8.5	9.5	10.5	11.5	12.5
W	0.5	0.13	0.22	0.13	0.03	0.00	0.00	0.00	0.00
a	1.5	1.65	2.89	2.40	1.14	0.34	0.05	0.02	0.00
w	2.5	1.50	5.87	7.67	5.70	2.32	0.62	0.15	0.00
e	3.5	0.54	3.46	4.29	7.78	4.42	1.63	0.44	0.10
	4.5	0.13	1.41	1.88	5.75	4.22	1.95	0.62	0.17
h	5.5	0.05	0.49	0.77	3.16	2.84	1.58	0.59	0.18
e	6.5	0.00	0.15	0.27	1.53	1.58	1.04	0.44	0.17
i	7.5	0.00	0.07	0.10	0.69	0.82	0.59	0.29	0.13
g	8.5	0.00	0.00	0.07	0.32	0.40	0.35	0.18	0.08
t	9.5	0.00	0.00	0.00	0.15	0.23	0.20	0.13	0.05
h	10.5	0.00	0.00	0.00	0.07	0.13	0.10	0.08	0.03

By the described procedure a weighted average value has been determined for each of the N_r considered responses relative to each ship.

At this point, in order to define a plausible rank index, partially following the methodology proposed by Bales, each one of the N_r responses has been normalized multiplying its reciprocal by the value relative to the ship for which the response presents a minimum. After this operation all the responses of a ship have been added together weighting them by an opportune factor α_k up to obtain an index R defined as follows:

$$R = \sum_{k=1}^{N_r} \alpha_k rms_k \tag{4}$$

The factor α_k is a real number varying between zero and one and may be chosen by the designer according to the importance he want to attribute to each response of the ship; this choice may be based on the experience of the designer himself. As an example, in the present work, α_k has been assumed equal to one everywhere, with the exception of the bending moment for the small ships category, where α_k was assumed equal to zero. In this case in fact, owing to the small lengths involved, the longitudinal strength is not a problem.

ANALYSIS OF RESULTS

The applicability of the seakeeping data set utilized in this study, has been already veryfied for some sample ships of different kinds in [18]. A further comparison between the responses retrieved from the series and those obtained by the direct application of a seakeeping computer code, has been performed for three Ro-Ro ships whose geometrical characteristics are in the range taken into consideration in this numerical example. The main dimensions of these ships are resumed in table VII while the results of the comparison are plotted in the diagrams of fig.2,3,4. It may be noted that an acceptable agreement exists between the trends of the two kinds of results, although the hull characteristics of the sample ships noticeably differ from those of the series.

Table VII - Main characteristics of the test ships.

Ship	L_{bp}	B	T	D	L/B	B/T	C_B
A	96.0	16.50	5.66	6228	5.82	2.92	0.678
B	163.6	27.00	9.50	25150	6.06	2.84	0.585
C	198.0	32.00	10.72	41645	6.19	2.99	0.598

Following the procedure described in the previous section, for each of the four groups of ships the relative indexes have then been computed for a fixed value of the velocity considered consistent with the displacement, on the basis of the collected data. The results are shown in the tables VIII, IX, X and XI. In these tables the weighted averaged responses by the statistics of the considered zone are reported together with the resulting rank indexes. They are, respectively, the heave, the pitch, the bending moment, the added resistance, the accelerations, the relative velocities and the relative motions at the forward (0) and after (20) perpendiculars and, finally, the rank index R.

Owing to the index definition, the seakeeping behaviour is better for higher values of the index itself, thus it results immediate to individuate the best solution among those considered for each group. In this case they are, respectively, the ship A8 for the displacement of 6000 t, the ship B5 for 10000 t, the ship C3 for 20000 t and the ship D1 for 35000 t. In addition, even if a parametric analysis was not the purpose of this work, the designer has the possibility to get interesting information from the results obtained by the procedure both in terms of global seakeeping qualities and for single responses. For example, in the present application, it seems evident the connection between the ratio B/T and the seakeeping index R, in the sense that to the highest values of the index generally correspond higher values of B/T, and to the lowest values of R, lower values of B/T correspond, with fluctuations in the intermediate fields. This fact is remarked for all the displacements considered. For what the remaining parameters are concerned, it is difficult to detect their influence on the index owing to the random nature of ships chosen for this numerical example; a parametric analysis would then be advisable.

The proposed procedure allows the designer both to compare even a large number of equivalent design solutions at his disposal, as shown in this paper, and to perform parametric investigations by systematically varying the main dimensional parameters of the starting design.

CONCLUDING REMARKS

In the present paper some calculations have been presented where the seakeeping responses systematically computed for a standard series have been used to evaluate rank indexes among a number of an early defined sets of ship dimensions and parameters. The data in the series are stored into a direct access file which can be quickly accessed so a large quantity of responses may be obtained in a very short time even by a small computer. This allows performance of calculations for a lot of environmental conditions and some ship dimensions. The environmental conditions are defined through the main period and significant wave height so the response may be evaluated knowing these two parameters and ship geometric characteristics. The data herein presented are to be used in the early stage of the design since they give a rough estimation of the desired seakeeping response in analogy with what is done for an early evaluation of still water performance by sistematic series or regressions formulas.

In the present application an example has been given where the best solution from the seakeeping point of view, is found among a number of ship dimensions randomly chosen. However the present methodology may be applied to study the effect of sistematic variations of hull forms parameters either for global seakeeping behaviour of the ship or for a single response in which the designer is particularly interested.

The purpose of this work has been that to show a possible application of the seakeeping data set just to choose the solution for the major ship dimensions. Herein only the seakeeping characteristics have been considered so it may happen that the best solution from this point of view may have not an optimal calm water performance. The best overall solution might be chosen as a compromise giving to the seakeeping behaviour the weight the designer believes most opportune. It would be advisable, as a further development, to implement the seakeeping file by extending the range of some variables and by taking into account additional parameters, as the wave direction.

Table VIII - Average responses and rank indexes for $\Delta = 6000$ t

SHIP	Heave	Pitch	B.M.	Add.Res.	Acc.0	Acc.20	Rel. Speed0	Rel. Speed20	Rel. Mot.0	Rel. Mot.20	R
A1	.866	1.568	1953.	20.789	1.832	1.377	2.366	1.331	1.992	.755	8.257
A2	.820	1.531	2379.	20.028	1.839	1.345	2.374	1.330	2.005	.772	8.340
A3	.862	1.595	2042.	20.541	1.827	1.412	2.359	1.363	1.958	.782	8.201
A4	.848	1.564	2211.	19.824	1.904	1.389	2.428	1.338	2.077	.787	8.162
A5	.800	1.492	2533.	18.544	1.911	1.322	2.444	1.326	2.123	.794	8.322
A6	.834	1.596	2279.	18.731	1.903	1.452	2.404	1.367	2.017	.823	8.147
A7	.757	1.424	2814.	19.393	1.817	1.223	2.377	1.297	2.041	.748	8.639
A8	.692	1.337	3351.	17.242	1.798	1.168	2.372	1.320	2.080	.819	8.831
A9	.846	1.540	2216.	20.594	1.876	1.345	2.415	1.320	2.069	.760	8.242
A10	.986	1.712	1577.	19.734	1.959	1.606	2.485	1.487	2.137	.940	7.535

Table IX - Average responses and rank indexes for $\Delta = 10000$ t

SHIP	Heave	Pitch	B.M.	Add.Res.	Acc.0	Acc.20	Rel. Speed0	Rel. Speed20	Rel. Mot.0	Rel. Mot.20	R
B1	.774	1.245	3746.	25.062	1.572	1.159	2.316	1.369	2.044	.799	9.175
B2	.722	1.199	4487.	24.470	1.545	1.107	2.302	1.366	2.023	.800	9.196
B3	.791	1.276	3791.	23.750	1.675	1.210	2.401	1.378	2.163	.827	8.949
B4	.718	1.264	4359.	23.502	1.491	1.201	2.237	1.429	1.872	.859	9.161
B5	.626	1.141	5539.	20.041	1.457	1.101	2.211	1.410	1.899	.896	9.422
B6	.646	1.155	5362.	21.329	1.479	1.098	2.235	1.395	1.927	.867	9.344
B7	.691	1.243	4666.	20.965	1.527	1.216	2.255	1.435	1.917	.899	9.147
B8	.642	1.111	5482.	20.900	1.532	1.048	2.287	1.372	2.052	.861	9.345
B9	.663	1.137	5243.	23.468	1.505	1.042	2.268	1.353	1.992	.803	9.357
B10	.710	1.247	4447.	24.071	1.468	1.167	2.221	1.413	1.852	.830	9.248

Table X - Average responses and rank indexes for $\Delta = 20000$ t

SHIP	Heave	Pitch	B.M.	Add.Res.	Acc.0	Acc.20	Rel. Speed0	Rel. Speed20	Rel. Mot.0	Rel. Mot.20	R
C1	.641	.884	7795.	31.654	1.165	.855	2.133	1.386	1.932	.801	8.905
C2	.765	1.151	5943.	28.204	1.214	1.224	2.164	1.682	1.883	1.207	8.071
C3	.503	.795	10834.	24.006	1.030	.828	2.018	1.449	1.748	.934	9.359
C4	.614	.856	8933.	29.330	1.205	.851	2.156	1.387	2.009	.839	8.817
C5	.692	.945	7172.	29.143	1.269	.959	2.213	1.416	2.053	.856	8.570
C6	.581	.901	9042.	25.958	1.119	.932	2.087	1.468	1.809	.925	8.880
C7	.597	.921	8549.	28.185	1.074	.936	2.060	1.488	1.742	.921	8.878
C8	.613	.865	8878.	29.464	1.188	.857	2.147	1.397	1.971	.840	8.828
C9	.730	1.027	6226.	31.195	1.231	1.036	2.195	1.487	1.954	.914	8.443
C10	.606	.878	8937.	28.658	1.173	.877	2.137	1.414	1.929	.858	8.821

Table XI - Average responses and rank indexes for $\Delta = 35000$ t

SHIP	Heave	Pitch	B.M.	Add.Res.	Acc.0	Acc.20	Rel. Speed0	Rel. Speed20	Rel. Mot.0	Rel. Mot.20	R
D1	.471	.648	15624.	28.750	.855	.702	1.935	1.436	1.719	.910	9.462
D2	.586	.776	11661.	31.786	.929	.823	2.023	1.517	1.796	.980	8.768
D3	.625	.831	10809.	28.453	.935	.928	2.021	1.617	1.801	1.147	8.536
D4	.590	.699	12246.	32.674	.966	.752	2.025	1.420	1.907	.865	8.965
D5	.611	.715	11360.	33.250	.965	.764	2.032	1.433	1.905	.873	8.925
D6	.670	.906	8846.	31.342	.906	.964	2.025	1.702	1.744	1.235	8.442
D7	.594	.701	12113.	32.782	.968	.754	2.026	1.420	1.910	.864	8.957
D8	.517	.705	13803.	33.039	.842	.728	1.963	1.485	1.689	.911	9.186
D9	.577	.698	12334.	33.636	.931	.735	2.011	1.435	1.853	.872	9.024
D10	.607	.814	10944.	30.540	.908	.879	2.008	1.589	1.757	1.089	8.652

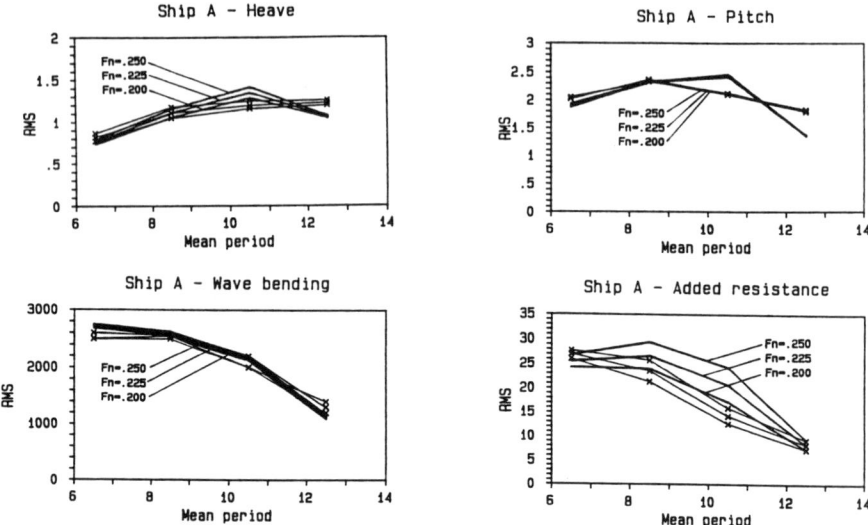

Fig. 2 - Ship A -Comparison between SCORES responses (thin lines with crosses) and series responses (bold lines) for three different Froude numbers.

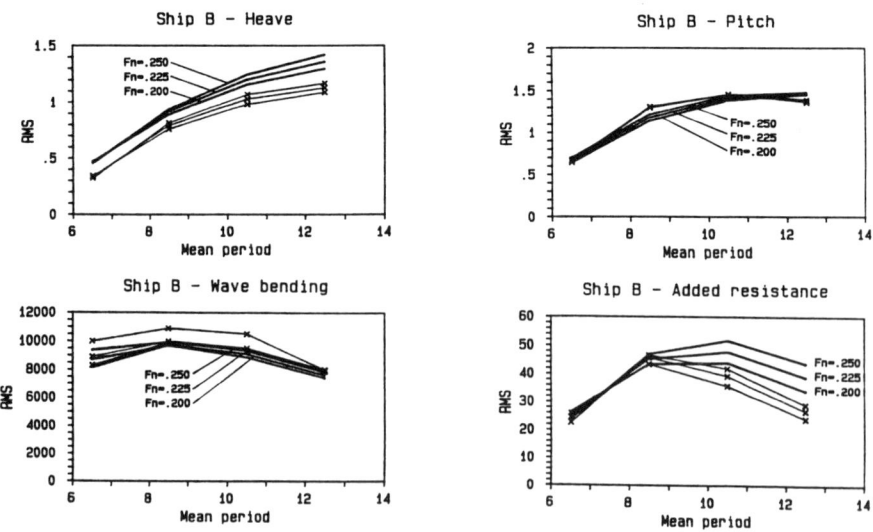

Fig. 3- Ship B -Comparison between SCORES responses (thin lines with crosses) and series responses (bold lines) for three different Froude numbers.

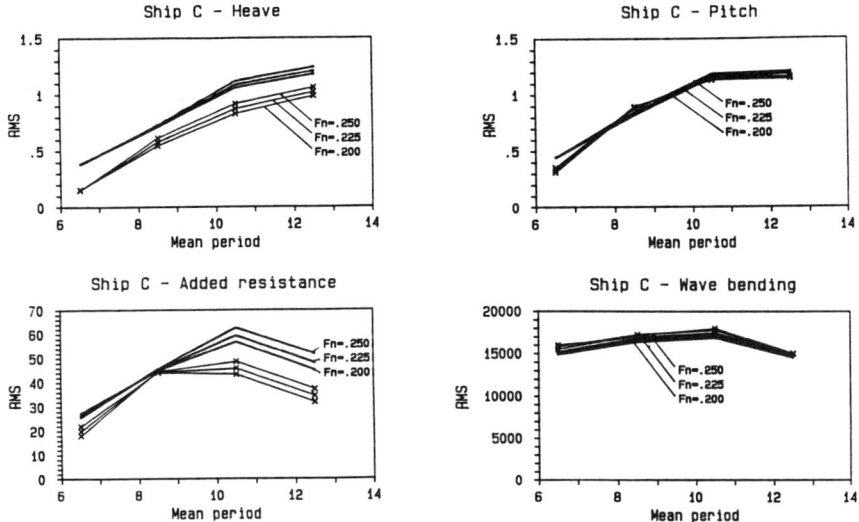

Fig. 4 - Ship C -Comparison between SCORES responses (thin lines with crosses) and series responses (bold lines) for three different Froude numbers.

REFERENCES

1. Bales N.K. and Cummings W.E., The Influence of Hull Forms on Seakeeping, Transactions SNAME, Vol. ,pp. 49-87, 1970.

2. Loukakis T.A. and Chryssostomidis C., Seakeeping Standard Series for Cruiser Stern Ships, Transactions SNAME, Vol. 83, pp. 67-125, 1975.

3. Wilson P.A., A Seakeeping Analysis of a Family of Merchant Ships, CADMO 1986.

4. Moor D.I. and Murdey D.C., Motions and Propulsion of Single Screw Models in Head Seas, Transactions RINA, Vol.110 ,pp. 403-446, 1968.

5. Moor D.I. and Murdey D.C., Motions and Propulsion of Single Screw Models in Head Seas, Part II, Transactions RINA, Vol.112 ,pp. 121-164, 1970.

6. Murdey D.C., An Analysis of Longitudinal Bending Moments Measured on Models in Head Waves, Transactions RINA, pp. 221-240, 1972.

7. Gregoropoulos G.J. and Loukakis T.A., On the Optimization of Hull Forms with Respect to Seakeeping, 5th IMAEM Congress, Athens, 1990. 8. Bales N.K., Optimizing the Seakeeping Performance of Destroyer-Type Hulls, 13th Symposium on Naval Hydrodynamics, pp. 479-503, Tokyo, 1980.

8. Bales N.K., Optimizing the Seakeeping Performance of Destroyer-Type Hulls, 13th Symposium on Naval Hydrodynamics, pp. 479-503, Tokyo, 1980.

9. Wijngaarden van A.M., The Optimum Form of a Small Hull for the North Sea Area, International Shipbuilding Progress, Vol. 31, pp. 181-187, 1984.

10. Zborowski A. and Sainsbury S.R., Small Vessel Hull Form Optimization for Heave and Pitch Performance, Marine Technology, Vol.25, pp. 293-303, 1988.

11. Nabergoj R. and Cipollini M., Ottimizzazione delle forme di carena dei pescherecci di altura in funzione della tenuta al mare, Tecnica Italiana, pp. 121-145, 1989.

12. Moor D.I., Parker M.N., Pattullo R.N.M., The BSRA methodical Series-An overall presentation, Transactions RINA, Vol. 103, 1961.

13. Bruzzone D., Caratteristiche di tenuta al mare per una serie sistematica, Proceedings NAV82 Congress, Naples, 1982.

14. Raff A.,Program SCORES-Ship Structural Response in Waves, Ship Structures Committee SSC-230, 1972.

15. Munro-Smith R., Merchant Ship Design, Hutchinson of London, 1964.

16. Watson D.G.M. and Gilfillan A.W., Some Ship Design Methods, Transactions RINA, Vol 119, pp 279-324, 1977.

17. Hogben N., Dacunha N.M.C. and Olliver G.F., Global Wave Statistics, British Marine Technology Limited, Feltham, 1986.

18. Bruzzone D. and Ferrando M., Influenza di alcuni Parametri Caratteristici di Carena sulla Tenuta al Mare, Inst. of Naval Architecture, Report n.8606, Genova, 1986.

Shipbuilding Software Technology for the Nineties

G. Marshall, W. Horsham and D. Catley

BMT CORTEC Limited,

Wallsend Research Station,

Wallsend, Tyne and Wear, NE28 6UY, U.K.

1   INTRODUCTION

The Froude Tank of the National Physical Laboratory was founded in 1911 and the The British Shipbuilding Research Association was founded in 1947. These two major U.K. Institutions were amalgamated in 1985 to become British Maritime Technology, now known as The BMT Group. It was not until the late 1950's that any serious attempt was made to use computer systems to automate the Naval Architecture and Marine Engineering routine calculations. The first computers used were very primitive by today's standards and had for input and output either paper tape or punched cards. It is understandable therefore that no real progress was made until the introduction of batch systems and dial up time-sharing systems in the mid 1960's. As the routine calculations were programmed and the user manuals written the software was made available to the British shipbuilding industry as a bureau service. This continues today, but to a limited extent, since most users have their own computer hardware on which the BMT CADCAM modules are used extensively.

In the 1970's and 1980's a decision to acquire a CADCAM system often involved the expenditure of large sums of money and major changes in organisation and methods. It was vital to consider carefully the benefits involved. The simple purchase of a computer-based system does not, of itself, solve problems. Effective use requires proper training of all appropriate staff and shop floor workers, and a proper integration of the management, computer and production functions. None of this was achievable quickly or cheaply, indeed the set up costs could in those early days far exceed the cost of the software, but the medium to long term benefits could be immense.

To be effective, the potential benefits to be realised must include:-

- a quicker response to requests for quotations and tenders comprehensive, accurate, presentable documents, and the ability to investigate modifications rapidly;

- an improved communication of information, e.g 3-D views and accurate readily accessable data;

- an effective management of design - resource planning, use of design standards, improved cost control;

- better designs through rigorous evaluation and selection and the provision of comprehensive functional analysis;

- increased productivity in the design office, quicker modifications, and elimination of tedious manual and repetitive calculations;

Software is one of the easiest form of technology transfer, and users need little knowledge of the contents of the "black box", but should have the ability to understand and interpret the results. The greatest requirement of software is that it be user-friendly and have a common external presentation or "house style".

The first release of BRITSHIPS was in 1972 [Reference 1]. This system was integrated to a limited extent, since it provided facilities for data transfer from the BSRA software into the IBM CADAM system. The software is continually being enhanced to encapsulate the latest technical innovations in computer hardware and software. This paper is intended to describe this evolutionary process, with emphasis on the impact of changes in analytical techniques, the increased power of computing hardware, plus system design technology, data handling techniques and the latest releases of the various systems.

The availability of computer graphics at affordable prices had a major impact on the utilisation of computers, especially in geometric visualisation and design, and the management of manufacturing drawings. Consequential increase in the size and complexity of the digital ship models being processed necessitated changes in software design to replace the simple input/output of serial files with random access datastores, and new interactive user interfaces were introduced to made systems more user-friendly. These features were used as a basis for the second release of BRITSHIPS [Reference 2] to give a computer graphics system for the definition of ship structural steelwork. None the less, the application programs tended to develop separately and piecemeal as the software tools and appreciation of the technology were utilized to a maximum extent by different groups of staff in a particular application areas.

As the software systems increased in size and complexity, leading to a larger number of programs, the management of system development and maintenance became increasingly difficult and more costly. There were two options available: continue modification of existing software to give the required improvements, or a re-design of the systems using the best available system engineering principles and quality control procedures. The first course of action was easily the cheaper in the short term, the consequential cost was falling behind technically; the second required a major commitment by BMT and a large investment of effort. After extensive discussions and deliberations the latter course of action was selected and has resulted in the development of the current software systems. The most important features of the BMT systems today may be summarised as follows:

- The system recognises and reflects the essential concepts of the ship design and detailing process.

- The system is based on a hierarchical principle where the design and manufacturing processes are laid out in a natural succession. At each hierarchical level the methods and technology employed are compatible and consistent.

- Each hierarchical level may contain one or more functions and each function is performed by a separate module.

- Each module operates in a well defined environment with its own datastore and can be used both as a part of the total system and as a stand-alone module. Integration of modules within the system environment is performed through the datastore accession routines.

- The user interface to the modules is user selectable depending upon the skill and familiarity of the user with the system.

- Each module has the same self-contained structure containing command, functional routines and graphics libraries with its own datastore and access routines.

- On-line help information is available.

- Each module is designed to be "utilitarian" instead of "totalitarian", i.e. the system supplies capabilities to the user, but decision making is left to the user. The system assists in the decision making process by indicating alternatives and by providing technically and economically acceptable solutions, but the user selects the suitable course of action.

- Where appropriate the modules are interactive, and maximum use is made of computer graphics to improve the user/machine communication.

It is believed that adoption of the principles outlined in this paper has led to the development of a disciplined and yet flexible system where replacement of modules and hence inclusion of new technology poses no serious problems. Furthermore, utilisation of organised datastores allows for integration of other third party proprietary software by the use of relatively straight forward interface routines and datastore access routines. The modular design of the system facilitates integration with a wide range of software and enables customisation without major expenditure.

This paper is a summary - a more detailed version is available from BMT.

2   SYSTEM DESIGN CONSIDERATIONS

2.1   General

During the late 60's and 70's the Shipbuilding Industry witnessed the development of a number of computer-aided design and detailing systems, including AUTOKON [3], STEERBEAR [4], SICEN [5], FORAN [6] as well as earlier versions of the BSRA developed system. With the exception of conventional hydrostatics and stability analysis, none of these systems contained design and analysis capabilities and the word 'design' was used mainly within the context of draughting. To match the design process used in current shipbuilding methods, the applications developed by BSRA required a change in methods and technology. The advancement of software engineering further facilitated this development.

The first consideration was the establishment of a rational interpretation of ship design and manufacturing processes and how the software to be developed would contribute to the outputs required by the shipbuilder. A simple overview of the ship design and construction process can be seen in Figure 1. So far as design and detailing is concerned, it is possible to distinguish three activity areas:-

● **Configuration design**

This phase essentially encapsulates pre and post contract design plus operational design. All of this activity is focused on the design office team. Pre-contract design comprises the evaluation of a wide range of alternative solutions to meet the shipowners' requirements. Post-contract design picks up from where the pre-contract design finishes and results in a final definition of the hull form with accurate predictions of its characteristics such as hydrostatics, loading conditions and powering with the prime objective of ensuring that contract terms will be achieved. The operational design then details the physical arrangement of the ship and its major components in terms of system diagrams, etc., such that the vessel will be able to meet its contractual performance requirements.

● **Engineering design**

This stage of the vessel design is primarily a drawing office activity concentrating upon complying with the Classification and Production requirements, i.e. engineering the major aspects of the design ready to commence product design. At the end of this stage adequate detail is available to progress into product design. The Classification design establishes a structural arrangement for each of the main areas of the ship with scantlings and ensures a scheme for Product Work Breakdown Structure (PWBS) is detailed, see Ref. 7. The Transitional design prepares all the detail for the production of conventional two dimensional working drawings.

● **Product Design**

The work is predominantly drawing office based with the single objective of creating production information consisting of workstation drawings, parts lists, and parts definition. Detailed design is commenced on each of the four major areas of hull, outfit, engineering, and electrical to produce parts definition, pipe geometry, and switchboards/consoles. Manufacturing information is likewise produced for each functional area resulting in parts lists, material requirements, numerical control tapes, and nesting drawings. Assembly and installation information for work packages and zones are the final output.

This potted version of the ship design and detailing process is important to appreciate how the application software is used to help the shipbuilder. Whilst the exact details of the process will vary from one ship production facility to another, the general principles are similar. A more detailed discussion of each design level is given below together with identification of the BMT CADCAM software modules which assist in the automation of the various processes (in parentheses {}). Other processes are sometimes automated using general engineering packages for two, and three-dimensional draughting, word processors, spreadsheets, etc. Additionally, other engineering software, for example finite element, or Classification requirements, are used in conjunction with the BMT CADCAM to supply all the needs of the shipbuilder.

2.2   Configuration design

The three sub-stages introduced in Section 2.1 are discussed below:

2.2.1 Pre-Contract design

The conventional design office carries out work in three functional areas:

Hull

The principal activities lie in the creation of the General Arrangement {CODES, COMPGEN}, the midship-section, the hull form {STARTER, CODES}, space allocation {CODES, COMPGEN, and SFOLDS}, accommodation arrangement, weight calculations {CODES and SFOLDS} and hull specifications. Figure 2 shows the use of surface manipulation points to control the form generated in HULLGEN.

Engineering

The essential calculations are those such as powering {STARTER, CODES, and SEHAM}, the preliminary machinery arrangement, preliminary main system diagrams, preliminary heat balance, auxiliary machinery loading's, machinery particulars and specification.

Electrical

The main requirement such as electrical loading for generator sizing {ELOAD}, short-circuit loading {SCIRC}, and the electrical specification are concentrated upon at this level.

From the authors' practical observations of this stage the conceptual design process is characterized by an ill-formulated array of activities superimposed on a general sense of direction. The practitioner will no doubt recognise this description. The process can be thought of as 'wicked' according to the definition of 'Rittel', see Ref 8 which contains a fuller discussion of this subject. The bottom line is that the tools used must allow multi-starts for alternative designa and with varying levels of available information.

2.2.2 Post Contract Design

Following on from the pre-contract work the designers continue to firm up the details to give a final hull form and arrangement:

Hull

The hull form characteristics and form are finalised {SFOLDS, B-LINES} and a faired lines plan is produced {B-LINES and/or HULLSURF}. For plate ordering purposes the shell expansion is derived {HULLSURF, SHELLDEF, BRITSHELL} complete with a total definition of all the seam and butt positions and stern frame/rudder particulars, etc. To ensure the ship will meet its contractual obligations the design is analysed for trim and stability, longitudinal strength, damaged stability, probability of survival, {SFOLDS}, seakeeping {SEDAS}, dynamic positioning {DPDAS}, and other important performance related characteristics.

Engineering

Together with the final hull form the final machinery, control room and shaft arrangement {SHAFT} are determined.

2.2.3 Operational Design

With a finalised hull/arrangement, and in order to progress to the next stage, the ship designers need to consider the on-board systems and to create system diagrammatics and materials/equipment lists. The latter being used for purchasing.

Hull

Arrangements for mooring {MODAS}, deck machinery, and all items of equipment are required, together with material lists.

Engineering

System and control loop diagrams are to be specified, together with, pumping plans {FABS} and material and equipment lists.

Electrical

The electrical equipment arrangement together with main cable diagrams {ELECTRO}, control loop diagrams, instrumentation, and equipment requirements are to be determined.

2.3. Engineering Design

The attributes of the design are to be assessed in order that the vessel built will comply with the design parameters and in order that the vessel can be production engineered. The two stages introduced in section 2.1 are now discussed below:

2.3.1 Classification Design

Hull

The overall structural requirements are to be assessed and the major structural plans which are required for Classification Society approval are produced, for example the profile and decks, detailing the scantlings of all the major and minor structure such as frames, girders and transverses {MINT and STRUC}. The scantlings will be assessed from programs such as LRPASS (Lloyds Register of Shipping) and the two-dimensional drawings will be created using a proprietary draughting system, such as CADDS4X (ComputerVision). Direct interfacing of the ship structure modelling system and the draughting systems have been sought as a premium and have provided a major advantage when using BMT software - this is discussed in more detail in Section 3.

Engineering

The major composite systems require to be engineered into the hull and internal definition { MINT and STRUC} using the preliminary system diagrams: module/pipe bank arrangements; floorplate arrangements; module/block interfaces; machinery and equipment diagrams; etc. Interference checking is carried out using the structural model and the details of the siting for each of the composite systems in the draughting system.

Electrical

The preliminary arrangement diagrams are to be overlaid with detailed cable routeing {ELECTRO}.

2.3.2 Transition from Design to Production

The detailed area structural drawings (e.g. machinery space, cargo space and fore end) are to be prepared giving preliminary datums, Classification drawings, material lists and requirements, holes, pads and penetrations, such that materials can be requisitioned and workstation plans are made available for full detailing of each of the main functional groups.

2.4 Product Design

There are three principal sub-stages identified as detailed design, manufacturing information, and assembly/installation information which are in four major areas - hull, electrical, engineering and outfit. The prime emphasis is on producing all the final workstation drawings, complete with production information to allow fabrication and assembly. Importantly, dimensional checking should be carried out to assure quality standards.

Hull

The final definition of all the parts of the structure are required, together with NC tape generation or optical drawings as appropriate, nesting drawings and identification on each steel part {STWKDES and BRITSHAPE}, (See Figure 3). Fabrication and assembly drawings defining work packages and parts lists, check dimensions, welded attachments, rolling lines, powder marking, parts list for work packages by unit, zone and block are required in order for the build sequence to proceed.

Engineering and Outfit

The pipe geometry and details are to be created together with parts lists for each work package at each unit, zone, and block. Typically the shipbuilder will use a proprietary system such as HICADEC - GRADE/P or ComputerVision application modules.

Module assembly drawings are produced for machinery, frames, seats and minor steelwork, parts lists and requisitions.

Furniture, ventilation, and cabin drawings are output.

For each of these elements zone composite drawings are required, in order to proceed with fabrication, assembly and pre-outfit, complete with parts lists per work package for each unit, zone, and block are produced including layout and compartment outfit drawings.

Electrical

From the routeing carried out previously the installation drawings and parts lists are finalised in order to produce a complete equipment and materials list {ELECTRO} for each unit, zone, and block. Two-dimensional drawings are obtained from the draughting system again using direct interfacing.

3.0   APPLICATION SOFTWARE

The shipbuilding process comprises many stop-go break points frequently associated with long periods of apparent inactivity on any one particular project by disparate groups of designers, loftsmen, draughtsmen, etc., using different computers, necessarily different operating systems, and application software licenced from different manufacturers/suppliers.

The naval architect must be able to evaluate, quickly and reliably, and technical and economic feasibility of alternative ships in the conceptual phase of the design process and thereby continue into full production to cutting steel without any time wasting manual transfer of design data. This is achieved through the use of an integrated CADCAM system, i.e. improving the efficiency of the input/output phase between systems [9]. Such a system reduces the amount of effort needed for routine calculations;

the time which can be devoted to creative design is correspondingly increased.

The truth is that there is no single company that can supply all the required functionality for the shipbuilder. However, some individual companies, because of their unique blend of experience and working relationships with the worlds major hardware/software suppliers, are able to supply a substantial element of the shipbuilder's needs. The development of software at BMT and the direct feedback from current users has created a software system that offers its users major advantages over competitors particularly in the critical area of information transfer between systems. Some of the salient points are now highlighted to illustrate this further.

The hull form definition generated from the configuration design stage can be instantly used for lofting, numerical fairing, shell plate development, framing, internal major/minor steelwork definition, piece parts, interference checking with composite systems, modelling for outfit,etc [10].

The internal definition generated from the configuration design stage can be used in a similar way. The full topology is retained enabling detailed changes to be effected rapidly without re-definition of the major/minor structural elements e.g. a bulkhead position may need to be moved to accommodate a design requirements change to say the next in a series of ships. The model can be regenerated without re-definition of any of the associated piece-parts.

Maximum usage of surface mathematics for definition of the production hull and its integrated appendages together with additional information for its Manufacture, e.g. sight lines for the shaping of plates (shell sets) and accurate jig assembly data. Figure 4 illustrates the most understood feature of a surface - the ability to be able to pick a point on the surface and the surface moves in the same way as a piece of plastic. Other systems merely tie splines together or make no attempt to keep the splines connected. Both alternatives have the same problem - to get intermediate data points the user has to interpolate across the gaps in the data. To illustrate the point further, a typical surface to surface intersection is shown in Figure 5.

Customised surface data files can be output e.g. sets of facet points, compartment details and other surface attributes to be used, for example, in advanced hydrodynamic or structural analysis calculations.

4   REGULATING AUTHORITY REQUIREMENTS

Some specific application areas where BMT has recently taken the lead and invested in software are ship manoeuvrability and probability of survival both incidentally being driven as a direct result over concern for safety.

Ship Manoeuvrability

For the last few years the International Maritime Organisation (IMO) has had the subject of ship manoeuvrability on the work programme of its maritime Safety Committee. In 1987 they approved and the Assembly adopted Resolution A601 for the "Provision and Display of manoeuvring Information on Board Ships", which include the descriptions of the pilot card, wheelhouse [poster and manoeuvring booklet]. Currently the subject of "Manoeuvring Standards" is on the agenda, and guidelines have been developed for the estimation of manoeuvring performance during the design stage of the ship, which would be verified during the initial full scale trials on the ship.

Guidelines have been developed stating that all ships should have manoeuvring qualities which permit them to keep course, to turn, to check turns, to operate at acceptably slow speeds and to stop, all in a satisfactory manner. Since most manoeuvring qualities are inherent in the design of the hull and machinery they should be consciously estimated during the design stage and as a result BMT has developed software for the ship designer to predict compliance with these guidelines {SMDP}. It is also stated that, full scale tests to confirm the manoeuvring performance of all new ships greater than 100 metres in length should be estimated using these guidelines.

Probability of Survival

As from 1 February 1992 all dry cargo ships, including tweendeckers, container ships, bulk carriers and ro-ro vessels greater than 100m in length will have to comply with a new set of damage stability regulations currently being incorporated into SOLAS 74 by IMO. Any ship sailing under the flag of a signatory to this convention having its keel laid after this date will have its internal subdivision assessed on the basis of a new probabilistic approach to the problem of damage stability.

Likely future developments include extension of the rules to cargo ships of less than 100m and re-introduction of the concept of probabilistic assessment of watertight subdivision to passenger ships. A previous attempt was made at this in IMO Resolution A265(VIII) in November 1973 where a probabilistic approach was presented as an alternative to the deterministic method of assessing damage stability in SOLAS 60. The use of these regulations was optional and for this reason they were largely ignored by designers.

Clearly assessing the probability of survival of a ship under the new regulations is fairly complex and to perform the calculations economically it is essential to use a computer program. Such software has now been produced as part of a joint-industry project. A software system, known as PASS, has been closely integrated with the widely-used SFOLDS suite of design analysis programs.

5   PARALLEL PROCESSING APPLICATIONS

Ship handling simulators have been used for a number of years as a training aid in Maritime Institutes. The main objectives were to offer experience to nautical students in bridge team training. However, in recent years a new role for the use of real time ship handling simulators in ship design has emerged. Here they can be used as a design tool to investigate certain specific tasks, such as handling of ships in harbours, under various ambient weather conditions.

A compact real time ship handling simulator called REMBRANDT (Real Time Bridge Aid to Navigation and Training) using 2-D and 3-D graphic displays has been developed by BMT. The simulator uses a standard P.C. plus transputers and generalised mathematical model, capable of simulating the ship motion together with the surrounding environment. This may be a harbour, or sector of an estuary, complete with berths, buoys, coastline, land marks, navigational lights. Also included are tidal currents, varying water depth, wind, mooring lines and fender forces.

The system has been used successfully to investigate the manoeuvring behaviour of ships after lengthening, to determine the optimum size of side thruster and rudder type. Experience has been given to many ship captains on the behaviour of their ships after modification, or on the behaviour of their ships in unfamiliar harbours. In addition, the system may be used to find the best ergonomic lay-out of the manoeuvring control panel.

6   CONCLUDING REMARKS

In this paper the authors have attempted to identify and discuss research projects and developments which have led to the availability of todays computer software packages at BMT. It is evident that no significant process was made until the mid 1950's. However, developments have been very rapid in the last decade. This has been due to customer demand, availability of advanced computer hardware and software operating systems which have allowed the execution of applications requiring high rates of computation and relatively large data storage.

The authors are of the opinion that the next decade will see the miniaturisation of computer hardware devices and further reductions in costs; to such an extent that it will become cost effective for every engineer to work with a portable machine i.e. speed, main memory and storage capacity, comparable with a typical workstation of today. Together with other improved means of communication and data transfer, this will encourage professional engineers to work at home for increasing proportions of their time rather than travel to the office just to use a workstation.

Software packages purchased now, should therefore be seen to be capable of adapting to this background of continual progress and projected further major changes in hardware. This forward thinking is an important aspect of the BMT software which has been briefly described herein. The days of machine dependent software systems are considered to be strictly limited.

7   REFERENCES

1. Parker, M.N. and Chadbund, J.E. "Recent Developments in the BSRA "BRITSHIPS" System", IESS, Vol. 117 (1974).

2. Forrest, P.D. and Parker, M.N. "Steelwork Design Using Computer Graphics" Transactions Royal Inst. of Naval Architects, Vol. 125, 1983.

3. A. Landmark "ALKON-A Language for Algorithmic Design" ICCAS 1973.

4. J. Valovirta and P. Laitaakari "Hull Structure Generation in the STEERBEAR System" ICCAS 1976.

5. B. Baret "SICEN A Computer-Aided Design for Preliminary Draft and Detailed Steel Structure Studies of a Ship " ICCAS 1976.

6. J.A. Belda "The FORAN System System"   Third International Conference, ICCAS, 1979.

7. Chirillo: Improving shipyard production with standard components and modules. Short course on Ship Production and Production Planning, Univ. of Michigan. Oct 26-31, 1980.

8. G. Marshall and D.J. Archer "Preliminary Ship Design A Rational Approach Using Microcomputers" CADMO '88.

9. Catley, D. et al: Design Optimization, Committee V.5 Report to International Ship and Offshore Structures Congress: Lyngby, Denmark, 1988. See also Executive Summary, J'. Marine Structures, 2ppll5-9, 1989.

10. Catley, D., Williams, E.A. and Romero J.L.: Interactive Definition of Steelwork for Production: Advanced Production Technology and Prediction of Boat Behaviour in Real Conditions, Milan TekNautic Fair, Italy, November 1989.

ACKNOWLEDGEMENT:

The authors would like to thank BMT CORTEC Ltd. for permission to publish this paper.

BMT CORTEC is a subsidiary of British Maritime Technology Limited, the independant international maritime research and technology organisation.

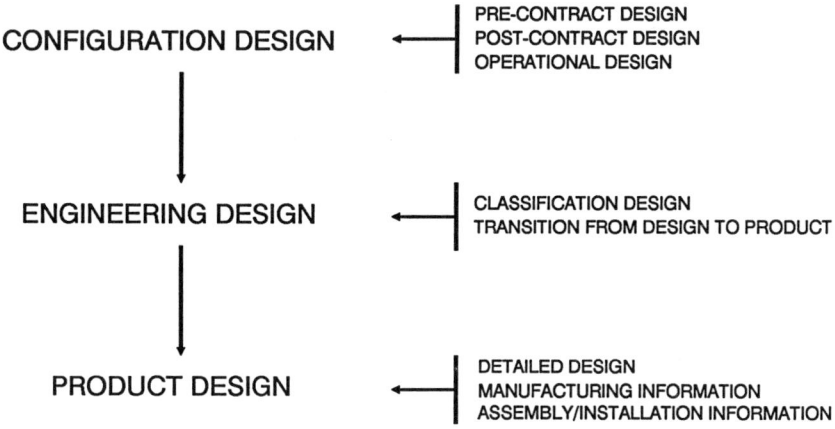

Fig.1   The Ship Design & Construction Process

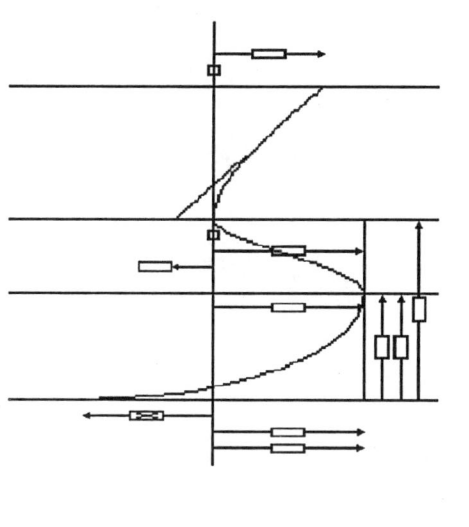

Fig.2   Interactive Graphical Manipulation of Curves Defining the Surface to be Generated (HULLGEN) - in this instance, the bow profile.

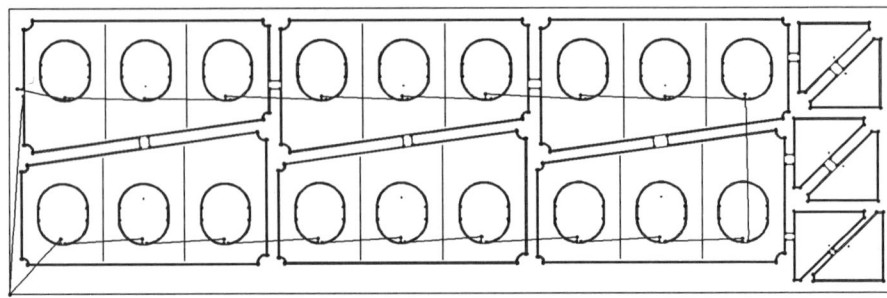

Fig.3   Typical plate nesting diagram showing cutting lines, bridges, marking lines, etc. (BRITSHAPE) above, with Steelwork unit assembly (STWKDES) below.

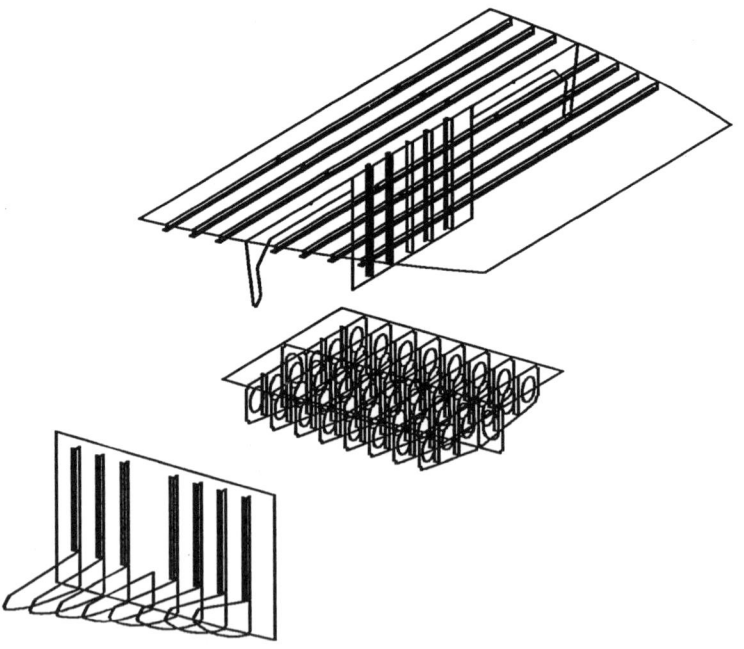

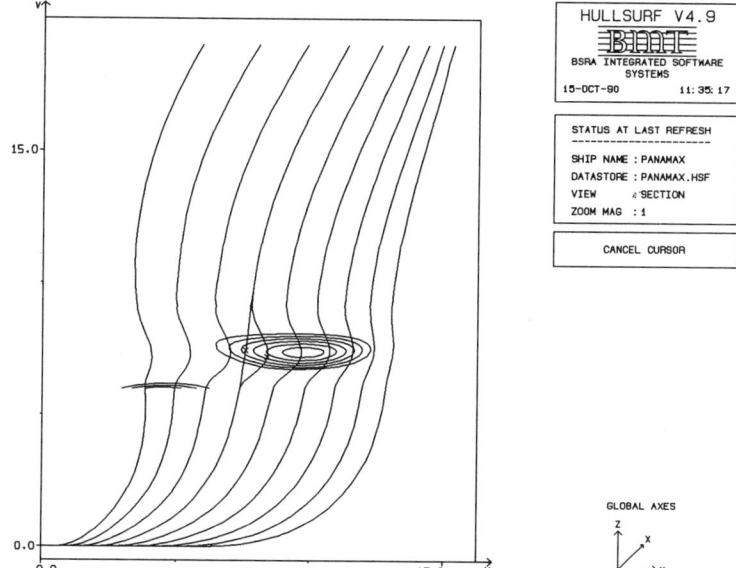

Fig.4   Typical surface definition characteristics (HULLSURF)
- Introduction of a bulge by pulling the surface and simplified
display of Gaussian curvature.

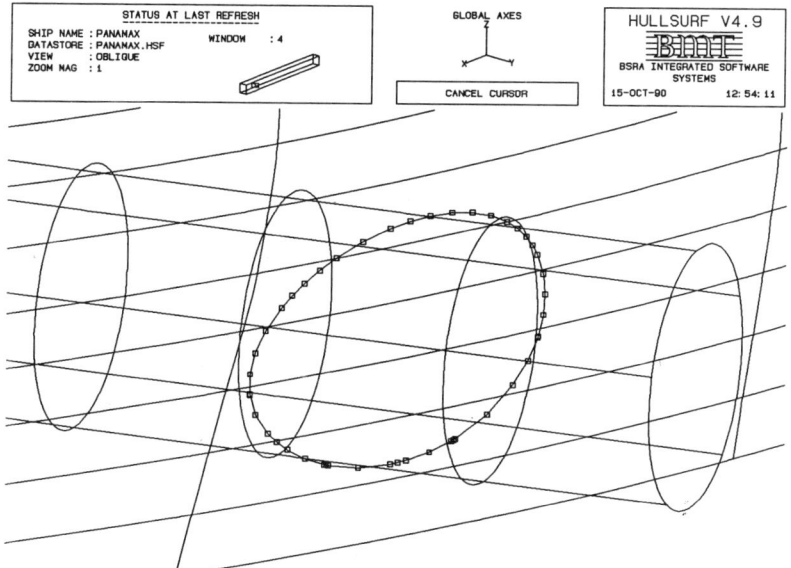

Fig.5   Typical surface definition characteristics (HULLSURF)
- Exact surface-to-surface intersection points for transfer to shell
plates, etc.

Computer Graphics in Warship Design

J.M. Duncan, P.H. Rutland and P.E. Gibbs

Sea Systems Controllerate, MOD(PE), Foxhill, Bath, U.K.

ABSTRACT

Computer visualisation transforms the symbolic into the geometric enabling scientists and engineers to observe their simulations and computations. It unifies the largely independent but convergent fields of computer graphics, image processing, and computer-aided design.

Interactive 3D graphics applications require a complete geometric or product model of an object's structure, behavioural characteristics and environment programmed into the computer to simulate the object's appearance and performance as realistically as possible. The model can be depicted on a display screen as a wireframe plot or as a complex solid or hollow image bounded by smooth shaded doubly curved surfaces. Realistic 3D graphic renderings significantly improve a user's ability to conceptualise, experiment, and gain insight into the results of complex synthesis and analysis.

Standardisation in computing is growing in significance with the arrival of open systems concepts such as PHIGS PLUS and PEX promising portability of applications across a wide variety of computing platforms.

This paper discusses the ways in which advanced graphics and user interface techniques may be applied in generating portable warship design software.

INTRODUCTION

Modern computer graphics workstations provide a powerful means for performing detailed design tasks and limited analyses, with faster computers accessed via networks providing additional analysis capability. Many analysis packages express their output in graphical form and a number of enabling technologies are required to permit analysis packages to run on one computer with graphical display on another.

As the cost of computer power reduces, system proliferation means that users are located on different machines separated from each other. Interoperation requires the connection of these systems to data networks. Single vendor networks are virtually impossible in government departments and the emerging networking standards are becoming more and more important for system integration.

In the past a variety of graphics standards have been proposed and these have tended to make a mockery of the word 'standard'. The argument in favour of standard graphics libraries is based on the notion that where new hardware supports the standard and where user applications use the standard, then moving existing code to new hardware should just require re-compilation. In recent years graphics hardware has evolved at a fast pace with significant increases in functionality whereas standards tend to succeed best in areas where there is a more moderate growth rate otherwise there is a tendency for them to lag behind the state of the art by a number of years.

The development of visualisation techniques for the commercial and entertainment marketplace, where the objective is to generate realistic looking images, is already a substantial field of investigation. Traditionally, scientific problems that required large-scale computing resources needed all the available computational power to perform the analyses or simulations. The ability to visualise results or guide the calculations themselves requires considerably more power. Today, workstations with impressive power fit on desktops and are well integrated into local networks. Advanced concepts are being developed and marketed for user interfaces.

The purpose of this paper is to describe some of the graphics tools which may be adopted to enable visual displays to be at the centre of the warship design process.

BRIEF HISTORY OF COMPUTER GRAPHICS

Interactive computer graphics applications were pioneered by automotive, aerospace and academic institutions during the 1960s. In many areas efforts at practical applications were thwarted by limited software and by prohibitively high hardware costs. By the middle 1960s there were around 20 manufacturers of commercial CRT graphics systems including Digital, Ferranti, IBM and Sanders supporting some 100 customers. Prior to the introduction of the storage tube terminals around 1968, many users were faced with investments of £100,000 or more for hardware alone and largely indeterminate software costs. Storage tube systems succeeded in opening computer graphics to thousands of new users and forced manufacturers with other technologies to offer their products at more competitive prices.

Computer graphics hardware incorporates many diverse technologies but two techniques are regarded as fundamental - raster scan and random position (calligraphic). Raster scan graphics provides excellent presentation and high flicker-free information content but has always been limited by resolution. Random positioning offers high resolution and line quality, but its drawbacks include limited colour capability and lower flicker-free information content. Early random position refresh systems could display about 2,000 vectors flicker-free, barely sufficient to display a warship body-plan.

Over the past few years 3D graphics workstations and high performance graphics terminals have evolved into systems that allow designers and engineers to work interactively with increasingly sophisticated 3D models.

First generation 3D workstations and terminals could produce typically 20,000 smooth shaded polygons or triangles per second with limited (entire screen) windowing capability and no anti-aliasing. Performance was generally limited by the rasterizing process rather than the rate at which the 3D geometric

data could be transformed. Later, manufacturers began designing their own rasterizing chips so that multiple pixels could be Z-buffered, smooth-shaded and painted
concurrently. Some of todays workstations and terminals will produce 100,000
smooth shaded polygons or triangles per second with windowing and full colour
graphics, but without anti-aliasing.

In parallel, supercomputing graphics workstations with fast multiple RISC
based processors and high performance vector pipelines have appeared, well suited
to floating point intensive applications such as fluid dynamics and finite element
analysis.

Today, hardware acceleration is available for real time anti-aliasing, texture
mapping, translucency and alpha compositing (combining anti-aliased images and
rendering translucent objects) allowing complex models of 1 million anti-aliased
vectors and up to 1 million shaded polygons or triangles to be manipulated in real
time.

There are a number of modelling bases available for describing warship geometry including; cellular or spatial occupancy (for example the octree method)
where a volume of space entirely containing the object is assumed to be divided
into discrete cuboidal cells which are labelled as to whether or not they are occu-
pied by material; constructive solid geometry (CSG) where an object is represented in terms of a tree of simple volumetric primitives with boolean operations;
boundary representation (Brep) where the objects boundary is stored using a topological structure defining the interconnections of faces, edges and vertices [1].
None of these is ideal for all applications and hybrid modellers are common.

The choice of geometric model is affected by the fundamental nature of the
data, the style and quality of rendering desired, and the costs of data storage and
communication. Surface modelling techniques are moving away from polygonal
primitives to NURBS (non-uniform rational B-splines) which can be meshed automatically during rendering.

WARSHIP DESIGN GRAPHICS

A wide variety of software has been written for use in the design of warships to
facilitate the description and variation of design proposals, auditing the weight,
space and volume requirements and performing analyses. The design process is
generally looked upon as a series of cycles. For each cycle, a design is proposed
and its characteristics determined and evaluated against the criteria that the design
is to satisfy. Where criteria are not satisfied, a modified design is proposed and a
new cycle commences. Design cycles move from low to high levels of detail.

Automated design optimisation is available for the most elementary level of
design and for well defined parts of the design. Ultimately it may become possible for the majority of the design process to be carried out by a computer, however
it is unlikely that innovation will be undertaken by computers in the foreseeable
future and current software requires designers to take the decisions while the computer is used to determine configurations and characteristics - the entire process
being under the control of the designer.

In contrast to the Army and Air force, the Navy has no prototype to test under realistic service conditions and this increases the need for accurate computer
modelling and prediction. Early design studies for warships are carried out by the
Ministry and detailed work by industry. They have been developing their separate

computer systems independently. New designs are produced to greater levels of detail, in shorter time-scales and with fewer people than hitherto. Most ship design organisations have access to software dealing with a wide variety of tasks [2].

Interactive 3D graphics facilities have been used by the Ministry for:

- Hull shape generation (also decks, curved bridge fronts etc) based on general 3D surface design techniques such as Coons, Bezier and B-Splines.

- Layout of hull and superstructures, decks and bulkheads using topological or symbolic methods.

- Placement of weapons, sensors and machinery.

- Production of drawings.

- Assessment of intact and damaged stability.

- Assessment of safe firing arcs for missile launchers and blind arcs for directors, taking account of ship motions.

- Mathematical modelling of ship motions to improve seakeeping ability, for example to allow operation of large helicopters in bad weather, and to study the effect of the shape of the hull on motions.

- Mathematical modelling of loading on the structure including 3D finite element analysis for detailed evaluation.

- Analysis of signatures, particularly radar cross section and infra red, to counter the use of missiles using these aspects on which to home.

- Analysis of vibration caused by waves, propellers and large machinery which affect strength of the hull, equipments and men.

- Computers used for controlling machinery and weapons and for assessing countermeasures to a threat where reaction time must be short.

3D GRAPHICS STANDARD

Phigs

PHIGS (Programmers Hierarchical Interactive Graphics System) [3] is a standard set by the ISO and ANSI for a program callable set of graphics functions. Important features of PHIGS are:-

- A 3-dimensional scene described by primitives and attributes can be built up into a display list which can be rendered and edited.

- The display list has a hierarchy of structures and coordinate transformations.

- User interaction is supported by the inclusion of input functions for devices such as locators and valuators.

A picture can be divided into parts and held in the display list as a number of structures which can then be built together into a complete scene. It is possible to reference a structure more than once in different positions and each part can be rotated by editing transformation matrices.

The basic PHIGS standard provides simple rendering of lines and filled polygon primitives specified in 3D coordinate space. By editing the display list it is possible to add or remove structures or change attributes of the primitives such as their colour.

When the display list is changed the whole picture is normally redrawn. If this can be done quickly enough then double buffering can be used to create animation.

User interaction is an important part of PHIGS. Input devices are defined to allow the user to pick elements of the display with a locator. Valuator devices attached to transformation matrices can be used to move and rotate parts of the scene or change the view point.

Because of its high level scene editing and user interaction features PHIGS is becoming an important standard in 3D computer aided design.

Phigs Plus

PHIGS PLUS [4] is an extension of PHIGS which includes more advanced rendering and more complex primitives. The rendering algorithms used can produce semi-realistic images of shaded surfaces. They are designed to be fast so that the interactive nature of PHIGS can be retained. No support is given for algorithms such as ray-tracing which produce cast shadows. This makes PHIGS PLUS suitable for geometric visualisation rather than producing photo-realistic images.

There are two main types of surface primitive available in PHIGS PLUS; Fill areas and NURBS surfaces. Fill area primitives are based on surfaces made from polygonal facets. These could represent a surface made up of flat sides like a box or pyramid, or they could be an approximation to a smoothly curved surface. The NURBS surface primitive provides a more accurate description of curved surfaces.

Surfaces can be given a range of attributes which affect their appearance in order to simulate various types of material. This includes a combination of ambient, diffuse and specular reflection properties. By varying the proportions and colours of the components it is possible to simulate matt, glossy and metallic surfaces. For example a glossy surface would have coloured diffuse reflection and white highlights while metallic surfaces have highlights similar in colour to the diffuse reflection.

PHIGS PLUS provides a lighting equation which supports multiple light sources of a variety of types including parallel, point and spot lights. The lighting equation is applied at a point on a surface using the surface normal at that point, the direction of view, the relative positions and directions of the light sources and the surface attributes.

In order to speed up the rendering the lighting calculation is performed only at the corners of a facet and the shade is then linearly interpolated across the rest of the facet during scan-line conversion. This method, known as Gouraud shading, is often performed by dedicated hardware graphics accelerators in the more powerful workstations. Z-buffer hidden surface removal can be performed at the same time. Speeds as high as 50,000 polygons a second are common and this is fast enough to allow complex models of warship structures to be animated in real time.

3D GRAPHICS APPLICATIONS

Surface Fairing

A good example of how 3D graphics can be used in computer aided warship design is in hull surface fairing and visualisation. Traditionally naval architects have used section curves taken at intervals along the hull to visualise its curvature. This method can be supported within PHIGS using wireframe rendering on a vector graphics screen or quality raster screen with high resolution and line antialiasing.

Most design systems use some kind of sculptured surface modelling technique such as bicubic NURBS surfaces. These are appropriate for producing faired surfaces because they are automatically curvature continuous except where discontinuity lines are specified. It is also possible to modify the shape with local changes by moving the surface control points.

With PHIGS PLUS the surface can be rendered either using the NURBS primitives or by dividing the surface into a mesh of facets and supplying surface normals calculated at the vertices. The shaded image shows details which might not be seen by looking at section curves alone. Bumps and defects in the surface can be seen provided the surface is viewed from the right angle and the lighting is from the right direction.

Surface Visualisation

To visualise the shape of a curved surface it is important to be able to rotate the surface on the display so that it can be viewed from different directions. The most convenient way to do this is with a dial box which allows rotation and translation in any direction and changes to zoom and perspective, although a mouse can be used to drag the surface around, but this does not provide as many degrees of freedom. A typical dial box is a device with eight valuator dials.

Light sources can also be moved to help reveal subtleties in the surface shape. Defects often show up best when the shadow bands and highlights move over them. Point light sources can be positioned near the surface to exaggerate shadowing around shallow bumps and dents. This is analogous to the old trick of shining a torch along a hull surface!

As the user moves the light sources with the dial box he will need some indication of the position of the light source with respect to the viewer. For point and spot light sources a small sphere or cone can be rendered at their location. Directional light sources are more difficult to show but one method that appears to work well is to illuminate a sphere rendered in a corner of the display.

Notice that it is not necessary for the surface to be rendered realistically. Cast shadows would probably just obscure detail. The most important requirement is that the surface can be smoothly shaded by fast methods. Furthermore it is not so important to have such high resolution or antialiasing as it is with a wireframe image because aliasing effects only appear along the edges of the surface.

Another powerful surface visualisation method is to use false colour plots to show curvature variations over the surface. The mean or gaussian curvature can be calculated anywhere on the surface and can be mapped onto a hue from a range of the spectrum. The PHIGS PLUS NURBS primitive allows such data to be mapped onto the surface. Alternatively if facet primitives are being used a colour can be associated with each vertex and this will be smoothly interpolated over the rest of the facet as part of the shading process. With experience this technique can show up areas where curvature is not varying smoothly, that would be hard to find by other methods.

The surface must be subdivided into facets small enough to ensure that shading artefacts such as mach banding do not give false impressions of the surface shape. Typically a hull surface may need to be divided into a thousand facets. If the user is to be able to rotate the surface without unacceptably long delays between image updates then it must be redrawn at least 3 times a second. This sets a minimum performance requirement of at least 3,000 Gouraud shaded polygons a second for this application.

Ideally the designer should be able to move the control points of a surface and have it redrawn interactively so that he can work the surface into a fairer shape. This is a more difficult process than rotating a surface since it requires recalculation of the shape but is less of a problem if the NURBS primitive is used.

Walkthroughs

A second application where PHIGS PLUS rendering can help a designer concerns the task of laying out equipments on or in a warship. It is necessary to do this in such a way that there is sufficient clearance between them for people and objects to move past. Controls on machinery will need to be accessible and visible and other more subjective aspects of ergonomic layout are important.

One way to tackle this problem is to perform a walkthrough. A faceted model of the warship including the relevant equipments must be loaded into the PHIGS display list. The designer can then use a dial box to move about in and around the model to observe the layout.

The dial box controls have to be set up differently than when normally viewing an object since it is necessary for the model to rotate about the viewer's frame of reference rather than the object's. This means that one of the translation dials will move the viewer in and out while the other two will move him from side to side and up and down. The rotation dials enable him to turn round. It is very difficult to do such a walkthrough without a dial box.

Once again it is not very important that the scene be rendered realistically. It is more helpful to colour code different types of object rather than give them correct colours.

Another feature of PHIGS which may prove useful in this application is the use of model clipping planes. A clipping plane passing through a model gives a cut away image which may be useful to check relative positions of equipments internal to the warship structures.

If walkthroughs are to be performed easily on complicated models then a very high graphics performance will be demanded. Such performance is now becoming available on a few of the most powerful workstations with dedicated graphics hardware.

OPEN SYSTEMS ASPECTS

To use CAD software, a designer needs access to a computer with a reasonable amount of compute power and a high resolution colour graphics device. This broad specification covers all machines from mainframes and minis with graphics terminals to workstations and desktops with integral graphics screens. To enable as large a number of designers as possible to have access to the software, it must be capable of running on a variety of computers under various operating systems through different graphics devices.

Until recently this has been achieved by producing several specific versions of the software to suit the different computer environments, but this has caused problems with maintenance, user interfaces, configuration control and distribution.

To ease problems of multiple copies of software, manufacturers and standards bodies have been developing and proposing standards to provide a single operating system and vendor independent environment to allow the generation of single versions of source compatible software capable of running under a variety of hardware and software configurations.

<u>Software Standards</u>
The first issue for standardisation is the programming language to be used. This is a well developed area giving a choice of languages such as FORTRAN 77, C and ADA. FORTRAN, being the oldest of these languages, is not as complete or flexible so the majority of vendors include extensions to the standard in their implementations. These extensions commonly include such things as data structure declarations, the IMPLICIT NONE and INCLUDE statements and a wider range of looping facilities. Given the large number of vendors supporting extensions to the standard it is possible to identify a common extension to FORTRAN 77 which may, perhaps, be included in future FORTRAN standards.

With a CAD system, graphics is as important as the programming language. To utilise a graphics standard in source compatible software it must be possible to rely on all implementations of the standard having the same interface in the chosen language. The interface between applications software and an implementation of the graphics standard is called a 'language binding' and defines, precisely, the subroutine names and parameter lists required to access all the graphics functionality. At the time of writing, PHIGS language bindings exist for FORTRAN and C and a draft language binding exists for ADA.

Although PHIGS defines standard graphics functionality, actual implementations may differ in areas such as PHIGS workstation initialisation and the availability of logical input devices. Therefore graphics software written for one PHIGS implementation may not necessarily be source compatible with all others.

Notwithstanding this, software written in a standard programming language and using the language binding to access PHIGS functionality will, with care, be source compatible from a graphics point of view with any computer that supports these standards.

However, software may also require to access operating system services such as file handling, terminal I/O and interrogation of environmental features. The drive for a standard operating system interface is leading towards POSIX [5], an operating system independent interface to system services. This proposed standard is based on UNIX like operating systems [6] and is intended to eliminate application level differences between the various system implementations. Currently only workstation vendors, who mostly use a UNIX like operating system, can provide POSIX compliant interfaces but other vendors are proposing to provide POSIX conformant interfaces to their proprietory operating systems. Thus, software that utilises the POSIX interface will be portable across all conformant operating systems.

Window Systems

With true portability across machines that support the above standards it only remains to determine how an application will run on any computer and graphics device in a network of disparate computers. The user must be able to run the software without 'needing to know' the CPU on which the software runs and without the software 'needing to know' the graphics device being accessed by the user.

One solution to this problem is the X Window system [7] which has been widely adopted and which has therefore become a 'de facto' standard for interfacing between application software and a user across a network containing different vendors hardware.

The X Window system is based on a Client-Server model, where the application software is the client, in some CPU, which sends data to the user's workstation or terminal (the server) using a standard network protocol (X-protocol). With the X-protocol it is also possible for the client and server to be on the same computer.

The server understands all X-protocol commands and operates under those commands to present information to the user in an area of the user's graphics device called a window. For each client connected to a server there is a separate window so that the server is able to display correct information appropriate to a particular application.

The client sends commands to the server via a library of routines (Xlib) that interfaces the applications software and system services to the X functionality. To display a graphical image in a window on a server, the PHIGS implementation on the client CPU must process all the commands in the client's display list and reduce them to actual displayable primitives such as lines and text and call the appropriate Xlib routines.

As well as processing output commands, the server also processes client requests for input from any of the available input devices (such as locators and valuators). The server will initiate a request for input and, on receipt of acknowledgement of completion, will return the input data to the requesting client.

With this system an application can run on any host and interact with the user on a terminal or graphics screen either local to the application CPU or remote via a network.

The 2D graphics features of the X Window System are not appropriate for 3D CAD software where the requirement is for fast interactive selection and transformation of structures. A proposed solution to this problem is a 3D graphics extension to X, called PEX [8], which is based on the PHIGS graphics standard although the PEX functionality will support any 3D graphics library.

The features of 3D graphics systems, such as generalised hierarchies, segment transformations, selection, highlighting, segment editing are provided by additions to the X client interface library, by PEX protocols which sit along side X protocols and by extensions to the server that allow server machines to understand and operate on the 3D graphics functions. Thus with PEX, it is possible for the PHIGS implementation to pass all graphics display list commands to the server and allow the server to process them whilst the client continues in parallel.

User Interfaces

To access the same application software running on different computers it is possible that the designer would need to know more than one operating system interface, as well as the application interface.

The designer can be shielded from the command line interface of an operating system with a Graphical User Interface (GUI) that would allow access to operating system functions via menus, icons and user generated windows.

A GUI can be generated using a 'toolkit', a library of routines that sits above Xlib and provides a high level applications interface for creating and processing menus and icons (widgets) within a windows environment. To reduce a designers learning requirement it is also possible to create an applications software GUI using the same toolkit that presents itself in the same manner as the operating system GUI (has the same 'look and feel').

To avoid different GUI's being implemented on different computers, efforts are underway to develop a standard 'look and feel' for interfaces using a standard set of widgets. Currently there are two contenders; Open Software Foundation's (OSF) MOTIF and Unix Software Operation's (USO) Open Look, although IEEE are developing a common toolkit which could be used for both MOTIF and Open Look but which may not allow access to all the features of each.

When using a toolkit with a PHIGS application it is important to consider the interaction between the toolkit and the graphics output and input devices. When an overlapping window, for menus, text etc, is closed then some implementations of PHIGS will automatically redraw the PHIGS window whilst others expect the application to control window updates. To do this it would require a message to the application to indicate that a redraw is necessary. PEX should provide this feature, and should also resolve the problem of allowing PHIGS events, such as pick and locator, to be handled simultaneously with other X events such as menu selection.

CONCLUSIONS

Proliferation of computers is a trend that is likely to continue through the 1990s and beyond. Developing software which is portable and capable of utilising significant numbers of network connected systems is the common goal. This paper has outlined a number of graphics programming tools which are likely to have a significant impact on the way in which software is developed for use in the warship design process.

REFERENCES

1. J.M. Duncan, I.M. Yuille, Representation of Compartmented Spaces for Computer Aided Warship Design, Computer Aided Design, Volume 16 No. 1, January 1984.

2. D.R. Pattison, R.E. Spencer, W.J. Van Griethuysan, The Computer Aided Ship Design System GODDESS and its Application to the Structural Design of Royal Navy Warships, Computer Applications in the Automation of Shipyard Operation and Ship Design IV, 1982.

3. Programmer's Hierarchical Interactive Graphics System (PHIGS), International Standard ISO/IEC 9592-1:1988(E), International Standards Organisation, Geneva, Switzerland, Oct. 1988.

4. PHIGS PLUS Functional Description, International Draft Proposed Standard

ISO/IEC 9592-4: 199X, International Standards Organisation, Geneva, Switzerland, March 1990.

5. POSIX (Portable Operating System Interface for Computer Environment), ISO/IEC 9945, International Standards Organisation, Geneva, Switzerland, March 1989.

6. UNIX is a registered trademark of AT&T in the USA and other countries.

7. Robert W. Scheifler, Jim Gettys, and Ron Newman, X Window System - C Library and Protocol Reference, Digital Press, Bedford, MA, 1988.

8. PEX Architecture Team, PEX Protocol Specification, PEX Version 5.0P, Sally C. Barry and Randi J. Rost, Document Editors, May 1990.

SECTION 2: SHIP HYDRODYNAMICS AND FLOW SIMULATION

Computing Wave Resistance, Wave Profile, and Sinkage and Trim of Transom Stern Ships

L. Cong and C.C. Hsiung
Centre for Marine Vessel Design and Research, Department of Mechanical Engineering, Technical University of Nova Scotia, Halifax, Nova Scotia, Canada

ABSTRACT

A numerical method for evaluating the wave resistance, wave profile, and sinkage and trim of a surface ship with a transom stern is presented in this paper. Based on the thin-ship and flat-ship theories, it is devised that a source system distributed on the longitudinal center-plane, a sink line along the bottom edge of the transom stern and a sink plane on the bottom of the overhang aftbody are used. An efficient treatment of the Green function is utilized to speed up the computation. An iterative procedure is adopted to determine the final sinkage and trim.

INTRODUCTION

In evaluating the sinkage and trim of ships, Yeung [1] has already obtained good results by using the first order thin-ship theory, but these calculations were limited to ships without transom sterns. In Yim's work [2], the influence of the transom stern was included, but no sinkage and trim were considered. It is intended to predict the wave resistance, wave profile as well as sinkage and trim for surface ships with transom sterns by using a simple numerical method in this paper. The hull without the transom stern is approximated by a source system distributed on the center-plane as the thin-ship theory, while the transom stern is approximated by a sink line along the bottom edge of the transom stern based on Yim's work [2]. The third singularity system is distributed on the bottom of the overhang aftbody which is approximated by the flat-ship theory locally.

The wave resistance and wave profile with sinkage and trim can be obtained by iterative computation. First, the wave profile, sinkage and trim are computed at the design draft and trim. Then, a new sink line corresponding to the computed sinkage and trim is determined. With this sink line, an additional wave system can be obtained. A final wave system corresponding to the free sinkage and trim can be determined by the superposition of the original wave profile along the hull and the additional wave system. The final results of the wave resistance, sinkage and trim under the condition of free sinkage and trim can also be obtained based on this superposed wave system. The iterative procedure can be continued, but two iterations are usually sufficient for engineering accuracy.

It is well-known that numerical calculation of the double integral in the Green function of ship flow is cumbersome and time-consuming. An improved numerical approximation of the Green function [3] which is based on Newman's work [4] has been adopted. The center-plane of the hull was discretized into a set of quadrilateral elements, with a few triangular elements if necessary. The hull form was depicted by the tent function [5]. Analytical integral formulas of wave resistance and free wave due to each element have been derived. Similar derivations have also been worked out for the transom sink line.

FORMULATION

The right handed $oxyz$ Cartesian coordinate system moving with the ship at a constant speed C is shown in Fig.1. The origin o is located on the undisturbed free surface with oz vertically downwards and ox parallel to the direction of the motion. Another coordinate system $o'x'y'z'$ fixed in the ship coincides with the system $oxyz$ at rest, but differs in the presence of sinkage and trim.

It is assumed that the fluid is inviscid and incompressible and the flow is irrotational. Furthermore, the surface tension is neglected. Then the flow of an inviscid fluid past a fixed ship can be described by a total velocity potential $\Phi(x, y, z)$ which is defined as:

$$\Phi(x, y, z) = -Cx + \phi(x, y, z)$$

where $\phi(x, y, z)$ is the perturbation potential which satisfies the Laplace equation:

$$\nabla^2 \phi(x, y, z) = 0 \tag{1}$$

subjected to the linearized free surface boundary condition:

$$C\phi_{xx}(x, y, 0) + g\phi_z(x, y, 0) = 0 \qquad z = 0 \tag{2}$$

the hull boundary condition:

$$\phi_n = Cn_x \qquad \text{on hull S and } z \geq 0 \tag{3}$$

the bottom condition for infinite depth:

$$\phi_z = 0 \qquad \text{as} \quad z \to \infty \tag{4}$$

and the radiation condition:

$$\phi = \begin{cases} o(\frac{1}{r}) & \text{if } x > 0 \\ \\ O(\frac{1}{r}) & \text{if } x < 0 \end{cases} \qquad \text{as } r \to \infty \tag{5}$$

The linearized free-surface elevation, ζ, is given by

$$\zeta(x, y) = \frac{C}{g}\phi_x(x, y, 0) \tag{6}$$

The solution of moving source under the free surface satisfying (1), (2), (4) and (5), the Green function to form the desired potential, has been investigated by Wehausen and Laitone [6]. With the coordinate system defined above, the Green function is

$$G(x, y, z, x_0, y_0, z_0) = -\frac{1}{r_1} + \frac{1}{r_2} + \frac{1}{\pi}\int_0^{2\pi}\int_0^\infty \frac{k_0 sec^2\theta}{k - k_0 sec^2\theta}e^{k[-(z+z_0)+i\omega]}\, dk\, d\theta$$

$$+ Re\, 2i \int_{-\frac{\pi}{2}}^{\frac{\pi}{2}} k_0 sec^2\theta e^{k_0 sec^2\theta[-(z+z_0)+i\omega]}\, d\theta \tag{7}$$

where,

$$r_1 = \sqrt{(x - x_0)^2 + (y - y_0)^2 + (z - z_0)^2}$$
$$r_2 = \sqrt{(x - x_0)^2 + (y - y_0)^2 + (z + z_0)^2}$$
$$\omega = (x - x_0)cos\theta + (y - y_0)sin\theta$$
$$k_0 = g/C^2$$

Then, the perturbation potential of a ship can be written as

$$\phi(x, y, z) = \int\int_S \sigma(x_0, y_0, z_0)\, G(x, y, z; x_0, y_0, z_0)\, ds \tag{8}$$

From the well-known thin-ship theory, the source strength $\sigma_M(x, z)$ distributed on the center-plane of a ship is

$$\sigma_M(x, z) = -\frac{C}{2\pi} f_x(x, z) \tag{9}$$

where $f(x, z)$, the hull function, defines the local half-beam of the hull surface; $f_x(x, z)$ is the derivative of the hull with respect to the ox-axis.

For a ship with a transom stern, $f_x(x, z)$ is usually undefined on the bottom around the overhang aftbody near the stern, thus the thin-ship approximation fails at this local area. Another first-order approximation, the so-called flat ship theory, could be introduced to handle this case. The hull bottom near the transom stern is quite flat and the stern draft, Z_T, is sufficiently small too, the local draft of the hull bottom of the overhang aftbody can be defined by

$$z = d(x, y)$$

Using the similar perturbation technique as in thin-ship theory, the boundary condition (3) becomes:

$$\phi_z = C\, d_x(x, y)$$

Then, the sink strength on the bottom of the overhang aftbody can be determined as

$$\sigma_F = -\frac{C}{4\pi} d_x(x, y) \tag{10}$$

To replace the transom stern itself, a sink line is distributed on the bottom adge of the stern. In terms of Yim's work [2], the strength of this sink line is:

$$\sigma_T = -\frac{C}{4\pi} Z_T \tag{11}$$

Thus, the whole ship can be represented by three singularity systems: (1) a source system distributed on the center-plane which is discretized as shown in Fig. 2; (2) a sink plane on the bottom of the overhang aftbody as shown in Fig. 3; and (3) a sink line distributed along the bottom line of the transom stern also as shown in Fig. 3.

Based on these analyses, the wave resistance formula can be written as

$$R = 16\pi\rho k_0^2 \int_0^{\frac{\pi}{2}} [P^2 + Q^2]\sec^3\theta \, d\theta \tag{12}$$

and the Kochin functions P and Q are defined as

$$
\begin{pmatrix} P \\ Q \end{pmatrix} = \int\int_{S_0} \sigma_M(x,z) \begin{pmatrix} \cos \\ \sin \end{pmatrix} (k_0 x \sec\theta) e^{-k_0 z \sec^2\theta} \, dz dx
$$

$$
+ \int\int_{S_F} \sigma_F(x,y,d(x,y)) \begin{pmatrix} \cos \\ \sin \end{pmatrix} [k_0(x\cos\theta + y\sin\theta)\sec^2\theta] e^{-k_0 z \sec^2\theta} \, dS_F
$$

$$
+ \int_{L_T} \sigma_T(x_T,y,Z_T) \begin{pmatrix} \cos \\ \sin \end{pmatrix} [k_0(x_T\cos\theta + y\sin\theta)\sec^2\theta] e^{-k_0 Z_T \sec^2\theta} \, dy \tag{13}
$$

where $K_0 = g/C^2$, S_0 is the center-plane of hull, S_F is the bottom plane of the overhang aftbody, and L_T is the bottom line of the transom stern.

Once the wave profile (6) is known, the sinkage and trim can immediately be found by solving the following equations by static equilibrium:

$$
\begin{pmatrix} h + x_w\alpha \\ x_w h + \frac{\nabla}{A_w}H_p\alpha \end{pmatrix} = \frac{2}{A_w} \int_{-\frac{L}{2}}^{\frac{L}{2}} \begin{pmatrix} \zeta(x)f(x,0) \\ x\zeta(x)f(x,0) \end{pmatrix} dx \tag{14}
$$

where h is sinkage which is positive as the ship raises; α is trim angle which is positive as the bow is up; x_w is the x-coordinate of the center of flotation; A_w is the water-plane area; ∇ is the ship displacement volume; and H_p is the longitudinal $\bar{GM}$ of the ship.

NUMERICAL METHOD

The tent function [5] for a ship hull $f(x_i, z_j)$ associated with a control point is defined as

$$
f(x,z) = \sum_{k=0}^{1}\sum_{\ell=0}^{1} y_{i+k,j+\ell}(1 - \frac{x_{i+k} - x}{\Delta x_k})(1 - \frac{z_{j+\ell} - z}{\Delta z_\ell}), \quad x_i \le x \le x_{i+1};\ z_j \le z \le z_{j+1} \tag{15}
$$

where i and j are index numbers of stations and waterlines, respectively, and

$$
\Delta x_k = \begin{cases} x_i - x_{i+1} & k=0 \\ x_{i+1} - x_i & k=1 \end{cases}
$$

$$
\Delta z_\ell = \begin{cases} z_\ell - z_{\ell+1} & \ell = 0 \\ z_{\ell+1} - z_\ell & \ell = 1 \end{cases}
$$

Then the derivative with respect to x on a quadrilateral can be written as

$$
f_x(x,z) = \sum_{k=0}^{1}\sum_{\ell=0}^{1} y_{i+k,j+\ell}\frac{1}{\Delta x_k}(1 - \frac{z_{j+\ell} - z}{\Delta z_\ell}) \tag{16}
$$

The simplest method of source distribution on a panel element is the monopole method which replaces the source panel by a point source at its centroid. Another choice is the panel method which distributes the source on the whole panel with uniform or linear density. A comparison of the calculated results of the Wigley model (in Fig. 4) by these two methods shows that the monopole method overestimates the wave resistance as Froude number F_n becomes lower. Thus, as the final choice, the panel method of source distribution has been adopted in this paper. For the purpose of numerical calculation, formula (12) can be rewritten as follows:

$$R = 16\pi\rho k_0^2 \int_0^\infty [P^2 + Q^2]/\sqrt{1 + u^2}\, du \tag{17}$$

where the Kochin functions can be written as:

$$\begin{pmatrix} P \\ Q \end{pmatrix} = \sum_{i=1}^{M} \sum_{j=1}^{N} \begin{pmatrix} P_{ij} \\ Q_{ij} \end{pmatrix} + \begin{pmatrix} P_F \\ Q_F \end{pmatrix} + \begin{pmatrix} P_T \\ Q_T \end{pmatrix} \tag{18}$$

The contribution to the Kochin functions from a quadrilateral panel is derived as:

$$\begin{pmatrix} P_{ij} \\ Q_{ij} \end{pmatrix} = \frac{C}{2\pi} \frac{1}{k_0^2(1 + u^2)} \left[\begin{pmatrix} sin \\ -cos \end{pmatrix} (k_0 x_{i+1}\sqrt{1 + u^2}) - \begin{pmatrix} sin \\ -cos \end{pmatrix} (k_0 x_i \sqrt{1 + u^2}) \right]$$

$$\times \left\{ [e^{-k_0 z_{j+1}(1+u^2)} - e^{k_0 z_j(1+u^2)}] \sum_{k=0}^{1} \sum_{\ell=0}^{1} \frac{y_{i+k,j+\ell}}{\Delta x_k} [1 - \frac{z_{j+\ell}}{\Delta z_\ell} + \frac{1}{k_0(1 + u^2)\Delta z_\ell}] \right.$$

$$\left. + [z_{j+1} e^{-k_0 z_{j+1}(1+u^2)} - z_j e^{-k_0 z_j(1+u^2)}] \sum_{k=0}^{1} \sum_{\ell=0}^{1} \frac{y_{i+k,j+\ell}}{\Delta x_k \Delta z_\ell} \right\} \tag{19}$$

The contribution from a triangular panel is derived as:

$$\begin{pmatrix} P_{ij} \\ Q_{ij} \end{pmatrix} = -\frac{C}{2\pi} \frac{f_x}{k_0^2(1 + u^2)} \left\{ e^{-k_0 z_j(1+u^2)} \left[\begin{pmatrix} sin \\ -cos \end{pmatrix} (k_0 x_{i+1}\sqrt{1 + u^2}) \right. \right.$$

$$\left. - \begin{pmatrix} sin \\ -cos \end{pmatrix} (k_0 x_i \sqrt{1 + u^2}) \right] + \left\{ e^{-k_0 z_a(1+u^2)} \left[tan\beta\sqrt{1 + u^2} \begin{pmatrix} cos \\ sin \end{pmatrix} (k_0 x_{i+1}\sqrt{1 + u^2}) \right. \right.$$

$$\left. + \begin{pmatrix} -sin \\ cos \end{pmatrix} (k_0 x_{i+1}\sqrt{1 + u^2}) \right] - e^{-k_0 z_b(1+u^2)} \left[tan\beta\sqrt{1 + u^2} \begin{pmatrix} cos \\ sin \end{pmatrix} (k_0 x_i\sqrt{1 + u^2}) \right.$$

$$\left. \left. + \begin{pmatrix} -sin \\ cos \end{pmatrix} (k_0 x_i \sqrt{1 + u^2}) \right] \right\} \frac{1}{1 + tan^2\beta(1 + u^2)} \right\} \tag{20}$$

where β is the angle between the base along the x-direction and the hypotenuse in a triangular panel, and

$$\begin{cases} z_a = z_{j+1} \quad and \quad z_b = z_j \quad if \ tan\beta > 0 \\ z_a = z_j \quad and \quad z_b = z_{j+1} \quad if \ tan\beta < 0 \end{cases}$$

The contribution from the sink line at the stern is:

$$\begin{pmatrix} P_T \\ Q_T \end{pmatrix} = 2\frac{\sigma_T}{k_0 u}sin(B_T k_0 u\sqrt{1+u^2})\begin{pmatrix} cos \\ sin \end{pmatrix}(k_0 x_T\sqrt{1+u^2})e^{-k_0 Z_T(1+u^2)} \qquad (21)$$

The wave profile can be calculated by using equation (6). In fact, the total wave system is a summation of the contributions from center-plane panel elements, the bottom sink plane and the transom sink line. The numerical calculation of the wave system has been separated into two parts: the local wave and the free wave. The local wave can be calculated as follows,

$$\begin{aligned} \zeta_\ell(x) = \frac{C}{g}\Bigg[&\int\int_{S_0} \sigma_M(x_0,0,z_0)G_{d_x}(x,0,0;x_0,0,z_0)\,dS_0 \\ &+ \int\int_{S_F} \sigma_F(x_0,y_0,z_0)G_{d_x}(x,0,0;x_0,y_0,z_0)\,dS_F \\ &+ \int_{L_T} \sigma_T(x_T,y_0,H_T)G_{d_x}(x,0,0;x_T,y_0,H_T)\,dy_0 \Bigg] \end{aligned} \qquad (22)$$

The free wave is

$$\zeta_f = \sum_{i=1}^{M}\sum_{j=1}^{N}\zeta_{fij} + \zeta_{fF} + \zeta_{fT} \qquad (23)$$

where free wave due to a quadrilateral panel is derived as:

$$\begin{aligned} \zeta_{fij}(x) = \frac{4C^2}{\pi g}\int_0^\infty \frac{du}{1+u^2} &\left[H(-\bar{x}_{i+1})sin\sqrt{1+u^2}\bar{x}_{i+1} - H(-\bar{x}_i)sin\sqrt{1+u^2}\bar{x}_i \right] \\ \times \Bigg\{ [e^{-k_0 z_{j+1}(1+u^2)} - e^{-k_0 z_j(1+u^2)}]\sum_{k=0}^{1}\sum_{\ell=0}^{1}&\frac{y_{i+k,j+\ell}}{\Delta x_k}[1 - \frac{z_{j+\ell}}{\Delta z_\ell} + \frac{1}{k_0(1+u^2)\Delta z_\ell}] \\ +[z_{j+1}e^{-k_0 z_{j+1}(1+u^2)} - z_j e^{-k_0 z_j(1+u^2)}]\sum_{k=0}^{1}\sum_{\ell=0}^{1}&\frac{y_{i+k,j+\ell}}{\Delta x_k}\frac{1}{\Delta z_\ell}\Bigg\} \end{aligned} \qquad (24)$$

the free wave due to a triangular panel element is derived as:

$$\begin{aligned} \zeta_{fij}(x) = \frac{4C^2}{\pi g}f_x(x_i,z_j)\int_0^\infty &\frac{du}{1+u^2}\Big\{ -e^{-k_0 z_j(1+u^2)} \\ \times [H(-\bar{x}_{i+1})sin\sqrt{1+u^2}\bar{x}_{i+1} - H(-\bar{x}_i)sin k_0\sqrt{1+u^2}\bar{x}_i] &+ \frac{1}{1+tan^2\beta(1+u^2)} \\ \times \{H(-\bar{x}_{i+1})e^{-k_0 z_a(1+u^2)}[tan\beta\sqrt{1+u^2}cos\sqrt{1+u^2}\bar{x}_{i+1} &+ sin k_0\sqrt{1+u^2}bar x_{i+1}] \\ -H(-\bar{x}_i)e^{-k_0 z_b(1+u^2)}[tan\beta\sqrt{1+u^2}cos k_0\sqrt{1+u^2}\bar{x}_i &+ sin k_0\sqrt{1+u^2}\bar{x} - x_i]\}\Big\} \end{aligned} \qquad (25)$$

The free wave due to the sink line is derived as:

$$\zeta_{fT} = -16k_0\frac{C}{g}\sigma_T H(-\bar{x}_T)\int_0^\infty e^{-k_0(z+Z_T)(1+u^2)}cos\sqrt{1+u^2}\bar{x}_T\frac{sin B_T k_0 u\sqrt{1+u^2}}{u}du \qquad (26)$$

where $H(x)$ is the Heaviside function, and

$$\bar{x}_j = k_0(x - x_j)$$

Based on the wave calculated above and equations (14), the sinkage h and trim angle α can be obtained.

The effect of sinkage and trim on the wave resistance and wave profile should not be ignored, especially in the case that Froude number F_n is higher then 0.3. An iterative procedure has been used to deal with this problem. The offsets of the ship with fixed sinkage and trim have been used in formulas (23), (24), (25) and (26) in order to determine the initial wave profile. Then, the initial sinkage h_1 and trim α_1 can be solved from equations (14) in terms of the initial wave system ζ_1. Thus, the new stern draft Z_{T1} can be found as:

$$Z_{T1} = Z_{T0} - [h_1 + (\frac{L}{2} + x_w)\alpha_1] \tag{27}$$

where subscript "0" denotes the parameters of the original condition, namely, the input data, while subscript "1" is for the parameters obtained from the first iteration. Then a new sink line with the strength, $-Z_{T1}C/4\pi$, will be used to calculate an additional wave system ζ_{T1} by (23) and (26) in an iterative procedure. By superpositioning ζ_1 and ζ_{T1}, the wave in the second iteration $\zeta_2 = (\zeta_1 - \zeta_{T0}) + \zeta_{T1}$ will be introduced to (14) again to obtain the sinkage h_2 and trim α_2. Moreover, the wave resistance coefficient c_{w2} can also be computed by including the effect of h_2 and α_2. This procedure can be continued until convergent results for h_k and α_k are obtained. In practical calculation, two iterations already satisfy engineering requirements, thus h_2 and α_2 may be taken as the final results.

COMPUTED RESULTS

A computer program based on the aforementioned method has been developed to predict the wave resistance, wave profile, and sinkage and trim for ships with transom sterns. DREA Model 316 [7] and ATHENA Model [8] have been chosen for calculations with this program. The computed wave resistance coefficients under the conditions of fixed and free sinkage and trim are shown in Fig. 5 and Fig. 6, and compared with experimental data and results calculated by other methods. The computed wave profiles are shown in Fig. 7 and Fig. 8, while the sinkage and trim are shown in Fig. 9. There are no published experimental data to show the sinkage and trim for Model 316. Thus the data of Model 316B, Model 316C and Model 316D [7] which are similar with Model 316 have been used for comparison.

Examination of all results indicates that the agreement between the present calculation and experiments is quite good. If the hull is discretized into 100 panels, the total cpu time for computing wave resistance, wave profile, sinkage and trim for 9 Froude numbers between 0.1 and 0.5 is about 6.7 minutes on a VAX 785 computer.

CONCLUDING REMARKS

The mathematical model WITH three singularity systems, which show the effects of the hull, transom stern and overhang aftbody, represents a ship with a transom stern to provide reasonable numerical results in predicting wave resistance, wave profile, sinkage and trim. Computing time can be significantly improved by the efficient numerical treatment on the Green function and on the derivation of analytical integrals of the wave system and resistance for each panel element. In computing wave resistance, the linearly distributed source strength on each panel element produces better results than the monopole source centered on the centroid of the panel. Especially in the case of low Froude number, the results calculated by the monopole source are usually overestimated.

Generally speaking, the computational results have shown a good agreement with the experimental results, but the discrepancy in the trim calculation, as shown in Fig. 9, increases as the Froude number becomes higher. It may be attributed to completely ignoring the ship wake where the vortex effect at high Froude numbers seems remarkable. The vortex effect in the wake behind the transom stern should be investigated.

ACKNOWLEDGMENTS

The authors wish to thank the Natural Sciences and Engineering Research Council of Canada and the Defence Research Establishment Atlantic for the research support.

REFERENCES

1. Yeung, R. W., "Sinkage and Trim in First Order Thin Ship Theory", Journal of Ship Research, Vol.16, No. 1, March 1972.

2. Yim, B., "Analyses of Waves and the Wave Resistance due to Transom-Stern Ships", Journal of Ship Research, Vol. 13, No. 2, June 1969.

3. Cong, L. Z. and Hsiung, C. C., "Numerical Treatment of the Green Fuction of a Moving Source under the Free Surface", Research Report of Technical University of Nova Scotia, in preparation.

4. Newman, J. N., "Evaluation of the Wave-Resistance of the Green function: Part 1- the Double Integral" Journal of Ship Research, Vol. 31, No. 2, June 1987.

5. Hsiung C. C., "Optimal Ship Forms for Minimum Wave Resistance", Journal of Ship Research, Vol. 25, No. 2, July 1981.

6. Wehausen, J. V. and Laitone, E. V., "Surface Waves", Handbuch der Physik, Springer-Verlag, Berlin, Vol. 9, 1960.

7. Murdey, D. C.,"Resistance, Propulsion, and Seakeeping Experiments with Model 316, and Propellers 53L and 53R", NRC Report LTR-SH-265, February 1980, Limited Distribution.

8. Proceedings of the Workshop on Ship Wave-Resistance Computations, Washington D.C., November 1979.

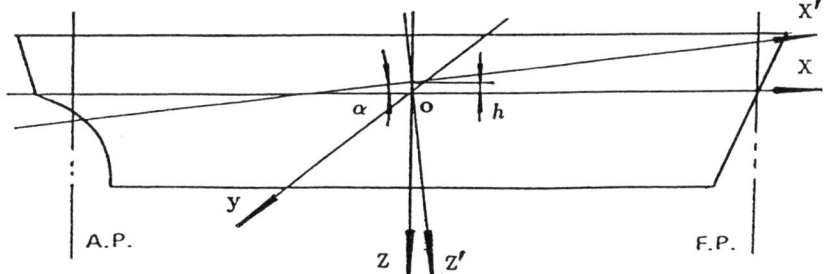

Fig. 1 Coordinate Systems

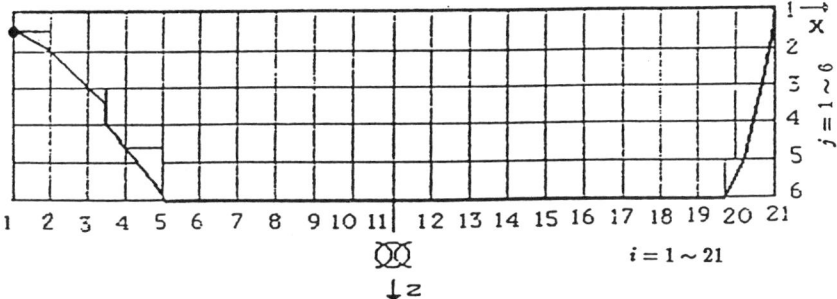

Fig. 2 Discretization of Center-Plane of DREA Model 316

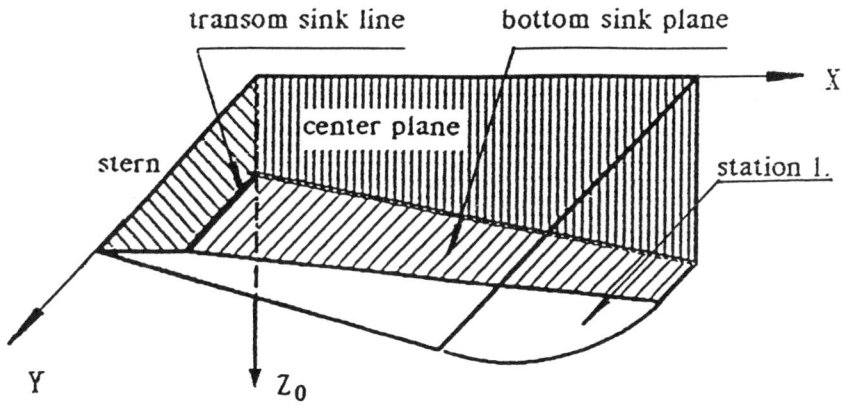

Fig. 3 Arrangement of Sink Line and Sink Plane

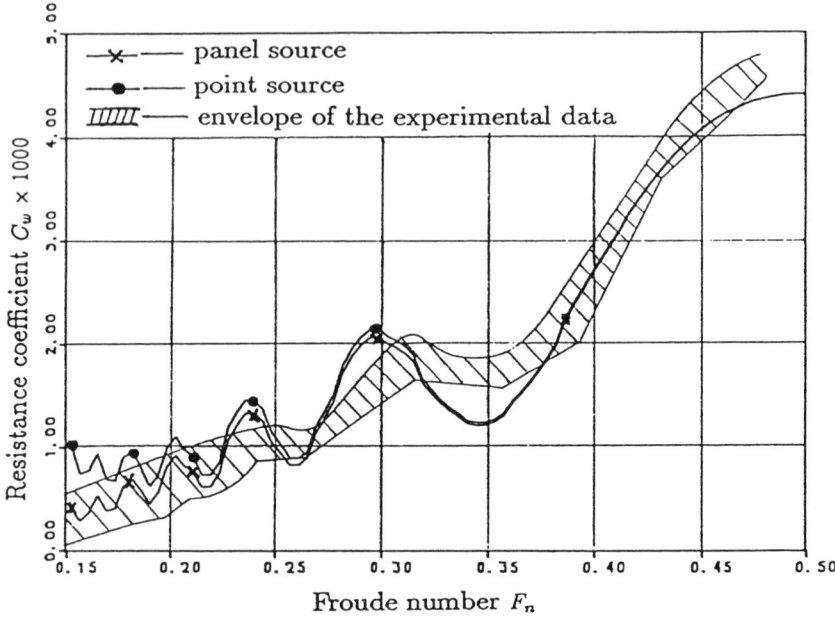

Fig. 4   Wave Resistance for Wigley's Model

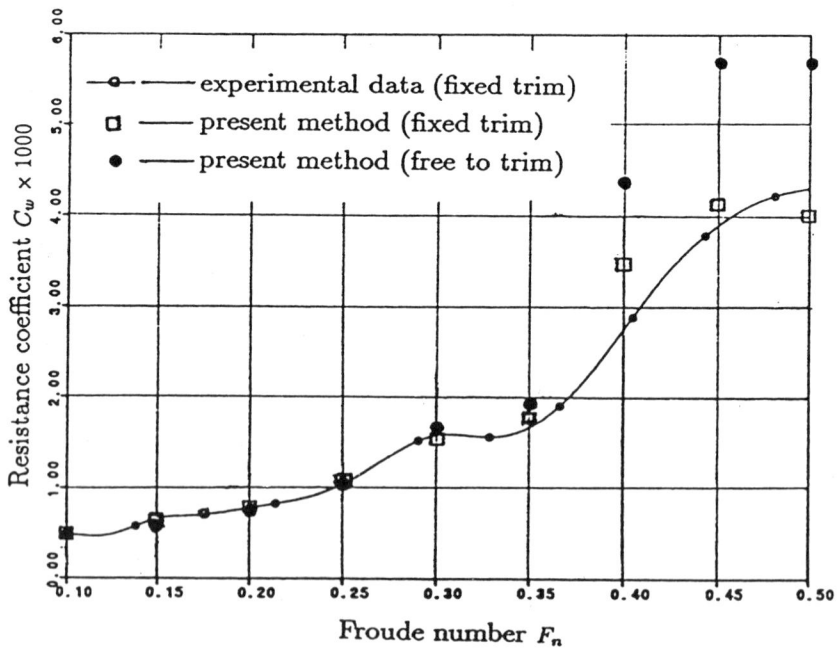

Fig. 5   Wave Resistance Coefficient of DREA Model 316

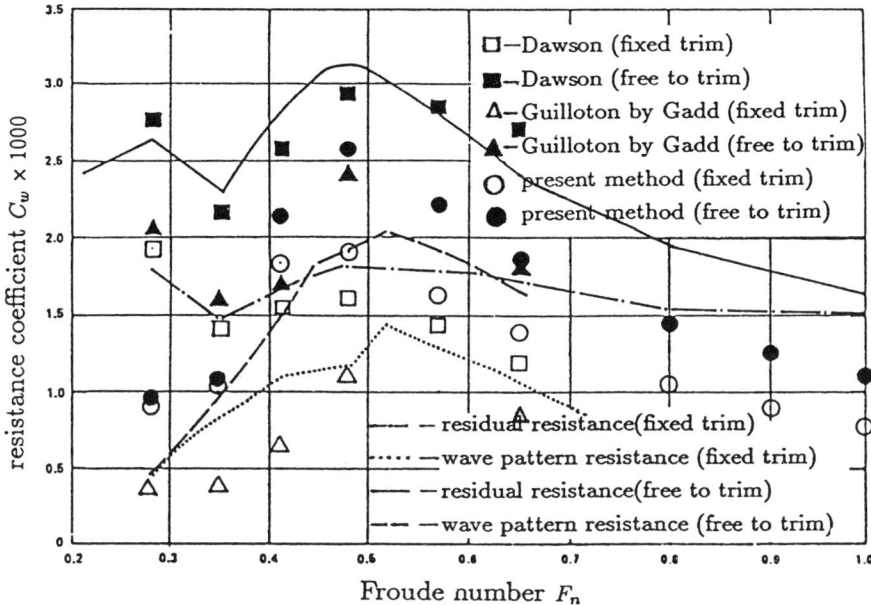

Fig. 6 Wave Resistance Coefficient of ATHENA Model

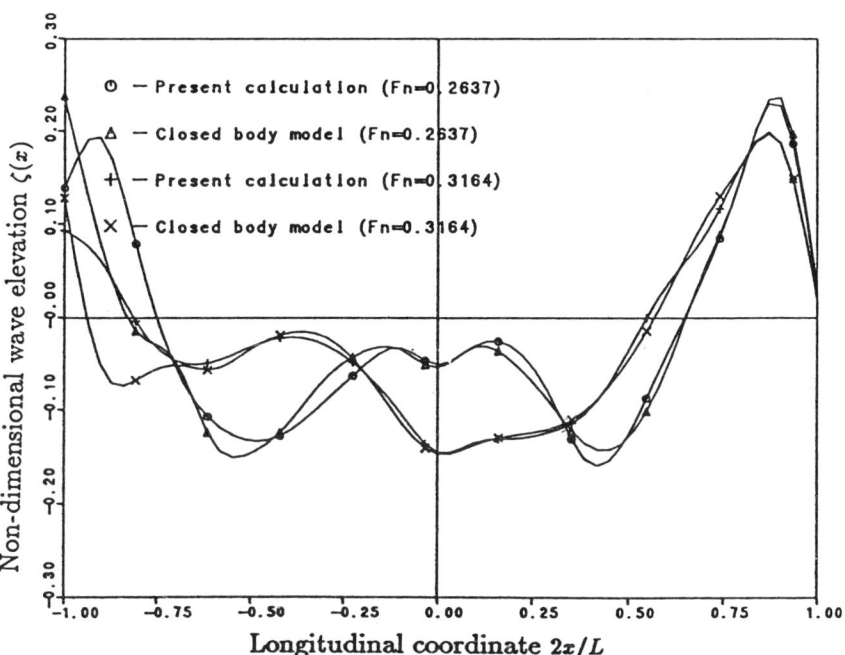

Fig. 7 Wave Profile for DREA Model 316

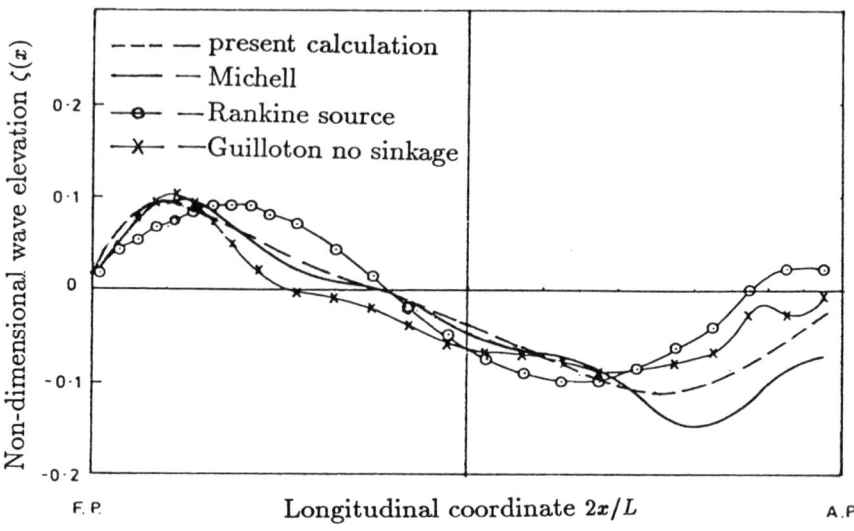

Fig. 8   Wave Profile for ATHENA Model ($F_n = 0.48$)

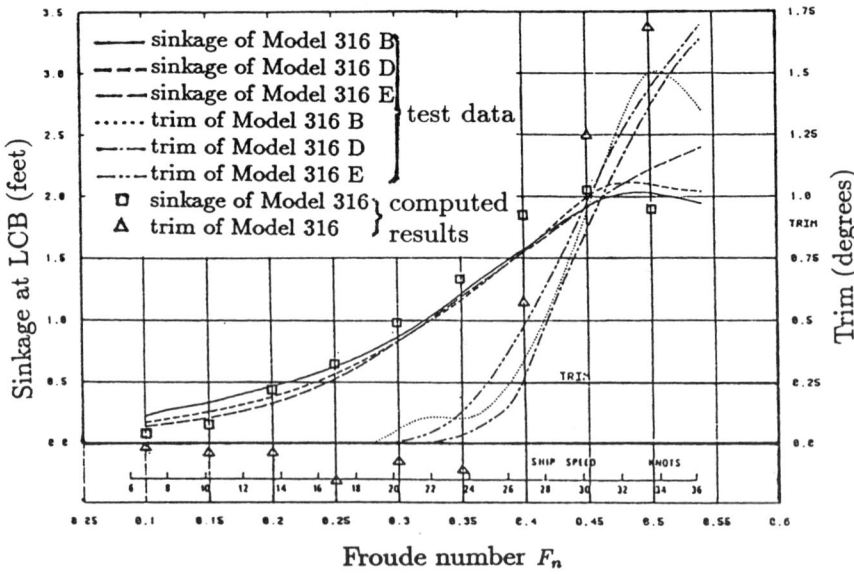

Fig. 9   Sinkage and Trim of DREA Model 316

A Computer Tool for Solving the Wave Resistance Problem for Conventional and Unconventional Ships

J.J. Maisonneuve and G. Delhommeau

SIREHNA and Laboratoire d'Hydrodynamique Navale, ENSM, Nantes, France

ABSTRACT

A computer code for solving the ship wave resistance problem has been developed since 1985 at the Laboratoire d'Hydrodynamique Navale of ENSM and Sirehna company, in Nantes. This program, based on a now classical Rankine source method, provides ample information on the flow around a ship moving at constant forward speed in calm water, and can be used efficiently by ship designers. Good results have been obtained for classical hulls, and the method has been extended to new ship shapes or new configurations. Some of these less usual applications are presented here.

INTRODUCTION

The wave resistance computation by Rankine source methods had first been proposed by G.E.Gadd [1] (1976) and C.W.Dawson [2] (1977). Several workshops have shown the efficiency and the versatility of the method on classical tests. After O.Daube [3] (1980) in France and L.Larsson [4] (1984) in Sweden, a computer program called REVA has been developed at the Laboratoire d'Hydrodynamique Navale ENSM and Sirehna in Nantes [5]. Some equivalent codes have then been written in the Netherlands by H.C.Raven [6] (NSMB), in Germany by G.Jensen and H.Söding [7] (HSVA), and in Japan by S.Ogiwara and A.Masuko [8]. Thorough analyses of the theoretical background of the method have been made, especially by P.D.Sclavounos and D.E.Nakos [9] (MIT) in 1988.

These computer codes are now being used by many institutions and give good results for the wave resistance of classical hulls. Besides, they allow a numerical comparison of flows around more complex hull shapes, at least in the areas where viscous effects are small.

REVA has already been carried out successfully in many cases, in collaboration with naval architects. Usually, an iterative procedure is used : the designer proposes a body plan; the computer code gives information about the flow that can be used to deduce a better form... This information typically is about velocities, pressures and streamlines around the hull, wave heights, hydrodynamic force versus longitudinal abscissa, dynamic position of the hull, and, the wave resistance of the ship. Graphical representations of these quantities are used to make the phenomenon easier to understand. This procedure has been applied to compare several forward shapes of fishing vessels, to the optimization of Swath ships hulls or to the location of bilge keels...

The versatility of the method made it applicable to less conventional applications. Some of these applications are presented here : the shallow water and channel effects on the flow and the wave resistance, the forces between two ships sailing side by side, the wave resistance of a Swath vessel and of a sailing-boat with heel and leeway, the effect of an air cushion.

NOMENCLATURE

The definition of the axes is shown in figure 1. The ship bow is pointing to negative abscissæ.

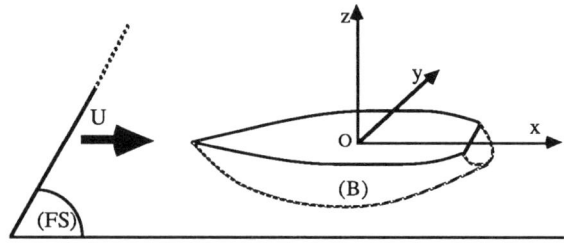

Figure 1 : coordinate system

(B)	: body
(FS)	: free surface
$(0,x,y,z)$	: coordinate system
U	: incident velocity
L	: length of the body
g	: gravitational acceleration
ρ	: density of water
Φ	: velocity potential
Φ_r	: double body velocity potential
H	: water depth
P	: pressure
Fn	: Froude number $= U/\sqrt{L\,g}$
Fh	: depth Froude number $= U/\sqrt{H\,g}$
Fh_c	: critical Froude number $= \sqrt{H/L}$
l	: curvilinear abscissa along streamlines

BRIEF THEORETICAL BACKGROUND

The fluid is assumed to be incompressible and the flow irrotational so that the velocity potential satisfies the Laplace equation in the fluid domain. On the hull, the normal velocity is set to zero : $\Phi_n = 0$. On the free surface, the boundary condition is linearized on the basis of the double-body flow :

$$(\phi_{r_l}^2 \, \phi_l\,)_l + \frac{g}{U^2}\, \phi_z = 2\, \phi_{r_{ll}}\, \phi_{r_l}^2 \bigg|_{z=0} \tag{1}$$

The method consists in distributing sources on the body and on a part of the free surface, so that integral equations can be written on these surfaces. The discretization of these equations lead to a linear system whose unknown values are the source densities. These densities allow us to compute velocities, and then pressures, forces, wave heights... The computed forces can be used to obtain dynamic trim and sinkage. The body is discretized again in its new position and another iteration can be processed. In order to build the linear system of equations translating the free surface condition, the double-hull problem has first to be solved.

A finite difference scheme is used to satisfy the free surface condition. The choice of this scheme is essential to get a good solution, and especially to satisfy the radiation condition, because in this kind of method, only the numerical damping prevents the waves from propagating upstream.

More details on the theoretical background of the method used in REVA can be found in [10].

THE PROGRAM : REVA

Special work was done to make the method easier to use and the results easier to understand (Figure 2).

The module called MAYA makes it possible either to read outputs from a CAD tool or to digitalize a body plan. The parameters of discretization can then be chosen to obtain the desired panelled geometry. Besides, other data like incident velocity, finite difference scheme, method for computing the forces, free surface grid parameters, control of outputs..., can be given with the help of the module DREVA.

The computation can then be done, in a batch mode if necessary (REVA).

The main standard results, which can be converted into graphical outputs, are the following:

      - velocities and dynamic or total pressures on the body. This can be used to understand the flow around the hull, in order to improve the shape.

      - integral of forces versus longitudinal abscissa along the hull. This makes it possible to see the longitudinal distribution of the wave resistance and is useful when comparing several hulls.

      - streamlines anywhere in the fluid. This is often used to locate correctly bilge-keels or other appendages.

      - wave elevation anywhere in the free surface discretized domain, and especially along the hull.

      - forces applied by the fluid on the body, in particular wave resistance.

      - dynamic displacements (trim and sinkage).

The computing time and storage capacity required by the program make it easier to be used on powerful computers. For six Froude numbers and for a case of 650 panels on the hull and on the free surface, the computation only takes about 80 seconds on a Cray XMP-14. So, when the cases are not too complex or too numerous, most existing personal workstations are sufficient. The adaptation to powerful micro-computers is possible and now being studied.

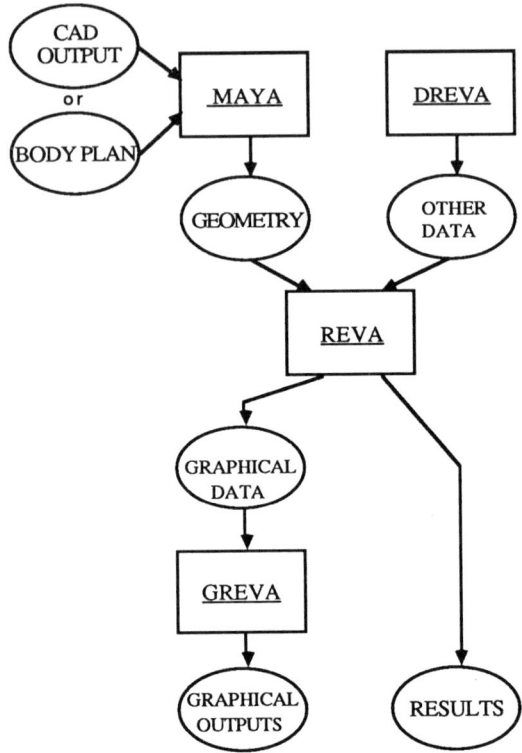

Figure 2: general organization REVA

LIMITATIONS AND STANDARD APPLICATIONS

The method is subject to some limitations. It is a potential flow method, and viscous effects cannot be taken into account. So the flow in the aft part of the ship may not be very well represented. In some cases such as transom sterns, a simple model has been developed to simulate viscous effects. This makes it possible to derive stable results on such ships, but the wave resistance itself is not always very accurate. On the other hand, differences between several shapes have always been well represented by calculation and confirmed by experiment. So, a comparative use of the program is recommended.

Low Froude numbers (< 0.2) can also be a problem for the method, especially when computing the wave resistance because of theoretical limitations [11]. Anyway, in this case, the wave resistance is very low compared with other resistances.

In spite of the basic assumptions of the method, results have been found to be acceptable for high Froude numbers (up to 1).

In all cases, the user must be very careful especially when defining the free surface grid, because results can be affected by small mistakes. The results must always be considered with care.

On these conditions, interesting applications can be made. After   many tests on classical hull shapes of the wave resistance literature, REVA has been applied to a number of industrial cases. Among them we can mention :
- the comparison between several bow shapes of trawlers and other fishing vessels.
- the location of bilge keels on several boats (fishing vessels...)
- the complete study of one hull of a SWATH vessel to minimize the wave resistance [12].
- the comparison of several bow shapes of a semi-submersible boat.
- the computation of the flow around multihull ships.
- the computation of the flow around specific bodies towed in a model basin.

Besides those common applications, it has been noticed that many people were also interested in the flow and wave resistance of new configurations. So several particular developments of the method have been made to handle with less conventional cases.

UNCONVENTIONAL APPLICATIONS

Flow around the ship in restricted water
The method used to take into account shallow water and channel effects consists in distributing sources on a part of these new surfaces, and to impose on them the same boundary condition as on the hull. The size of linear systems of equations to solve is simply increased by the number of panels on the boundary surfaces. This method allows us to deal with the case of a boat sailing in a channel or in a model basin with a transverse section of any shape.

This method was applied on a classical 124 m long Series 60 ship, first in infinite depth, then with a bottom for two depths (H = 12.70 m and   H = 19.84 m), and finally with walls 2H apart, with H = 12.7, 19.84, 30 and 40 m. The hull was discretized into 230 panels, the bottom and walls into 320   panels, and the free surface into 400 panels.

In figure 3, the wave resistance coefficient in finite depth is compared with the coefficient in infinite water and with the results of the simplified Schlichting theory [10]. The results are coherent with this theory, except when approaching the critical Froude number $F_h=1$ ($F_n = 0.32$ for H = 12.70, $F_n = 0.4$ for H = 19.84). It can be noticed that the wave resistance is increased in shallow water. On the other hand, in the figure 4 the walls seem to counteract this effect, and the coefficient decreases when the size of the basin decreases, except for high Froude numbers and depths. This phenomenon is also observed on the dynamic trim and sinkage. The presence of a bottom increases the "squat" phenomenon (Figure 5), but the addition of walls decreases it (Figure 6).

The pressures on the bottom and walls can easily be computed as well as the wave pattern.

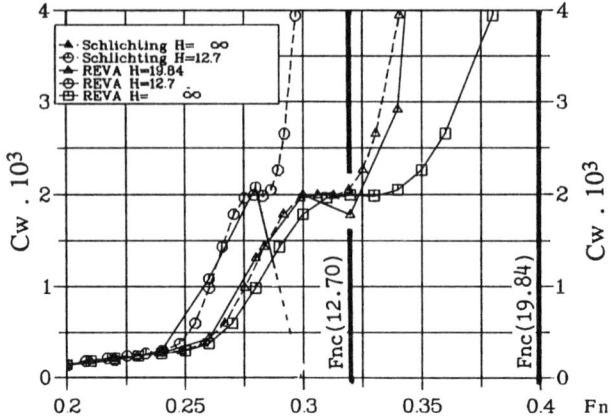

Figure 3 : Wave resistance coefficient for the Series 60 with a bottom

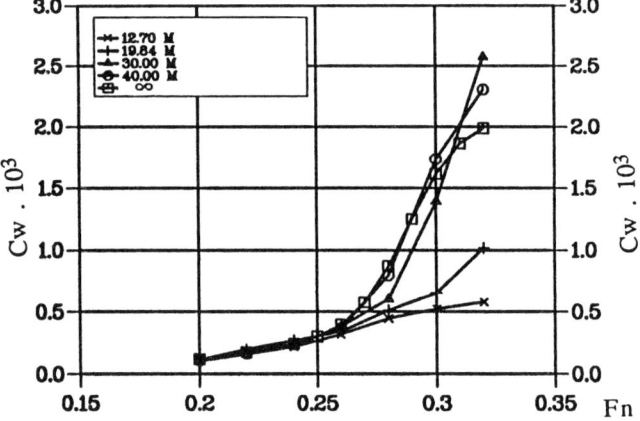

Figure 4 : Wave resistance coefficient for the Series 60 with a bottom and walls

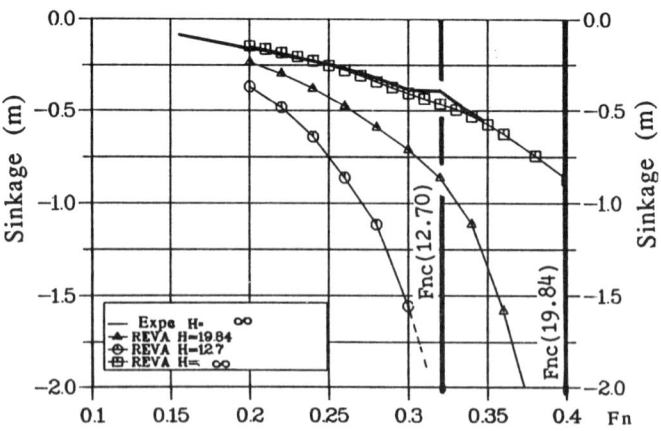

Figure 5 : Sinkage for the Series 60 with a bottom

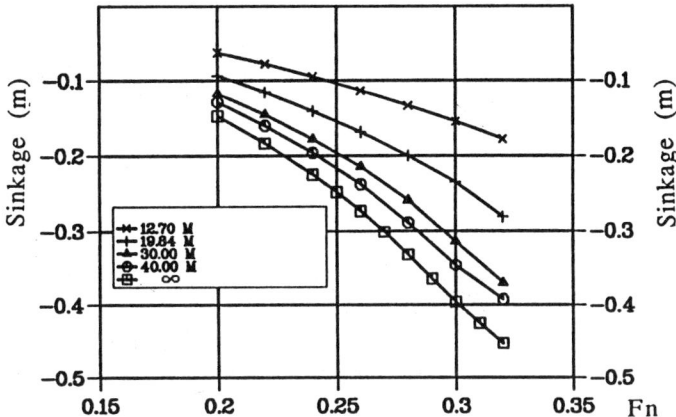

Figure 6 : Sinkage for the Series 60 with a bottom and walls

Ships sailing side by side

REVA is now able to deal with symetrical or dissymetrical boats with one, two or three hulls. One interesting application is the case of two ships sailing at the same speed, side by side.

We have considered here the same ship as in the previous paragraph (Series 60), and a 80 m long Wigley mathematical hull. The total number of panels is about 1380, because of the dissymetry of the configuration (460 on the Series 60, 260 on the Wigley, 660 on the free surface). The computation was made for two relative positions of the hulls : both bows or both sterns at the same abscissa.

In both cases, figure 7 and 8 show the longitudinal forces exerted on each hull, with and without the other one. The Froude number is here relative to the length of the Series 60. It can be noticed that the wave resistance of the rear hull is reduced to the detriment of the other. The lateral force on each hull is also computed.

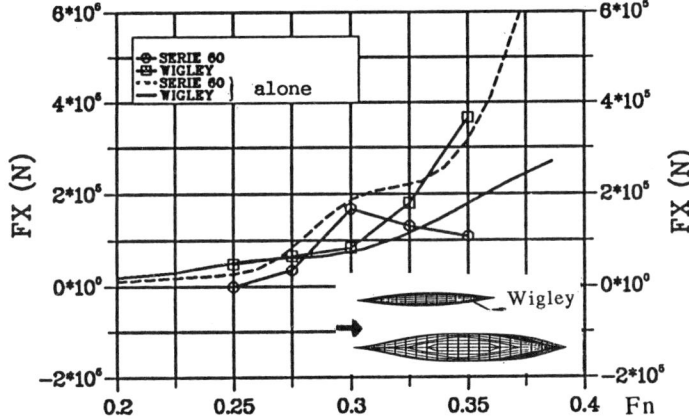

Figure 7 : Ships side by side, longitudinal force, bows at the same abscissa

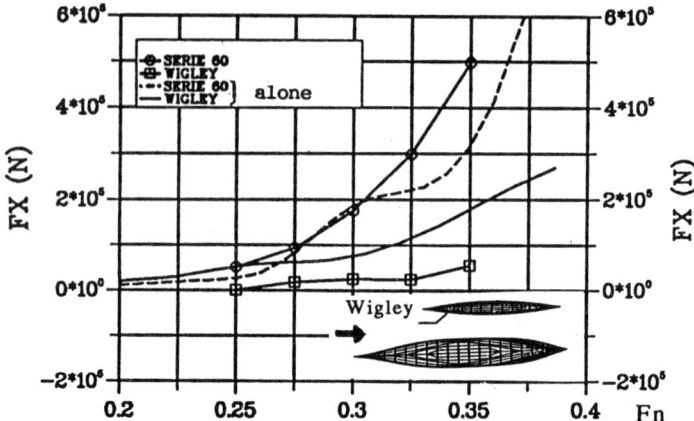

Figure 8 : Ships side by side, longitudinal force, sterns at the same abscissa

Sailing boats

Lifting effects are of prime importance for sailing-boats, therefore a modelisation of lifting surfaces was introduced in the program. Each profile is divided into lifting strips which support a dipole distribution on its centerplane in order to create an additional potential representing lifting effects. To get a unique solution to this problem, a Kutta-Joukowski condition must be applied at the trailing edge. The number of unknown quantities of the whole problem is increased by the number of unknown circulations, i.e. by the number of strips.

The method was first tested successfully in infinite water (without a free surface) on an RAE wing. It was then applied to a sailing-boat on which full scale experimental results were available : the 5.5 J.I. , 7.4 m long ANTIOPE yacht [13]. The hull was divided into 460 panels, including 240 on the keel, and the free surface into 800 panels.

Figure 9 shows the computed total resistance coefficient, using an ITTC57 formula to estimate friction, compared with experimental results obtained by HSMB and DTRC. The computed and experimental lifting coefficients are compared in figure 10, at a Froude number of 0.30. The accuracy is quite good. Figure 11 shows the effect of the heel on the total resistance for three Froude numbers.

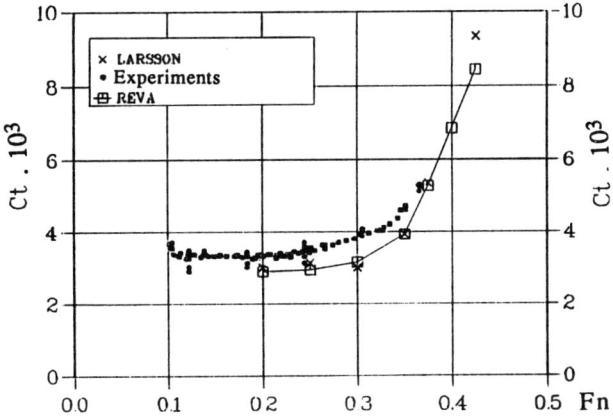

Figure 9 : Sailing boat without heel, total resistance coefficient

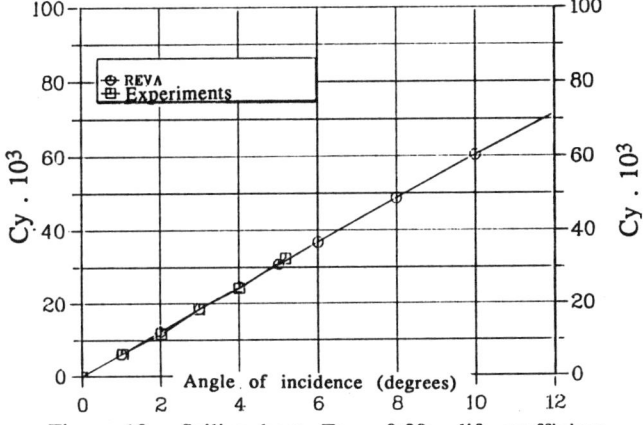

Figure 10 : Sailing boat, Fn = 0.30 , lift coefficient

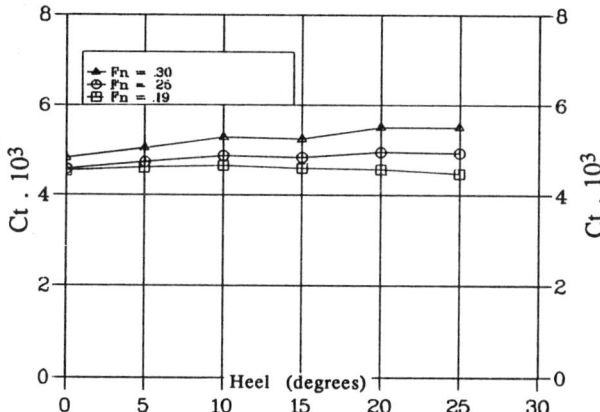

Figure 11 : Sailing boat, 5° of incidence, influence of the heel on the total drag

Surface effect ships

A pressure distribution different from the atmospheric pressure, located on a finite area of the free surface can simply be taken into account by adding a term in the free surface condition (1) which becomes :

$$(\phi_{\eta_1}^2 \phi_1)_1 + \frac{P_1}{\rho} \phi_1 + \frac{g}{U^2} \phi_z = 2 \phi_{\eta_{11}} \phi_{\eta_1}^2 \Big|_{z=0} \tag{2}$$

This possibility was introduced, and first tested successfully in the two-dimensional case by comparison with Havelock analytical results [10]. A three-dimensional rectangular air-cushion was then tested (Figure 12). The results were compared with those of Doctors [14] (Figure 13). The accuracy is quite good, especially at high Froude numbers, the free surface grid probably being a little too coarse at low Froude numbers.

The computation can easily be extended to surface effect ships (SES). As an example we have considered a ship made of two 80 m long Wigley hulls (biwigley). The Figure 14 compares the resistance of this boat to the resistance with a pressure of 1000 Pa between the hulls, and to the resistance of the air cushion itself. The total resistance with the cushion is almost the same as without it, but the displacement of the ship is increased by about 10 %.

A real surface effect ship was then tested (NES, [15]). The wave pattern created by this ship at a Froude number of 0.4 is shown in figure 15.

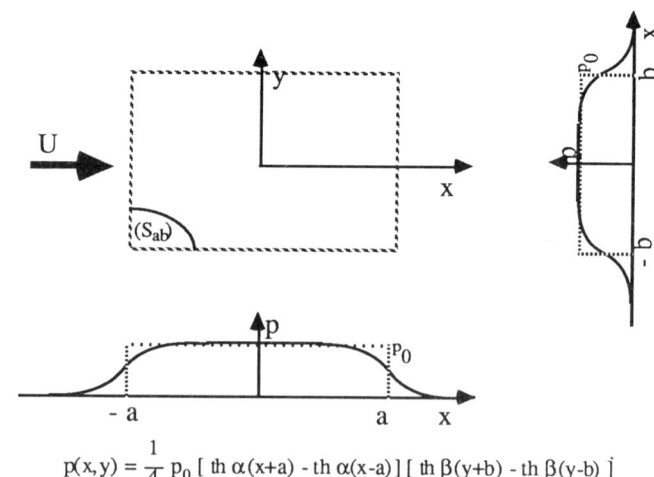

$$p(x,y) = \frac{1}{4} p_0 [\, th\, \alpha(x+a) - th\, \alpha(x-a)] [\, th\, \beta(y+b) - th\, \beta(y-b) \,]$$

Figure 12 : Definition of the air cushion

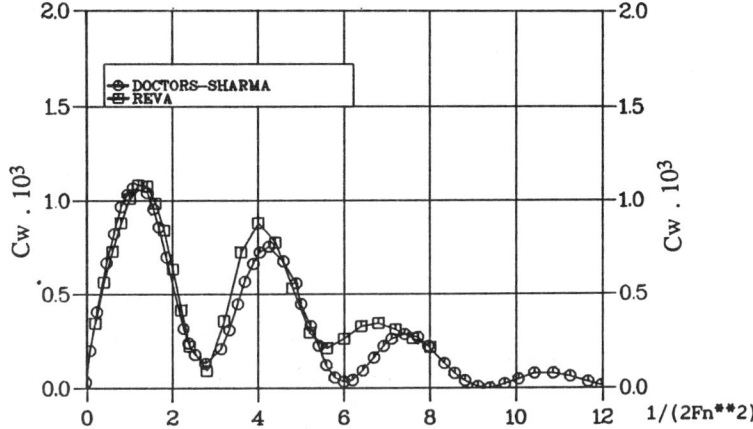

Figure 13 : Rectangular air-cushion, wave resistance coefficient

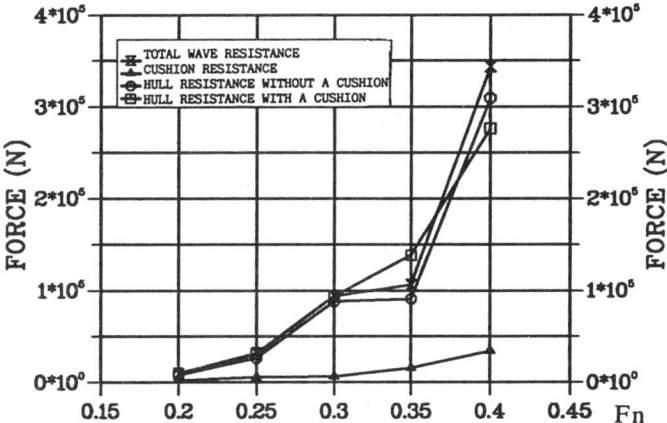

Figure 14 : wave resistance of a simplified Surface Effect Ship (biwigley)

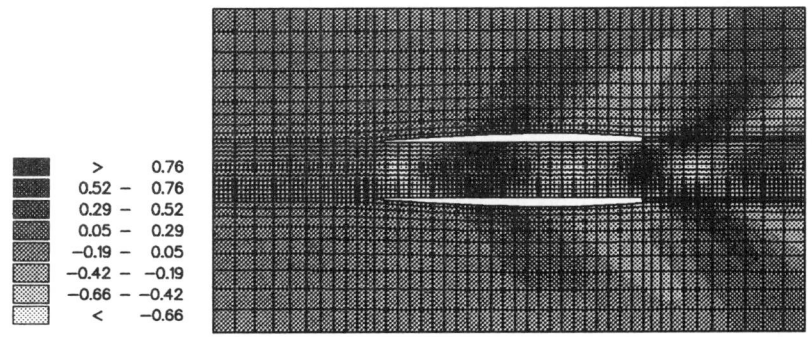

Figure 15 : wave pattern of a real Surface Effect Ship, Fn = 0.4

Semi-submersible vessel

As a last example of computation on unconventional ships, the case of a three hull semi-submersible vessel was studied. Figure 16 presents the wave pattern around the ship, at a Froude number of 0.76 (this Froude number is relative to the length of one hull). Some problems may occur with this kind of ships, because of strong interactions between hulls at high speeds, but the computed wave resistance curves are quite similar to the experimental ones.

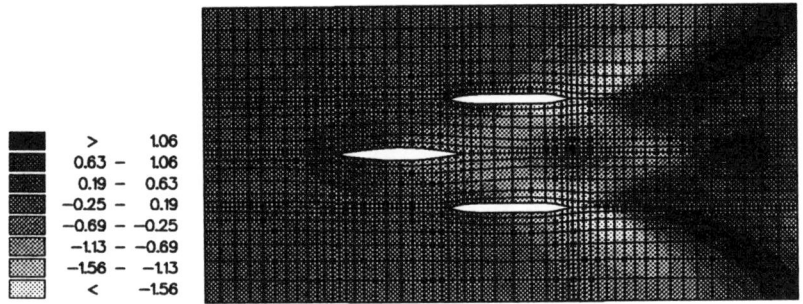

Figure 16 : wave pattern of a Swath ship. Fn = 0.76

CONCLUSION

The basic program, REVA, founded on a Rankine source method is already used in many classical cases to solve the wave resistance problem. Some specific developments have been made to apply this method to unconventional configurations. Some of these developments are now beginning to be used (multihull ships, sailboats...), others require further validation in real conditions (surface effect ships...).

Small improvements will be made about problems like flow at transom sterns...The refinement of the method itself, for example taking into account the exact nonlinear free surface condition, are foreseen, but are not considered as a necessity to obtain good results in most cases. So further developments will rather concern new applications of the method : far-field flow computation, other unconventional configurations...During the next few years, a method for automatic optimization of ship hull shapes will also be developed.

ACKNOWLEDGEMENT

This work was financially supported by the Direction des Etudes, Recherches et Techniques of the French Defence Ministry, under contracts n° 88/455/DRET and 89/456/DRET.

REFERENCES

[1]   GADD.G.E. Improves Methods for the Calculation of Ship Resistance , pp.
      139-156, International Conference CADMO , Washington ,1986

[2]   DAWSON.C.W. A Practical Computer Method for solving Ship-Waves
      Problems, pp30-38, Second International Conference on Numerical Ship
      Hydrodynamics , University of California, Berkeley, 1977

[3]   DAUBE.O. Contribution au Calcul Non Linéaire de la Résistance de
      Vagues d'un Navire, Thèse de Doctorat ès Sciences, Université   de   Paris
      VI, 1980

[4]   LARSSON.L , XIA.F. Calculation of Potential Flow with a Free Surface" ,
      Swedish Maritime Research Center SSPA , report No 2912-1, 1984

[5]   DELHOMMEAU.G , MAISONNEUVE.J.J. Application de la méthode des
      singularités de Rankine au calcul de la résistance de vagues, ATMA No
      86 , Paris, 1986

[6]   RAVEN.H.C. Variation on a Theme by Dawson, pp 151-172, Proceedings
      of the 17th Symposium on Naval Hydrodynamics , The Hague , 1988

[7]   JENSEN.G , MI.Z.X , SÖDING.H. Rankine Source Method for Numerical
      Solutions of the Steady Wave Resistance Problem, pp 575-582, 16th
      Symposium on Naval Hydrodynamics , Berkeley, 1986

[8]   OGIWARA.A , MASUKO.A. A Method of Computation for Steady Ship
      Waves by means of Rankine Sources and its Application to Hull Form
      Design, pp 97-116, International Conference CADMO, Washington, 1986

[9]   SCLAVOUNOS.P.D , NAKOS.D.E. Stability Analysis of Panel Methods for
      Free-Surface Flows with Forward Speed, pp 173-193, Proceedings   of the
      17th Symposium on Naval Hydrodynamics , The Hague, 1988

[10] MAISONNEUVE J.J. Résolution du problème de la résistance de vagues
      des navires par une méthode de singularités de Rankine, Thèse de
      Doctorat, ENSM, Nantes, 1989

[11] RAVEN H.C. Adequacy of Free Surface Condition for the Wave Resistance
      Problem, 18th Symposium on Naval Hydrodynamics, Michigan, 1990

[12] MAISONNEUVE.J.J , NAHON.J.C. Optimisation de la résistance de vagues
      de carènes semi-submersibles, pp 189-204, Deuxièmes Journées de
      l'Hydrodynamique , Nantes , 1989

[13] KIRKMAN K.L., PEDRICK.D.R. Scale Effects in Sailing Yacht
      Hydrodynamic Testing" , pp 77-125, Trans. SNAME   , 1974

[14] DOCTORS.L.J , SHARMA.S.D. The Wave Resistance of an Air-Cushion
      Vehicle in Steady and Accelerated Motion, pp 248-261, Journal of Ship
      Research ,Vol 16, No 4, 1972

[15] MAISONNEUVE J.J. Résistance de vagues des navires - Logiciel REVA,
      Rapport Sirehna 88/20/RF

2D-respective 3D-Flow Field Measurements in Open Channel Flow compared with Numerical Models

W. Bechteler, H. Sattel, K. Schätz, A. Tasdemir

Institute for Hydraulics

University of the Armed Forces, Munich, D-8014 Neubiberg, Germany

ABSTRACT

One obtains control data by means of various refined measurement methods in order to validate complex computer programs. The calculated forecast shows a good agreement with the measured results . The experiments were carried out in a 30 meter long tilting flume with a width and a height of 1 meter each [7].

NOMENCLATURE

ρ density of the fluid
g   gravity
b   width of the channel
h   waterdepth
ζ defined wavestructure in a final range $x_1 < x < x_2$ and $y_1 = -b/2 < y < y_2 = b/2$
$\Phi(x,y,z)$ velocity potential
Φ_x, Φ_y, Φ_z velocity components
A_v, B_v amplitude of a wave symmetrical or antisymmetrical to the channel
k_v propagation factor

INTRODUCTION

The first series of measurements determined the 3D-flow field around a vertical cylinder with a diameter of 10 centimeters in the middle of the channel. The cylinder consisted of several discs with the probes mounted in one of the discs and placed at the level required by the measurement. In addition, the entire cylinder may be rotated around its vertical axis in order to adjust a probe at desired angles.

The second series of measurements was the determination of the flow field around a pontoon ferry which has a length and a width of 1 meter each, i.e. there was no flow around the side of the ferry because it touched the channel walls. Due to this obvious 2-dimensional hydrodynamic problem the flow field was determined 2-dimensionally.

MEASUREMENT TECHNIQUES AND MEASURED QUANTITIES

In order to obtain various control data for validating the computer program PHOENICS, which is supposed to be designed not only for velocity and pressure distribution, but also for determining the deformation of the free surface, the following quantities were measured in this project:

- 2 resp. 3D-velocity distribution and turbulence intensity
  in the flow field around the body
- pressure on the surface of the body
- friction velocity on the surface of the body
- 2D-deformation of the free surface

The velocity distribution around the circular cylinder was determined first 2-dimensionally and then 3-dimensionally with a 3D-Laser-Doppler-Velocimeter by utilizing fiber optic probes. The velocity distribution around the ferry was measured 2-dimensionally due to reasons mentioned above. For all LDV measurements the coincidence mode was used in order to be sure that only signals were accepted, that were derived from the same particle in the flow, i.e. the signals had to occure simultaniously for each of the 2 resp. 3 channels before being stored. By this method the same number of data was obtained for each velocity component. One condition for the application of the coincidence mode is a good data rate and also a high data density. The 3D-Laser-Doppler-Velocimetry measurements therefore were carried out by using particles being dispersed in the water about 5 meters upstream of the cylinder. The best signal to noise ratio was achieved by means of Titaniumdioxide particles.

For the investigations of the flow around the cylinder the fiber optic probe carrier could be adjusted by a rack and pinion gear for changing and adjustiing the radius of the measurement grid. The velocity was determined around the cylinder for each of the 10 radii in 15 degree steps ranging from 0 degrees (luff of the cylinder) to 180 degrees (wake of the cylinder). For the Reynolds numbers 3.0E+04 and 1.0E+05, the velocity distribution was determined for 6 different levels (1.5, 8.0, 15.0, 25.0, 33.0 and 36.0 cm) at a flow depth of 40 centimeters. For each of the 6 levels velocity, turbulence and standard deviation were registered at 130 positions, i.e. the grid for the velocity measurements had altogether 780 knots for each of the Reynolds numbers.

The velocity distribution around the pontoon ferry was determined for 2 different drafts (loaded and unloaded) and for 3 separate water heights at critical Froude numbers. The grid ranged from 1.7 meters upstream to 2.0 meters downstream of the middle of the ferry and, due to the water heights had 200 to 350 positions. First tests with fine grids with up to 650 knots showed that one could dispense with a courser grid without impairing information on the flow. The LDV fiberoptic probe could be adjusted at desired levels, and the pontoon ferry was attached to a traversing system on top of the tilting flume, that could move the ferry up- and downstream automaticly by means of computer. For this traversing method it was necessary to have the same water height all over the tilting flume in order to have the same draft during a series of measurements. Therefore the inclination of the tilting flume was adjusted that way, that the surface of the flow was parallel to the ground of the flume. In order to have the opportunity to measure the velocities close to the surface of the ferry a new adjustable mirror for the LDV fiberoptic probe had been developed in our laboratory.

Until recently the wall shear stress or the skin friction could only be calculated from the measured velocity distribution close to the wall. In recent years friction velocity vector probes have been developed by G.Gust [4]. Once calibrated these flush mounted probes provide the opportunity of a direct determination of the skin friction without disturbing the flow. The friction velocity was determined according to the conditions required by the Laser-Doppler-Velocimetry.

The resistance of a body in a flow consists of the friction resistance, the viscous pressure resistance, and the wave resistance. The wave resistance can be calculated out of the deformation of the free surface, which was determined by means of a photogrammetic survey.   Four cameras - synchronized within 1/500s - were used to obtain a complete stereophotogrammetric survey of the deformation of the free surface 1.0 meter upsteam and 1.2 meters downstream of the object in the flow. The exact position of 5000 points was determined. A computer interpolated these points on an equidistant grid. The wave resistance was calculated out of the shape of the vertical profiles of the deformation of the free surface with a program developed at the institute.   For better comparison the wave resistance was determined for various Reynolds numbers.

Compared to the Laser-Doppler-Velocimetry the advantage of the photogrammetic survey is the simultanious coverage of a whole measurement field. The disadvantage is that one can only determine a momentary steady situation of a nonsteady problem.

Finally the measured data were compared with the results achieved by the computer program PHOENICS.

RESULTS

For the first series of measurements - the vertical cylinder in open channel flow - the comparison of the horizontal profiles of the velocity distribution showed, that the separation point of the flow moves upstream for higher measurement levels (fig. 1 and 2). Due to spaceproblems only the profiles of the lowest level and the 25 centimeter level is shown for   Re = 1.0E+05. For this Reynoldsnumber exists a two-phase mixture in the wake of the cylinder for the upper two levels, therefore the profile of the lowest level is compared with the one at the height of 25 centimeters. With increasing Reynolds numbers the separation point moves downstream; also the Karman vortex street becomes narrower, because the boundary layer becomes turbulent.

The vertical profile shows - as expected - that the z-component is very small (fig. 3), nevertheless especially in the wake it is obvious that the flow around a vertical cylinder is a 3-dimensional problem. The vectors of 36 centimeter level were deleted for Re = 1.0E+05 because of the mentioned two-phase mixture in the wake. The profile of the free surface, determined by the photogrammetic survey, can also be seen by figure 3.

The friction velocity of course increases with higher Reynolds numbers (fig. 4); From 0° to 180° measured values of the friction vector decrease roughly at the point of separation. Also the standard deviation increases in this turbulent wake region. At the stagnation point the friction vector has a value close to zero as expected.

Figure 5 shows the isoheights and the deformation of the free surface for Re = 1.0E+05 measured by the stereo photogrammetic survey. Figure 6 shows an example of the deformation of the free surface in the wake of the pontoon ferry. The following equation provides the theoretical base for calculating the wave resistance by means of the law of conservation of energy:

$$R_w = (\rho/2g) \int_{-b/2}^{b/2} \varsigma^2(x,y) \, dy + (\rho/2) \int_{-h}^{0} \int_{-b/2}^{b/2} (\Phi_y^2 + \Phi_z^2 - \Phi_x^2) \, dy \, dz$$

The wave propagation velocity is determined by the wave structure [8]. By this the next equation provides the practical base for calculating the wave resistance:

$$R_w = (\rho g b/8) \left[\sum_0^n \epsilon \, A_\upsilon^2 \, (2 - k_0/k_\upsilon) + \sum_{1/2}^{n+1/2} B_\upsilon^2 \, (2 - k_0/k_\upsilon) \right]$$

The results of the wave resistance calculation for the 5 different Reynolds numbers can be seen in figure 7 together with the results of a theoretical research [1]. Figure 8 shows the influence of the number of profiles on the results of the calculation.

The results of the second series of measurements - the 2-dimensional flow around a pontoon ferry - show an obvious influence of the trim, especially for lower water levels: E.g. for a water level of 100 mm and a draft of 36 mm (unloaded status) the flow field around the ferry with no trim is compared to that with a trim of +2 degrees (front up). The flow around the ferry with no trim separates where the front slope meets the bottom of the ferry (fig. 9), whereas the flow around the ferry with a trim of +2 degrees separates where the bottom meets the rear slope of the ferry (fig. 10). Both plots show a velocity vector with a value around zero in the separation zone. The results for the measured pressure distribution on the surface of the pontoon ferry also show this phenomina. Further the limits for the operation of this ferry in shallow water became obvious for a water height of 100 mm and a loaded draft of 72 mm: the deck of the ferry was flooded and no measurements could be carried out for this status. The results of figures 9 to 11 show the expected acceleration and figures 12 and 13 the increased standard deviation of the velocity of the flow under the ferry. Figures 12 and 13 compare this increase of the standard deviation for different trims: The standard deviation of the velocity increases more for a negativ trim.

As mentioned previously, the measurements are made to check the validity of the results obtained by commercial computer programs, which are already in existence or in development. The program is installed on a CONVAX C 202 with 128 MBytes of RAM. The standard K - ϵ model is applied as a closure model for the highly turbulent fluid motion.

First the program PHOENICS was developed to calculate the flow field around the cylinder. It was installed on a MicroVax 2000 with 3 MBytes of RAM for calculation that time. The calculation was performed in a cylindrical-polar grid with 100 cells in radial, 50 cells in azimuthal and 15 cells in vertical direction. Due to the limited RAM the extension of the grid was 0.25 m in radial and 0.25 m in vertical direction. The calculation was performed around one half of the cylinder taking advantage of the symmetry of the flow field in steady flow. The undisturbed velocities at the inlet were 0.3, 0.5 and 1.0 m/s according to the Reynolds numbers of 0.3E+05, 0.5E+05 and 1.0E+05 based on the diameter of the cylinder and the laminar viscosity of 1.0E-06 of water at a temperature of 20° C. The inflow region was extended above the total height to avoid free surface effects which are to be considered in future calculations and about 1/3 of the entire domain in azimuthal direction.

Figs. 14 and 15 show the comparison between the measured and computed flow field around the cylinder. The layer we took for this comparison was the fourth from the bottom of the channel at z = 8.0cm. The reason therefore lies in the assumption that in this layer neither boundary layer effects from the bottom of the channel nor gravity effects from the top of the computational domain have to be taken into account. There is a good agreement between both velocity fields. The differences which can be stated fairly are of two kinds. First, the differences in the upstream zone (dashed lines) lack any explanation. On the other side the divergence in the recirculation-zone might be traced back to the used turbulence model. As reported in [9] the model used underpredicts the recirculation zone by about 20% of the real zone found by measurements. Both the dependency of the expansion of the recirculation field on the inlet velocity and the dependency on the layer level show the same tendency like the results of the LDV measurements: The separation zone spreads with lower Reynolds numbers and increasing height.

The computed and measured results of the flow field around the pontoon ferry were compared by means of vertical profiles. Both the tendency of the velocity distribution and the values show very good agreement (fig. 16 to 18).

CONCLUDING REMARKS

When presented, the velocity distribution calculations of the chosen reference bodies - the vertical circular cylinder and the pontoon ferry - already showed fairly good conformity with measured results. The measurement techniques have a remarkable consistency which, of course is an important factor for obtaining reliable control data used to validate complex computer programs. The determination of the flow field around other ships will follow in order to make the program a powerful tool for an expert system .

ACKNOWLEDGEMENTS

This research is financed by the Department of Defense, West-Germany, contract T/R325/H0019/H2312. The authors are grateful to O.Dr. B. Hug for his support.

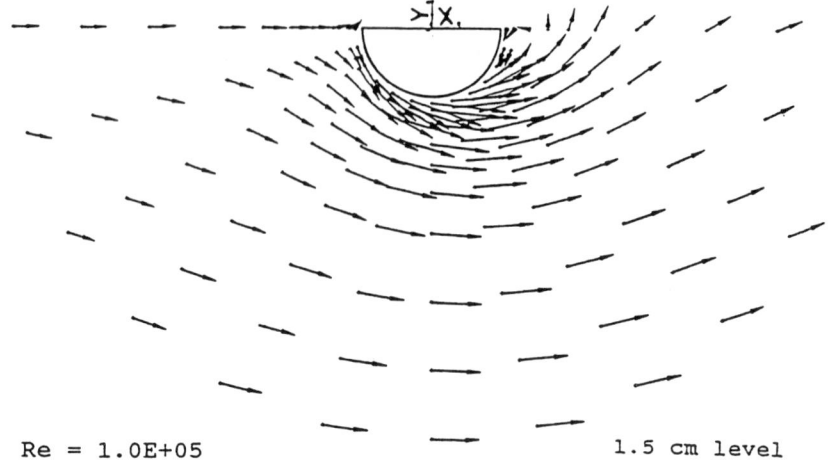

Re = 1.0E+05   1.5 cm level

Fig. 1:   Measured Horizontal Velocity Profile
          around the Cylinder

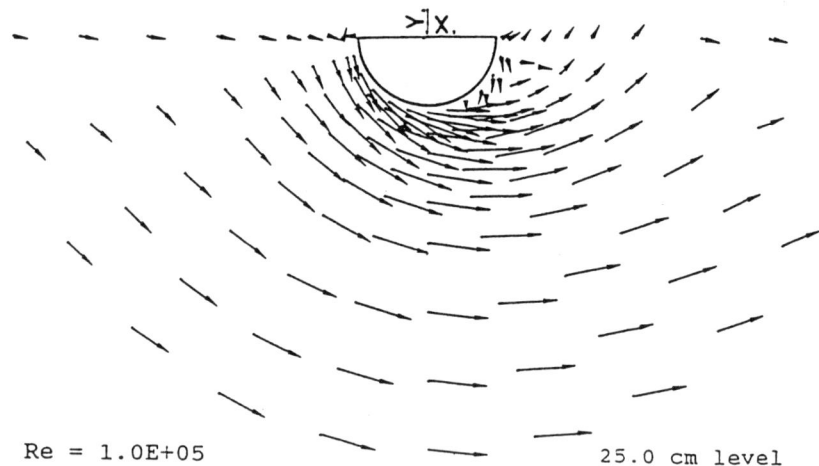

Re = 1.0E+05   25.0 cm level

Fig. 2:   Measured Horizontal Velocity Profile
          around the Cylinder

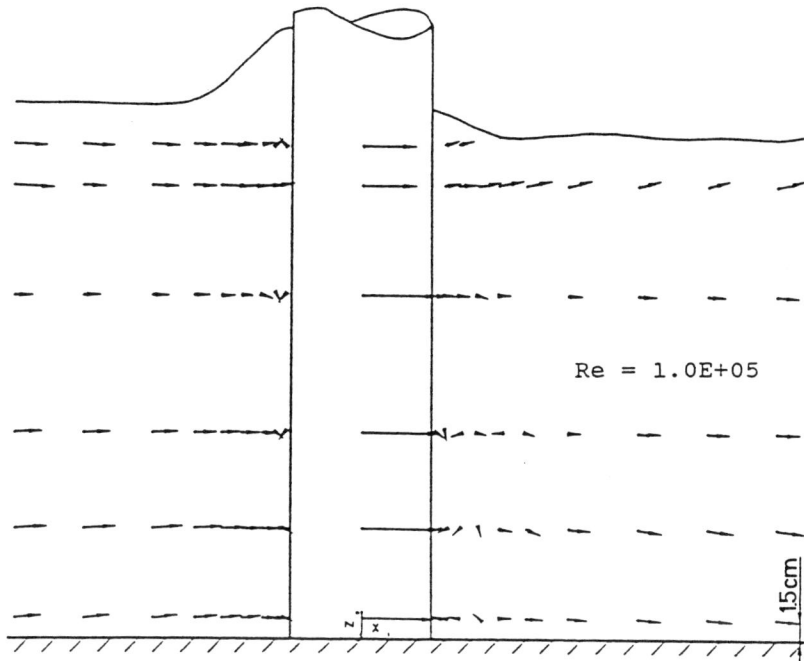

Fig. 3:   Measured Vertical Velocity Profile
           around the Cylinder

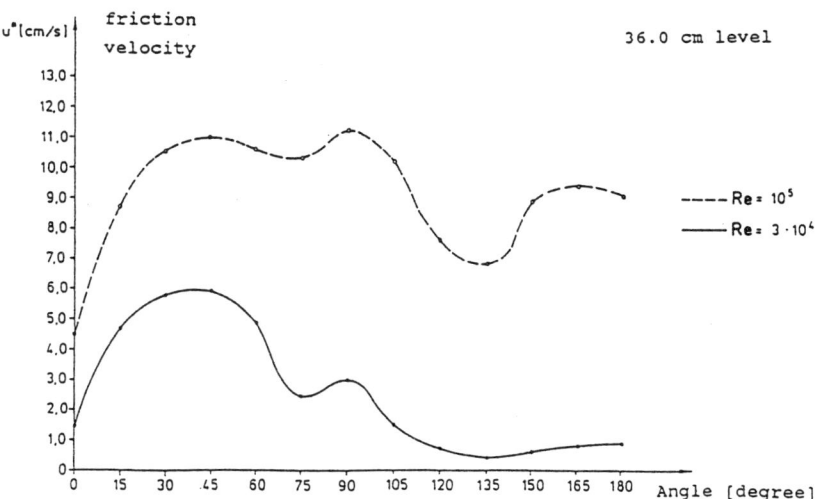

Fig. 4:   Measured Values for the Skin Friction
           on the Surface of the Cylinder

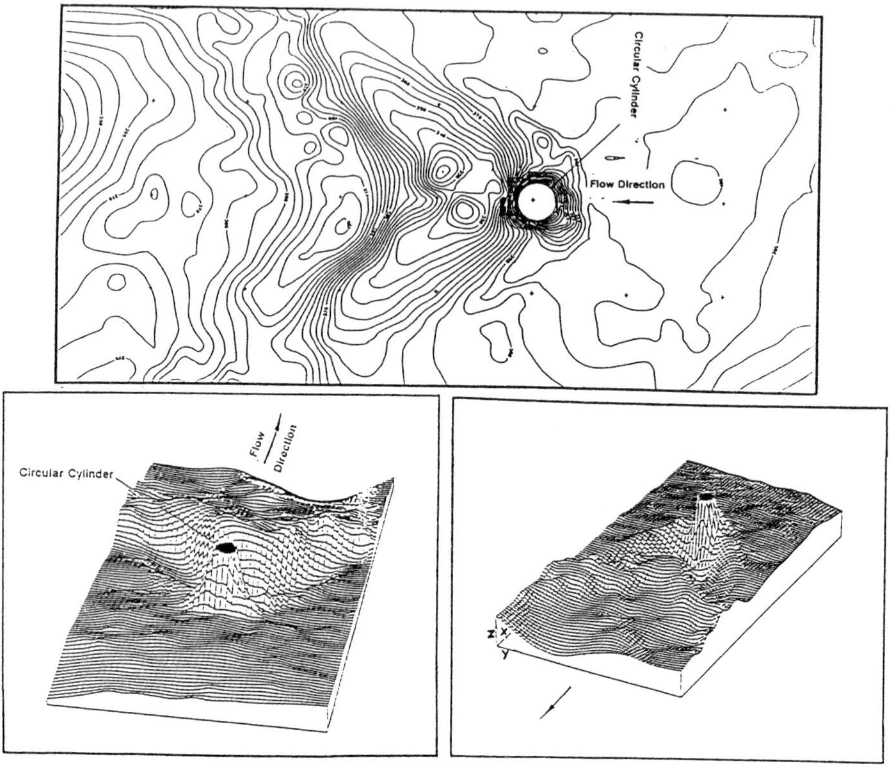

Fig. 5:   Isoheights and Deformation of the Free Surface
around the Vertical Cylinder

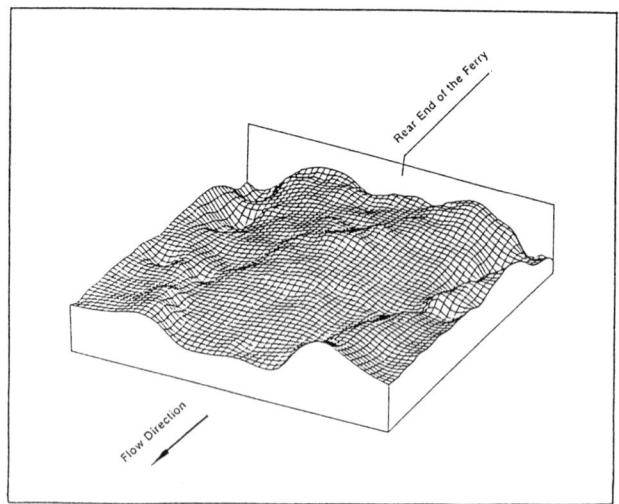

Fig. 6:   Deformation of the Free Surface in the Wake of
the Pontoon Ferry

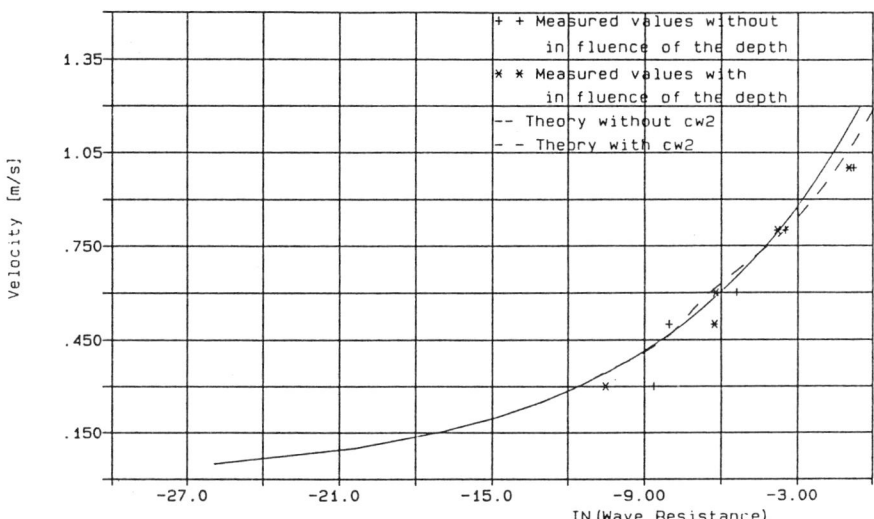

Fig. 7:   Theoretical and Experimental Results for the
          Wave Resistance of a Cylinder

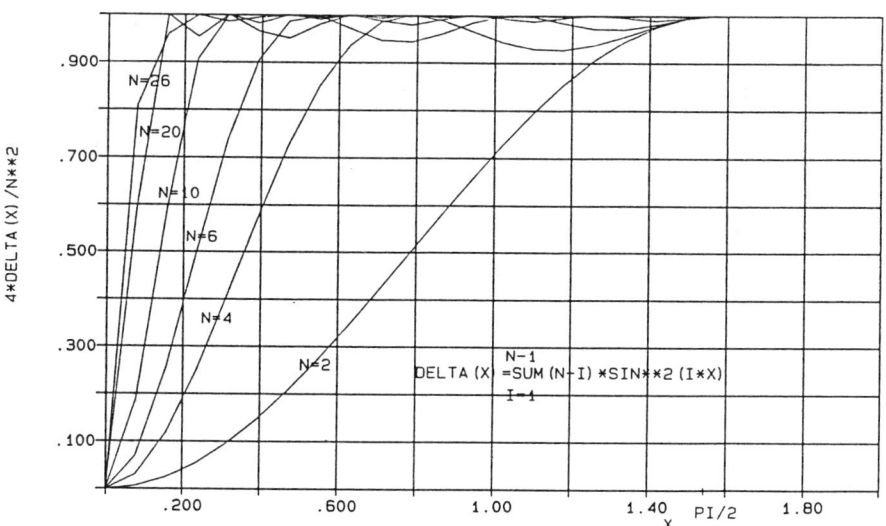

Fig. 8:   Dependency of the Results of the Wave Resistance
          Caculation on the Number of Profiles (N)

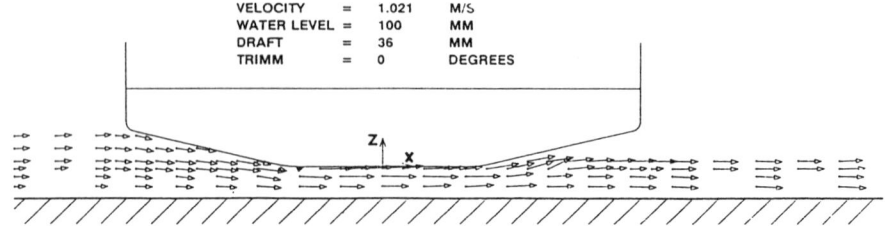

Fig. 9:   Measured Velocity Profile around the Pontoon Ferry

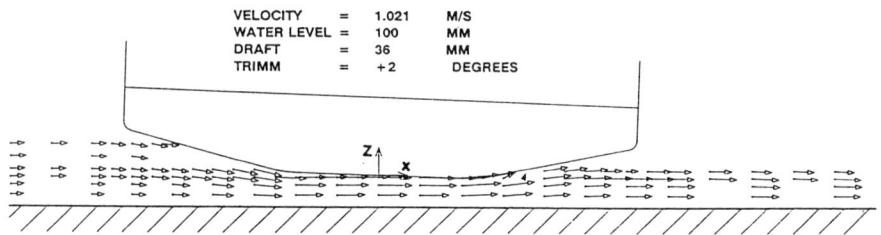

Fig. 10:   Measured Velocity Profile around the Pontoon Ferry

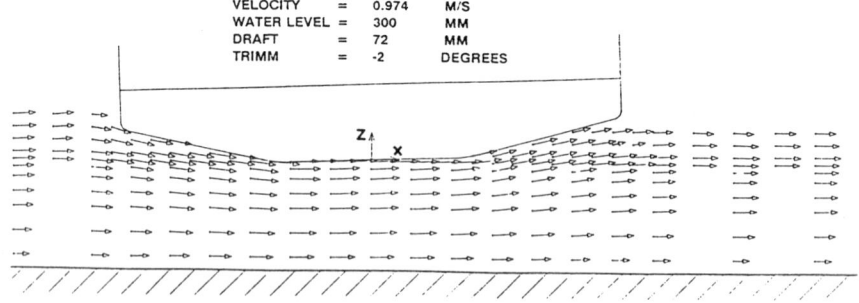

Fig. 11:   Measured Velocity Profile around the Pontoon Ferry

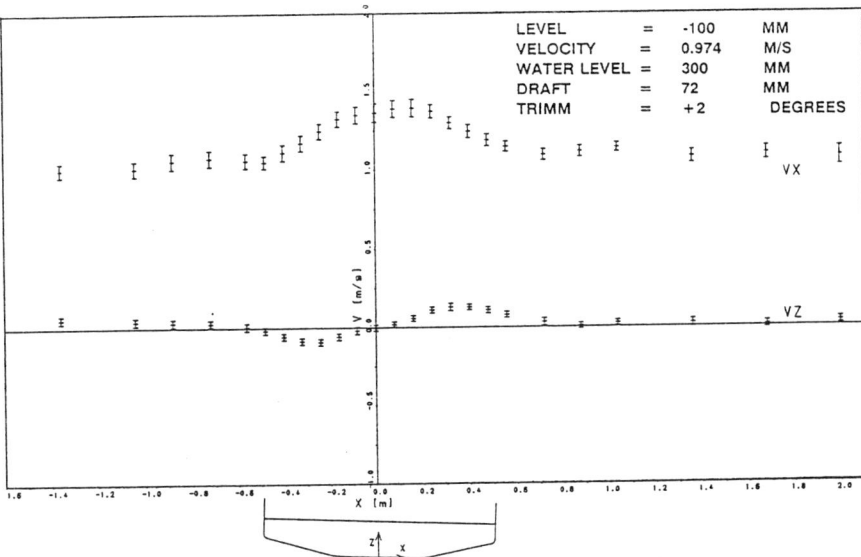

Fig. 12:   Standard Deviation of Measured Velocities

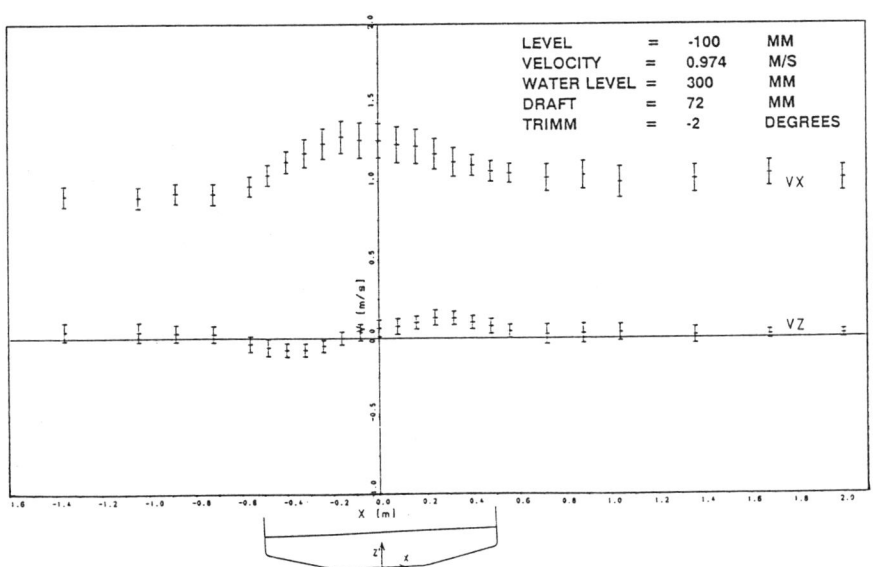

Fig. 13:   Standard Deviation of Measured Velocities

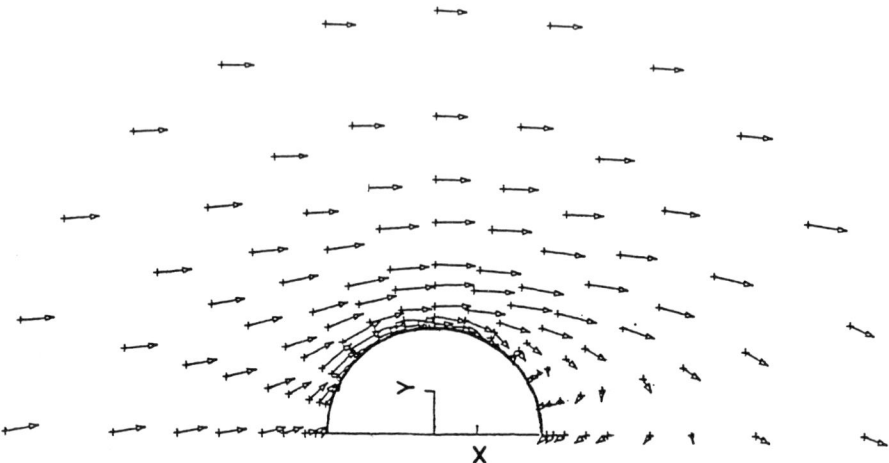

Fig. 14:   Measured Flow Field at Re = 1.0E+05

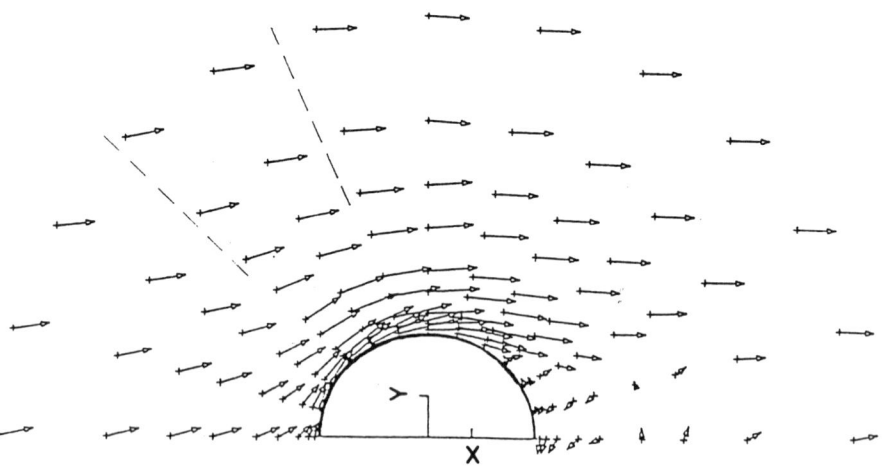

Fig. 15:   Computed Flow Field at Re = 1.0E+05

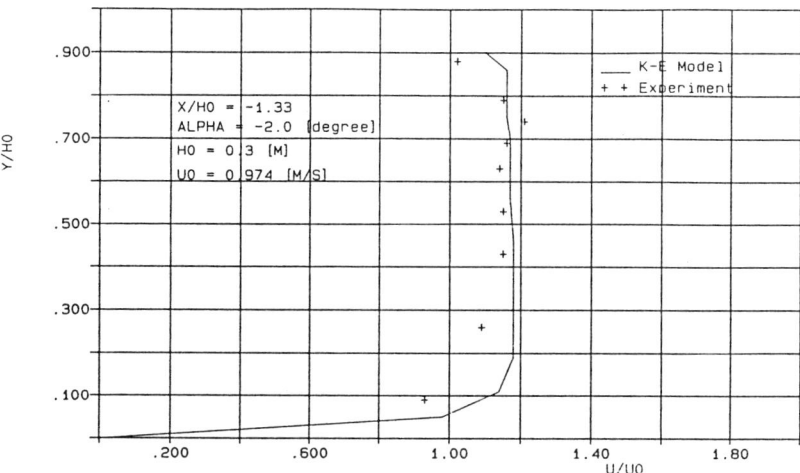

Fig. 16:   Comparison of Computed and Experimental Results

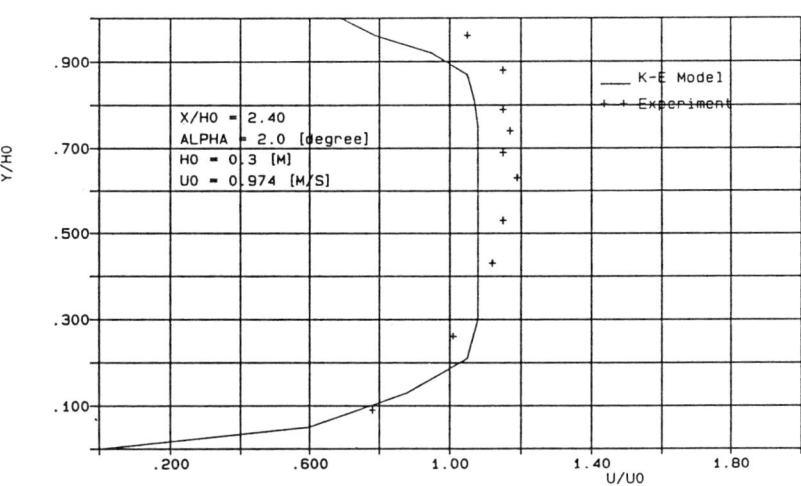

Fig. 17:   Comparison of Computed and Experimental Results

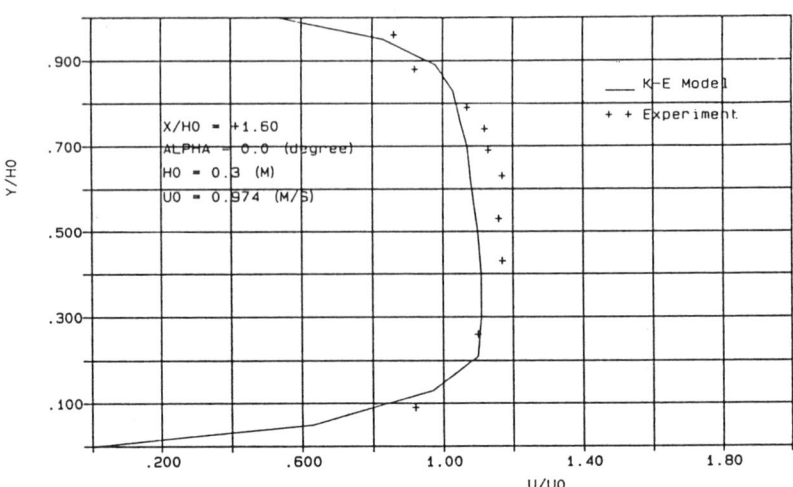

Fig. 18:   Comparison of Computed and Experimental Results

REFERENCES

[1] A.I. Aldogan, Theory and use of the nonlinear Wave Resistance of slow Ships (turkish), Technical University of Istanbul, 1983

[2] D. Durst, A. Melling, J.H. Whitelaw, Theorie und Praxis der Laser - Doppler - Anemometrie, G. Braun Karlsruhe, 1987

[3] K. Eggers, Über die Ermittlung des Wellenwiderstandes eines Schiffsmodells durch Analyse seines Wellensystems, Schiffstechnik Vol. 9 - 1962 Paper 46

[4] G. Gust, Skin Friction Probes for Field Applications, Journal of Geophysical Research, Vol.93 No. c11, pp. 14,121-14,132, 1988

[5] H. Kronewetter, Gleichzeitige Messung von drei Geschwindigkeitskomponenten, special Edition of " Jahrbuch Laser "

[6] G. Lammers, J. Laudan, Ermittlung des effektiven Zustromfeldes der Propeller von Wasserfahrzeugen unter Verwendung eines Laser - Doppler - Anemometers, The Hamburg Ship Model Basin, Report No. 1521, 1981

[7] Institute for Hydraulics at the Armed Forces University Munich, D-8014 Neubiberg, Vol. 27, 1988

[8] S.D. Sharma, Untersuchungen über den Zähigkeits- und Wellenwiderstand mit besonderer Berücksichtigung ihrer Wechselwirkung, Insitut for Ship building, University of Hamburg Report No. 138

[9] W. Rodi, Turbulence Models and Their Application in Hydraulics, 1981

[10] H.I. Rosten, B. Spalding, PHOENICS Reference Manual, 1988

Numerical Computation of the Flow Around a Diving Plane in Water Waves

S. Fontaine(*), S. Huberson(**) and J.L.Montagné(*)
(*) Bassin d'Essai des Carènes, Centre du Val de Reuil, 27100, France
(**) L.I.M.S.I. B.P.133 91403 Orsay cedex, France

ABSTRACT

The unsteady incompressible flow of an inviscid fluid around a diving plane of a submarine in water waves is computed. The distance to the free surface is a few chord length. The method used is an integral method to account for the lifting surfaces, and a particle grid free method for the wakes. The numerical results are compared to experimental data.

INTRODUCTION

We want to compute the unsteady incompressible flow of an inviscid fluid around a diving plane of a submarine in water waves. The distance between the diving plane and the free surface is assumed to be large enough so that the free surface perturbation due to this plane remains small. So, the free surface can be modelized as a sinusoidal surface and the incoming undisturbed flow is given by Stokes's explicit formula for linearized water waves. The motion of the submarine is also simplified and assumed to correspond to a constant velocity. Due to the unsteady flow around the diving plane, its hydrodynamic characteristics are unsteady and its wake can not be approximated by a plane wake as in a classical lifting surface problem. We present a method for solving this nonlinear problem.

THE NUMERICAL METHOD

The outline of the method
It results from the above mentioned assumptions that the velocity field can be expressed in the frame moving with the submarine as the sum of three terms: The submarine velocity, the velocity due to the waves and the perturbation due to the presence of the submarine and its diving planes.

$$\underline{U} = \underline{U}_\omega + \underline{U}_\infty + \underline{U}_w$$

Only the first term has to be computed. This is achieved by using an integral formulation which allows us to compute it as a linear combination of the influence of the diving plane, the influence of the submarine hull, and the influence of the vortical wakes of the lifting surfaces. The first two terms are obtained with boundary integral methods and the third one is computed with the particle method which will be decribed in the next section.

The particle method

the vortex particles concept has already been used to compute three dimensional unsteady flows around wings and helicopter rotors [1]. Particle methods have been introduced as early as 1931 [2] in order to compute two dimensional flows of incompressible inviscid fluids involving a singular vorticity distribution. The method consists in discretizing the vorticity support with a set of particles and to compute their motion This is achieved by using a lagrangian coordinate system to express the Euler equations.

$$\frac{d\mathbf{X}}{dt} = \underline{\mathbf{U}}(\mathbf{X}, t) \tag{1}$$

$$\frac{d\omega}{dt} = (\underline{\omega} \cdot \underline{\nabla})\underline{\mathbf{U}}$$

The vorticity distribution is discretized in particles. The particles are defined by two quantities: their circulation which is the amount of vorticity initially contained in the particle and a point which can be the geometric center of the particle, the baricentric center for the vorticity contained in the particle or any other point representative of the particle location.

$$\underline{\Omega_i} = \int\int\int_{P_i} \underline{\omega} d\sigma \tag{2}$$

$$\underline{\Omega_i} = \int\int\int_{P_i} \underline{\omega} d\sigma \tag{3}$$

$$\underline{\Omega_i} \wedge \underline{\mathbf{X}_i} = \int\int\int_{P_i} \underline{\omega} \wedge \underline{\mathbf{x}} d\sigma$$

The velocity field is obtained from the vorticity distribution by using the Biot-Savart integral law:

$$\underline{\mathbf{U}}(\underline{\mathbf{x}}) = \frac{1.}{4\Pi} \int\int\int_{R^3} \frac{\underline{\omega}(\underline{\mathbf{y}}) \wedge (\underline{\mathbf{x}} - \underline{\mathbf{y}})}{|\underline{\mathbf{x}} - \underline{\mathbf{y}}|^3} d\sigma \tag{4}$$

Thanks to the previous definitions, the discrete equations to be solved are:

$$\frac{d\mathbf{X}_i}{dt} = \underline{\mathbf{U}}_i$$

$$\frac{d\underline{\Omega}_i}{dt} = 0. \tag{5}$$

$$\underline{\mathbf{U}}_i = \frac{1.}{4\Pi} \sum_{j \neq i} \frac{\underline{\Omega}_j \wedge (\underline{\mathbf{X}}_i - \underline{\mathbf{X}}_j)}{|\underline{\mathbf{X}}_i - \underline{\mathbf{X}}_j|^3}$$

This method has been applied to many different problems. Particularly, Chorin [3] proposed an extension of the method to account for viscous effect both within the flow field and along the boundaries. Rehbach [4] extended the method to three dimensionnal flows. In that case, the vorticity transport equation has a non-zero right hand side which is called the deformation term. This term can be obtained through an integral relation by derivating the Biot-Savart law:

$$\frac{d\underline{\Omega}_i}{dt} = (\underline{\Omega}_i \cdot \underline{\nabla})\underline{U}(\underline{X}_i) \tag{6}$$

$$(\underline{\Omega}_i \cdot \underline{\nabla})\underline{U}(\underline{X}_i) = \frac{3}{8\Pi} \sum_{j \neq i} \frac{1}{|\underline{X}_i - \underline{X}_j|^5}[((\underline{X}_i - \underline{X}_j) \cdot \underline{\Omega}_i)((\underline{X}_i - \underline{X}_j) \wedge \underline{\Omega}_j) +$$

$$(((\underline{X}_i - \underline{X}_j) \wedge \underline{\Omega}_j) \cdot \underline{\Omega}_i)(\underline{X}_i - \underline{X}_j)]$$

The solid walls

As it has been mentioned in the previous section, this algorithm can be used to compute the wake of a moving body. The solid walls are taken into account by using an integral equation. This technique was developped first by Giesing. The velocity field computed through the Biot-Savart relation is added to a velocity field derived from a potential specially constructed in order to satisfy the boundary conditions. In this work, these boundary conditions have to be satisfied on two kinds of boundaries: those which belong to thick bodies, for example the submarine hull, and the lifting surfaces. We will split the potential into two parts with respect to these two kinds of boundaries and denote these two potentials ϕ_σ for the thick bodies, and ϕ_{ls} for the lifting surfaces. Both are obtained by solving a Poisson equation:

$$\begin{cases} \Delta\phi_\sigma = 0. \\ \frac{\partial\phi_\sigma}{\partial n} = -(\underline{U}_\infty + \underline{U}_\omega + \underline{\nabla}\phi_{ls}) \cdot \underline{n} \end{cases} \tag{7}$$

$$\begin{cases} \Delta\phi_{ls} = 0. \\ \frac{\partial\phi_{ls}}{\partial n} = -(\underline{U}_\infty + \underline{U}_\omega + \underline{\nabla}\phi_\sigma) \cdot \underline{n} \end{cases} \tag{8}$$

These two problems are coupled. Thanks to the unsteadiness of the flow, they are solved sequentially. The right hand side of the boundary conditions are computed by using the solution at the previous time step. Both potential are computed through an integral formulation of equations (7) and (8). The first one is derived from a source density distributed on the walls of the thick bodies, and the second one with a doublet distribution. This last kind of singularity makes the account for lifting effects easier.

The vorticity creation

The vorticity creation model establishes the link between the solid walls discretization and the particle wake. Particles are created along given lines in order to satisfy a condition formaly derived from the Joukovski condition. We use Hess relation between the doublet distribution and vorticity to obtain a vorticity distribution on the solid walls $\underline{\omega}_{sw}$ and this vorticity is assumed to be emitted within the fluid at a mean velocity computed along the separation lines. These separation lines are discretized in small segments σ_e which are represented by their mid point $\underline{X}_e$. At each point $\underline{X}_e$ and at each time step Δt , a particle is created according to the following relations:

$$\underline{X}_i = \underline{X}_e + \Delta t/2 \underline{U}_e \tag{9}$$
$$\underline{\Omega}_i = \int_{\sigma_e} \omega_{sw} dx \Delta t \, | \, \underline{U}_e \, |$$

The wave model

As it has been mentioned in the introduction, we assume that the free surface is not disturbed by the submarine and the deepth is infinite so the wave model is a simple two dimensional Stokes model. We just recall hereafter the explicit formula which are used to express the velocity potential. The free surface is assumed to be sinusoidal, according to the following equation:

$$y = a \sin(mx - nt) \tag{10}$$

It is a progressive wave with a wavelength $\lambda = 2.\pi/m$ and a period $\tau = 2.\pi/n$. The expression of the velocity potential is

$$\phi = \frac{acm}{\text{sh}(mh)} \text{sh}(my) \sin(mx - nt)$$

THE NUMERICAL RESULTS:

The external conditions

The diving plane has been discretized in a 12×5 mesh. This mesh is illustrated on figure 2.

The numerical conditions are specially designed to be representative of the conditions experienced by a submarine. The plane is several chord length deep, and the wave lengths are large compared to the size of the plane. The results are the global forces on the foil, and the chordwise distribution of the pressure jump at different sections of the foil. Experimental results used in the validation have been obtained in the towing tank of Paris for different wave conditions [5]. The speed of the submarine was constant and the wave caracteristics are given in table 1. h_{cc} is the total height between crest and trough, w is the pulsation and λ is the wave length.

$$h_{cc} = 1.76$$
$$w = 1.5321$$
$$\lambda = 28.31$$

An unsteady flow results from the composition of the wave induced velocity field with the submarine velocity. The time evolution of the angle of incidence of the diving plane of the submarine with this flow is:

$$\theta = \tan^{-1}((| \, \underline{U}_\infty \, | + u_w)/v_w)$$

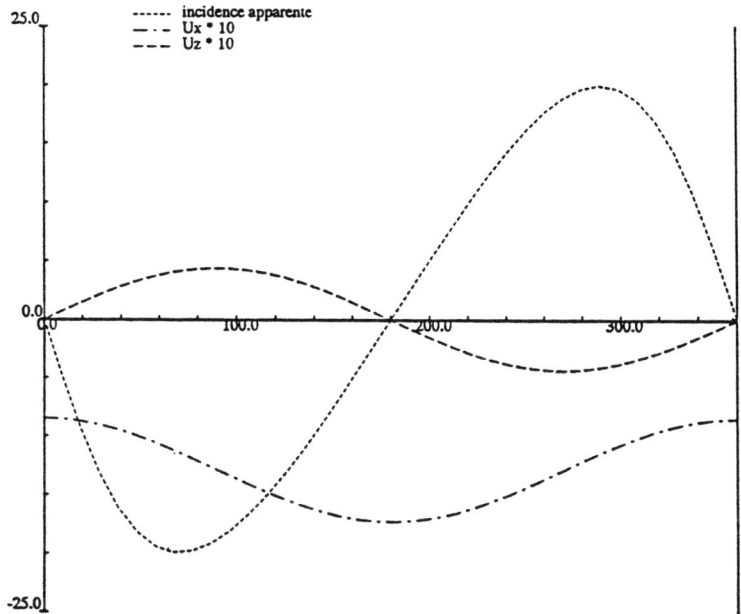

Figure 1: *angle of incidence and velocity component as a function of the phase of the waves*

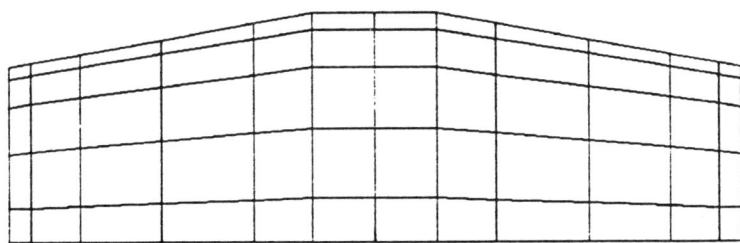

Figure 2: *Diving palne discretization: a 12 × 5 mesh is used.*

Figure 1 gives the variation of the angle of incidence and the two components of the unperturbed velocity with respect to the wave phase. It can be pointed out that the maximum angle of incidence experienced by the diving plane is always over 20^0. This, added to the unsteadiness of the flow, indicates that the wake of the diving plane must be generated at least by both trailing edge and tips. The vortical wake pattern is represented on figure 3 Under more severe conditions, leading edge separation can also occur during a part of the cycle. This last case will not be considered hereafter.

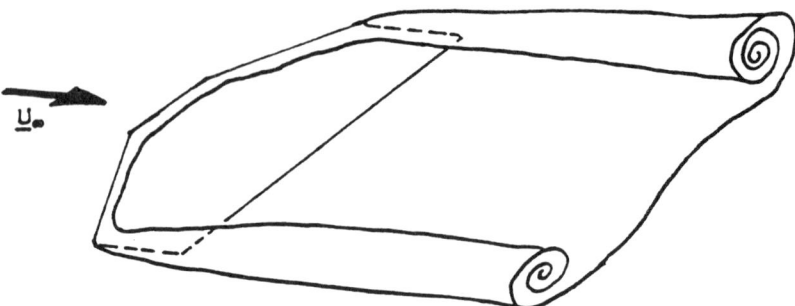

Figure 3: *wake pattern for a diving plane at an angle of incidence lower than 20 degrees*

The tips separation

In order to estimate the relative importance of the lift gap associated to the rolling up of the vortex sheet constituting the wake of the side edges, two different models were used. For the first one, only the trailing edge wake was discretized by means of vortex particles. The tips vortex is assumed to be well represented by the singularity lying on the edges of the lifting surface when no Joukovski condition is satisfied. For the second one, all the vorticity shed at the edges of the plane is discretized except at the leading edge. The results obtained with wave number 1 are presented on figure 4. It can be observed that the lift gain due to the presence of the two vortex cores emanating from the tips represent about 30 % of the total lift. This is of some importance because it indicates that a significant improvement of the efficacity of the diving plane in those conditions could be obtained by a modification of the diving plane shape.

The submarine hull

We have also tried to estimate the influence of the submarine hull. This was achieved by assuming that the perturbation caused by the presence of the hull could be modelized by the potential formulation (7). The hull of the submarine is discretized into a set of panels carrying a source density. because the wake of the diving plane still remains far from the hull, except in a small domain on the rear part of the tower, a coarse panelling is used (figure 5). The numerical results (figure 6) showed that this influence was very small, at least under the conditions considered here. These conclusions must be seen keeping in mind the limitation of the model which did not account for the possible separation from the submarine hull. However, it indicates that, within the limitation of the present model, the representation of the submarine hull is not crucial. This is an appreciable result because of the large saving of computing time.

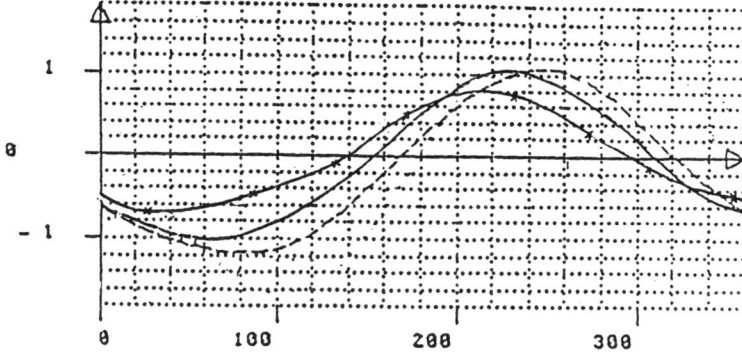

Figure 4: *Time evolution of the lift during one period. continuous: with tip separation, dashed: only trailing edge separation*

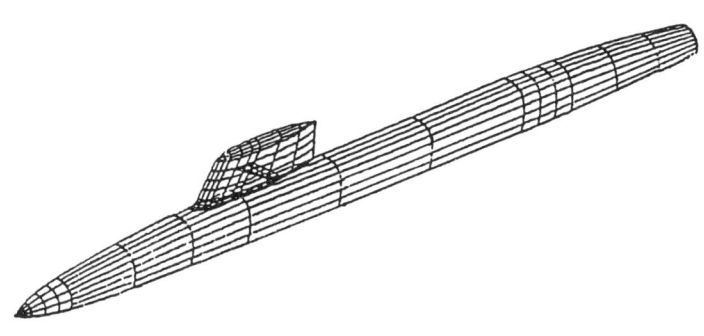

Figure 5: *Discretization of the submarine hull , 960 panels.*

The nonlinear effects

We can identify two principal nonlinear effects. Both are due to the large angle of incidence and to the resulting vortical wake. The first effect is the wellknown loss of lift induced by the separation. Although this loss is widely balanced by the lift gain due to the presence of the rolled up wake above the diving plane,

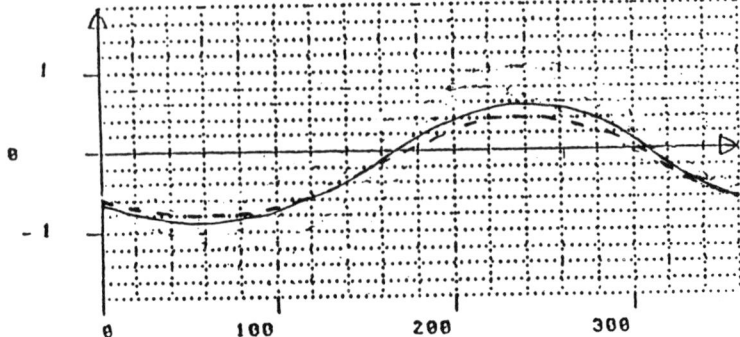

Figure 6: *Time evolution of the lift during one period. continuous: isolated diving plane, dashed: with the submarine hull.*

this loss is still observed on figure 7. The second effect is a memory effect. It is also caused by the presence of the vortex core above the plane. The vorticity emitted along the edges of the plane at time t_0 is convected downstream and will remain above the wing for a time running from 0 for the trailing edge to the ratio (Chord-length/submarine-speed) for the region close to the leading edge. As a result, the vorticity within the vortex core is for a part due to conditions which can drastically differ from the present ones. To illustrate this phenomena, we have plotted on figure 8 the numerical results obtained with the particle discretization of the wake and those extrapolated from figure 7 by considering that the unsteady lift is only a function of the angle of incidence θ and of the mean velocity on the diving plane $\bar{U}$.

Higher angles of incidence

We present in this section results obtain when the initial angle of incidence of the diving plane (when $\underline{\mathbf{U}}_w = 0$.) is not 0. In that cases the maximum or minimum angle of incidence are more than 20^0 and it is very probable that a leading edge separation occurs. Although this phenomenon has not been modelized in the present work, the discrepancy between the numerical results obtained for $\theta = \pm 10^0$ and the experimental results remain small. This may indicate that the leading edge separation remain limited to a small time interval. These results are presented on figures 9 and 10 below.

CONCLUSION:

The first conclusion of this study is quite agreeable: The particle method did not fail to predict the lift on diving planes in waves. More interesting is the fact that a large amount of this lift can be attributed to the flow pattern near the tip. This indicates that computational fluid dynamics could be of some help for the design problem of the tip. Vortex cores have a positive effect on lift. However, they induced a non linear response of the foil, and are noisy because of the cavitation filament which develops in the vortex core depression. A compromize has to be

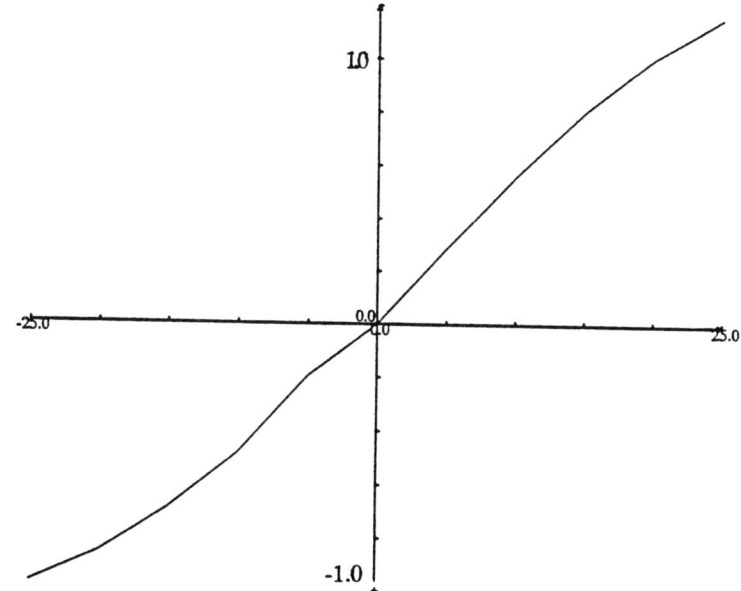

Figure 7: *Lift of the diving plane with respect to the angle of incidence.*

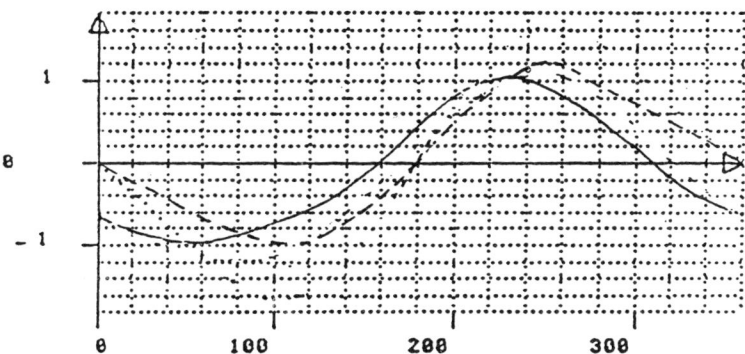

Figure 8: *Time evolution of the lift during one period. continuous: Particle method, dashed: extrapolation from figure 7*

found among all these different properties.

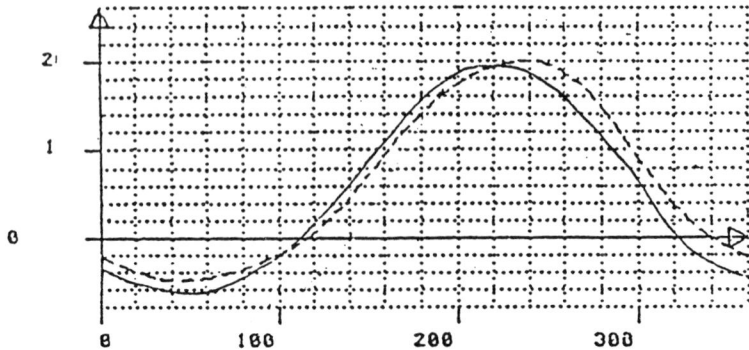

Figure 9: *Time evolution of the lift during one period.* $\theta = 10^0$

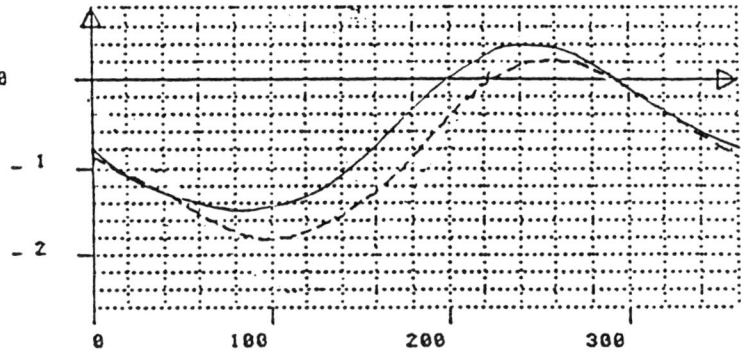

Figure 10: *Time evolution of the lift during one period.* $\theta = -10^0$

References

[1] Cantaloube B. and Huberson S. A new approach to compute the flow around helicopter rotors, $9^t h$ European Helicopter Forum, Stresa, 1983.

[2] Rosenhead. The formation of vortices from a surface discontinuities, Proc. of the Roy. Soc, Vol A 134, pp 170,192, 1931.

[3] Chorin A.J. Numerical study of slightly viscous flow, J. Fluid Mech., Vol57, 1973, pp 785-796

[4] Rehbach C. A numerical calculation of three dimensionnal unsteady flows with vortex sheets, AIAA 16th aerospace Sciences Meeting, Huntsville, 1978.

[5] Lefebvre. Efficacitié de la barre de plongée avant, Rapport interne du Bassin d'Essai des Carènes de Paris, 1988.

First and Second Order Wave Forces on Floating Bodies From Far-Field Relations

A.C. Fernandes and L.A.P. Levy

PETROBRÁS/CENPES/DIPREX/SEDEM

Rio de Janeiro, RJ 21910, Brazil

ABSTRACT

Some wave forces (radiation damping and horizontal drift forces) on floating bodies may be expressed in terms of far–field properties, specifically the Kochin function which is proportional to the wave amplitude at infinite. Several of these results depend on azimuthal integrations, which have been performed numerically. The present work, however, shows some analytical results concerning the latter integration which improve the computational method in terms of both precision and speed.

INTRODUCTION

The evaluation of waves forces on floating bodies by potential theory is a subject that is well established in theoretical terms but has required a lot of attention recently concerning numerical evaluations. This necessity has been caused by the increasingly complex shapes that have been built. The floating bodies evolved from the mono–hull ships to, for instance, the semi–submersibles and theirs pontoons, columns and bracings requiring therefore more precise geometrical definitions and computational methods. In this context, any improvements such as the one presented here is welcomed.

For the first order problem (linear wave–amplitude dependence), an alternative way of calculating the damping coefficient is presented here. The same is made for calculation of the mean drift forces on the horizontal plane, properties from the second order problem (quadratic wave–amplitude dependence)

BASIC FORMULATION

Consider a floating body subject to the action of ocean waves. Applying the conservations laws for the ideal fluid it can be shown that potential theory holds (see for instance Wehausen [1]). In this case the flow may be described by a velocity potential which satisfies Laplace equation. Within the linear problem formulation, it is possible to decompose the problem in the radiation and diffraction problems and from these calculate in a easier manner several properties. The ones that matters for the present work are the damping coefficient and the mean drift horizontal forces. The expression to be presented by the former is based on energy conservation laws within the radiation problem while for the latter momentum conservation for both diffraction and radiation problems must be applied. Please refer to Wehausen [1] for the general formulation.

For the present work which factors out the time dependence by considering only harmonic motions, the pertinent Kochin function (which is proportional to the wave amplitude in the far–field) may be written as follows

$$H_j(\theta) = \frac{-k}{D} \int\int_{S_B} [\frac{\partial\varphi_j}{\partial n} - \varphi_j \frac{\partial}{\partial n}]$$

$$\frac{\cosh k(z+h)}{\cosh kh} e^{-ik(x\cos\theta + y\sin\theta)} dS \qquad (1)$$

where k is the wave number, θ is the azimuthal angle, S_B is the mean body wetted surface, h is the water depth, φ_j is the complex velocity potential corresponding the unit amplitude body motion in direction j ($j = 1,2,3,4,5,6$) and

$$D = \tgh(kh) + kh \, \sech^2(kh) \text{ (a constant)}$$

$$\frac{\partial}{\partial n} \equiv n_1 \frac{\partial}{\partial x} + n_2 \frac{\partial}{\partial y} + n_3 \frac{\partial}{\partial z} \equiv \hat{n} \, \nabla$$

$\hat{n} = (n_1, n_2, n_3)$ is the normal towards the fluid and

$$\nabla \equiv -(\frac{\partial}{\partial x}, \frac{\partial}{\partial y}, \frac{\partial}{\partial z}).$$

Applying Green's Theorem and Method of the Stationary Phase, it can be shown (see for instance Mei [2]) that as the horizontal distance from the body vicinity (R) increases indefinitely, then, the asymptotic behavior of φ_j is such that:

$$\varphi_j \sim \frac{i}{(2\pi kR)^{1/2}} e^{ikR} e^{\frac{-i\pi}{4}} \frac{\cosh k(z+h)}{\cosh kh} H_j(\theta) \qquad (2)$$

shown that the Kochin function is proportional to the wave amplitude at infinite. This allows a relation between the energy radiated away in the form of waves with the damping coefficient.

FIRST ORDER

It is possible to show that (Mei [2]) the damping may be expressed as

$$B_{ij} = \frac{\rho}{4\pi wk} D \int_0^{2\pi} H_i^*(\theta) \, H_j(\theta) d\theta \qquad (3)$$

(* indicates complex conjugate). This shows that if the Kochin functions are known, then the damping coefficient follows provided the azimuthal integration is performed.

SECOND ORDER

Applying momentum conservation theorem, Maruo [3] and Newman [4] have shown that it is possible to express the mean drift horizontal forces only in terms of properties of the first order problem. These latter are the appropriate Kochin functions as shown next(see Wehausen [1])

$$\overline{F}_x = -\frac{\rho}{8\pi} \int_0^{2\pi} |H(\theta)|^2 (\cos\beta - \cos\theta) d\theta \tag{4}$$

$$\overline{F}_y = -\frac{\rho}{8\pi} \int_0^{2\pi} |H(\theta)|^2 (\mathrm{sen}\beta - \mathrm{sen}\theta) d\theta \tag{5}$$

$$\overline{K}_z = \frac{-\rho}{8\pi k} \mathrm{Im}\{ \int_0^{2\pi} H(\theta) \, D^*(\theta) d\theta \} + $$
$$+ \frac{1}{2k^2} \rho\omega \, |A| \, \mathrm{Re}\{D(\beta)\} \tag{6}$$

here $\overline{F}_x$ is the mean drift force in x– direction, $\overline{F}_y$ is the mean drift force in y–direction and $\overline{K}_z$ is the mean drift moment with respect to the z–axis (x and y are on the horizontal plane and z is vertical). Note also that

$$D(\theta) = \frac{d \, H(\theta)}{d\theta} \tag{7}$$

In this case the Kochin function may be decomposed as a summation of more elementary ones.

The total velocity potential may be written Wehausen [1]

$$\varphi_t = \varphi_w + \varphi_d + \sum_{j=1}^{6} \eta_j \varphi_j \tag{8}$$

where φ_w corresponds to the incoming wave, φ_d corresponds to scattering and φ_j to radiation (η_j is the displacement in generalized direction j). Hence, introducing φ_7 such that

$$\varphi_d = A\varphi_7 \tag{9}$$

it is easy to show that

$$H(\theta) = \sum_{j=1}^{7} \eta_j H_j(\theta) \tag{10}$$

and therefore

$$|H(\theta)|^2 = \sum_{i=1}^{7} \sum_{j=1}^{7} \eta_i \eta_j^* H_i(\theta) H_j^*(\theta) \tag{11}$$

AZIMUTHAL INTEGRATION

Observing the expressions (3), (4), (5) and (6) (also (11)) one notes that the azimuthal integration of products of Kochin functions are necessary. The first alternative is to calculate these numerically. However, another alternative is possible by integrating the referred products of Kochin functions analytically. In order to make it, consider

$$I_{ij} = \int_0^{2\pi} H_i(\theta) H_j^*(\theta)d\theta \tag{12}$$

Rewriting (1) such that

$$H_j(\theta) = \frac{-k}{D} \int\int_{S_B} [\alpha_j(P) + \alpha_j^c(P)\cos\theta + \alpha_j^s(P)\text{sen}\,\theta]$$

$$\frac{\cosh k(z+h)}{\cosh kh} e^{-ik(x\cos\theta + y\,\text{sen}\,\theta)}dS \tag{13}$$

where

$$\alpha_j(P) \equiv \frac{\partial\,\varphi j\,(P)}{\partial n_P} - kN_3(P)\varphi_j(P)$$

$$N_3(P) = \text{tgh}k(z+h)\, n_3(P)$$

$$\alpha_j^c(P) \equiv ikn_1(P)\varphi_j(P)$$

$$\alpha_j^s(P) \equiv ikn_2(P)\varphi_j(P)$$

one has

$$I_{ij} = (\frac{-k}{D})^2 \int\int_{S_B} dS_P \int\int_{S_B} dS_Q\, T_{ij}(P_j,Q) \tag{14}$$

where

$$T_{ij}(P,Q) = \int_0^{2\pi} \{[\alpha_j(P) + \alpha_i^c(P)\cos\theta + \alpha_i^s\,\text{sen}\,\theta]$$

$$[\alpha_j^*(Q) + \alpha_j^{c*}(\theta)\cos\theta + \alpha_j^{s*}(\theta)\text{sen}\,\theta]$$

$$\frac{\cosh k(z+h)}{\cosh kh} \cdot \frac{\cosh k(z'+h)}{\cosh kh}$$

$$e^{-ik[(x-x')\cos\theta + (y-y')\text{sen}\,\theta]}\}d\theta \tag{15}$$

with $P \equiv (x,y,z)$ e $Q \equiv (x',y',z')$ are points on the body. Multiplying and integrating, the result is

$$T_{ij}(P,Q) = \frac{\cosh k(z+h)}{\cosh kh} \cdot \frac{\cosh k(z'+h)}{\cosh kh}\{\alpha_j(P)\alpha_j^*(Q)I(P,Q) +$$

$$+ [\alpha_i(P)\alpha_j^{c*}(Q) + \alpha_i^c(P)\alpha_j^*(\theta)]\,I^c(P,Q) +$$

$$+ [\alpha_i(P)\alpha_j^{s*}(Q) + \alpha_i^s(P)\alpha_j^*(Q)]\,I^s(P,Q) +$$

$$+ [\alpha_i^c(P)\alpha_j^{s*}(\theta) + \alpha_i^s(P)\alpha_j^{c*}(Q)]\,I^{cs}(P,Q) +$$

$$+ \alpha_i^c(P)\,\alpha_j^{c*}(Q)\,I^{cc}(P,Q) +$$

$$+ \alpha_i^s(P)\,\alpha_j^{s*}(\theta)\,I^{ss}(P,Q)\,\} \tag{16}$$

The I's in equation (16) may be expressed in terms of Bessel functions (J_o,

J_1 e J_2)(see Appendix 1). With $\rho \equiv (\xi^2 + \eta^2)^{1/2}$; $\xi \equiv x-x'$; $\eta \equiv y-y'$ the result is

$$I(P,Q) = 2\pi\, J_0(k\rho) \tag{17}$$

$$I^c(P,Q) = i2\pi\, k\xi\, \frac{J_1(k\rho)}{k\rho} \tag{18}$$

$$I^s(P,Q) = i2\pi\, k\eta\, \frac{J_1(k\rho)}{k\rho} \tag{19}$$

$$I^{cc}(P,Q) = 2\pi\,[\,-(\frac{k\xi}{\rho})^2\, J_2(k\rho) + \frac{J_1(k\rho)}{k\rho}] \tag{20}$$

$$I^{cs}(P,Q) = -\,2P\,(\frac{k\xi}{k\rho})(\frac{k\eta}{k\rho})\, J_2(k\rho) \tag{21}$$

$$I^{ss}(P,Q) = 2\pi\,[-(\frac{k\eta}{k\rho})^2\, J_2(k\rho) + \frac{J_1(k\rho)}{k\rho}] \tag{22}$$

On the other hand the mean drift forces may be expressed as

$$\overline{F}_x = -\frac{\rho}{8\pi}\sum_{i=1}^{7}\sum_{j=1}^{7}\eta_i\eta_j^{*}[\cos\beta\, I_{ij} - I_{ij}^c] \tag{23}$$

$$\overline{F}_y = -\frac{\rho}{8\pi}\sum_{i=1}^{7}\sum_{j=1}^{7}\eta_i\eta_j^{*}[\text{sen}\beta\, I_{ij} - I_{ij}^s] \tag{24}$$

$$\overline{K}_z = \frac{-\rho}{8\pi}\,\text{Im}\{\sum_{i=1}^{7}\sum_{j=1}^{7}\eta_i\eta_j^{*}I_{ij}^d\} +$$
$$+ \frac{\rho\omega|A|}{2k^2}\text{Re}\{D(\beta)\} \tag{25}$$

where

$$I_{ij}^c \equiv \int_0^{2\pi} H_i(\theta)H_j(\theta)\cos\theta\, d\theta \tag{26}$$

$$I_{ij}^s \equiv \int_0^{2\pi} H_i(\theta)H_j(\theta)\text{sen}\theta\, d\theta \tag{27}$$

$$I_{ij}^d \equiv \int_0^{2\pi} H_i(\theta)D_j^{*}(\theta)d\theta \tag{28}$$

Proceeding analogously as in the development of I_{ij} (see (15) + (16) and Appendix 1) one has

$$I_{ij}^c = (-\frac{k}{D})^2 \iint_{s_B} dS_P \iint_{s_B} dS_Q\, T_{ij}^c(P,Q)$$

$$I_{ij}^s = (-\frac{k}{D})^2 \iint_{s_B} dS_P \iint_{s_B} dS_Q\, T_{ij}^c(P,Q)$$

with the expression for $T_{ij}^c(P,Q)$ following from (16) taking the I variables added by ( )$^{..c}$, that is, $I^c(P,Q)$ instead of $I(P,Q)$; $I^{csc}(P,Q)$ instead of $I^{cs}(P,Q)$; etc.

Analogously for the expression for $T^8_{ij}(P,Q)$ one should take from (16) the I variables added by $(\quad)^{\cdot\cdot 8}$, etc.

The remaining I functions are given by

$$I^{ccc}(P,Q) = 2\pi i\ (\tfrac{k\xi}{k\rho})\{-[3-4(\tfrac{k\xi}{k\rho})^2]\tfrac{2\,J_2(k\rho)}{k\rho}$$

$$- k\xi(\tfrac{k\xi}{k\rho})\tfrac{J_1(k\rho)}{k\rho}\} \tag{29}$$

$$I^{cc8}(P,Q) = 2\pi i\ (\tfrac{k\eta}{k\rho})\{-[1-4(\tfrac{k\xi}{k\rho})^2]\tfrac{2\,J_2(k\rho)}{k\rho}$$

$$- k\eta(\tfrac{k\eta}{k\rho})\tfrac{J_1(k\rho)}{k\rho}\} \tag{30}$$

$$I^{88c}(P,Q) = 2\pi i\ (\tfrac{k\xi}{k\rho})\{-[1-4(\tfrac{k\eta}{k\rho})^2]\tfrac{2\,J_2(k\rho)}{k\rho}$$

$$- k\eta(\tfrac{k\eta}{k\rho})\tfrac{J_1(k\rho)}{k\rho}\} \tag{31}$$

$$I^{888}(P,Q) = 2\pi i\ (\tfrac{k\eta}{k\rho})\{-[3-4(\tfrac{k\eta}{k\rho})^2]\tfrac{2\,J_2(k\rho)}{k\rho}$$

$$- k\eta(\tfrac{k\eta}{k\rho})\tfrac{J_1(k\rho)}{k\rho}\} \tag{32}$$

The expression for I^d_{ij} in (29) requires a further treatment of the function $D_j(\theta) = \dfrac{dH_j(\theta)}{d\theta}$.

Deriving and rearranging, one has:

$$D_j(\theta) = \frac{ik^2}{D} \iint_{S_B} \{\beta^c_j(P)\cos\theta + \beta^8_j(P)\mathrm{sen}\,\theta + \beta^{c8}_j(P)\cos\theta\,\mathrm{sen}\,\theta$$

$$+ \beta^{cc}_j(P)\cos^2\theta + \beta^{88}_j(P)\mathrm{sen}^2\theta\}\frac{\coshk(z+h)}{\coshkh}\,e^{-ik(x\cos\theta + y\mathrm{sen}\,\theta)}\,dS$$

where

$$\beta^8_j(P) \equiv -x\,\frac{\partial\varphi_j}{\partial n} - \varphi_j(P)(-n_1 - kx\,N_3)$$

$$\beta^c_j(P) \equiv -y\,\frac{\partial\varphi_j}{\partial n} - \varphi_j(P)(-n_2 - ky\,N_3)$$

$$\beta^{c8}_j(P) \equiv ik\,(xn_1 - yn_2)\varphi_j(P)$$

$$\beta^{88}_j(P) \equiv ik\,n_2\,\varphi_j(P)$$

$$\beta^{cc}_j(P) \equiv ik\,n_1\,\varphi_j(P)$$

Hence

$$I_{ij}^d = (\frac{-k}{D})^2 ik \iint\limits_{S_B} dS_P \iint\limits_{S_B} dS_Q \; T_{ij}^d(P,Q) \qquad (33)$$

with

$$T_{ij}^d(P,Q) = \frac{\cosh k(z+h)}{\cosh kh} \frac{\cosh k(z'+h)}{\cosh kh} \{\alpha_i(P)\beta_j^{s*}(Q)I^s(P,Q) + $$

$$+ \; \alpha_i(P) \; \beta_j^{c*}(Q) \; I^c(P,Q) + $$

$$+ [\; \alpha_i(P) \; \beta_j^{cs*}(Q) + \alpha_i^c(P) \; \beta_j^{s*}(Q) + \alpha_i^s(P) \; \beta_j^{c*}(Q)] \; I^{sc}(P,Q) $$

$$+ [\; \alpha_i(P) \; \beta_j^{cc*}(Q) + \alpha_i^c(P) \; \beta_j^{c*}(Q) \;] \; I^{cc}(P,Q) $$

$$+ [\; \alpha_i(P) \; \beta_j^{ss*}(Q) + \alpha_i^s(P) \; \beta_j^{s*}(Q) \;] \; I^{ss}(P,Q) $$

$$+ [\; \alpha_i^c(P) \; \beta_j^{cs*}(Q) + \alpha_i^s(P) \; \beta_j^{cc*}(Q) \;] \; I^{ccs}(P,Q) $$

$$+ [\; \alpha_i^c(P) \; \beta_j^{ss*}(Q) + \alpha_i^s(P) \; \beta_j^{cs*}(Q) \;] \; I^{css}(P,Q) $$

$$+ \; \alpha_i^c(P) \; \beta_j^{cc*}(Q) \; I^{ccc}(P,Q) $$

$$+ \; \alpha_i^s(P) \; \beta_j^{ss*}(Q) \; I^{sss}(P,Q)\} \qquad (34)$$

SUMMARY

In summary the results are the following: for the damping coefficient, from (3) and (12) one has

$$B_{ij} = \frac{\rho}{4\pi w k} \; D \; I_{ij}$$

with I_{ij} calculated by (14) and (15). On the other hand, for the drift forces one should use (23) and (24) with the I expression given by (17)–(22) e (29)–(32). For the moment, the expression is (25) with I_{ij}^d obtained from (33) e (34). This last expression requires the same I's as before.

RESULTS

The expressions cited in the Summary are easy programmable from any code that calculate the velocity potentials φ_j, j=1,2,...7 in points through the body. From a code developed by Levy [5] this has been made for the present work. As example of application two classical results have been chosen: the MacCamy and Fuchs slender cylinder [6] and the hemisphere as discussed in Kokkinowrachos [7]. In the Figures 1 and 2 the mean drift horizontal force are present respectively. In these figures the non—dimensional coefficients have been produced with D as the water depth and a the radius of the cylinder or the hemisphere. In Figures 3 and 4 one may observe the discretization of the bodies under considerations.

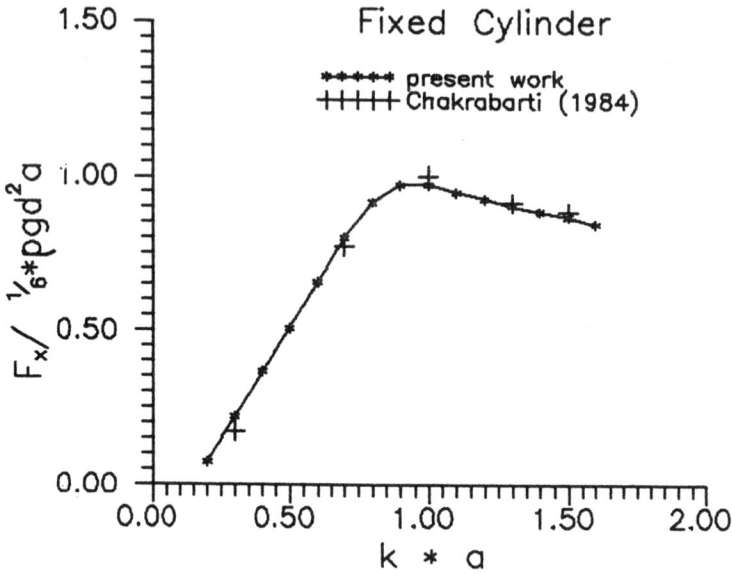

Figure 1 Mean Drift Horizontal Force, Cylinder

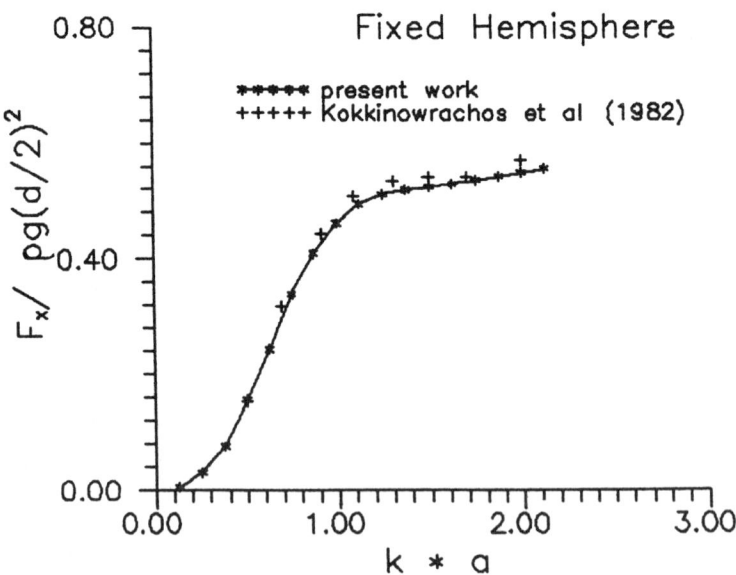

Figure 2 Mean Drift Horizontal Force, Hemisphere

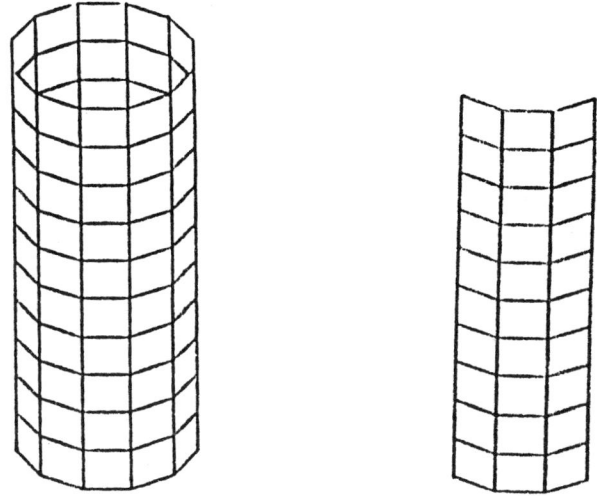

Figure 3 120 panels for full cylinder (30 for one quadrant)

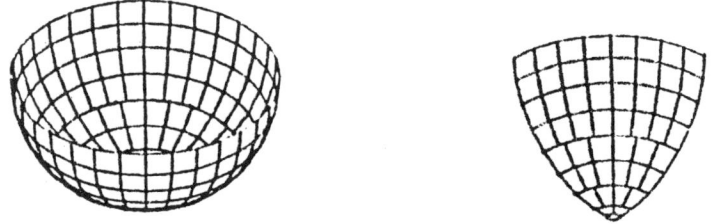

Figure 4 204 panels for full hemisphere (51 for one quadrant)

REFERENCES

1. Wehausen, J.V. The Motion of Floating Bodies, Annual Review of Fluid Mechanics, Vol. 3 pp.237–268, 1971

2. Mei, C.C. The Applied Dynamics of Ocean Surface Waves; John Wiley and Sons, 1983.

3. Maruo, H. The Drift of a Body Floating on Waves, Journal of Ship Research; December, 1960

4. Newman, J.N. "The Drift Force and Moment on Ship in Waves", Journal of Ship Research; March, 1967

5. Levy, L.A.P. Análise de Movimentos de Corpos Flutuantes no Domínio do Tempo, MSC thesis; Engenharia Oceânica; Universidade Federal do Rio de Janeiro, COPPE, 1989.

6. Chakrabarti, S.K. Steady Drift Force on Vertical Cylinder — Viscous vs. Potential; Applied Ocean Research, Vol. 6, No. 2, 1984

7. Kokkinowrachos, K., Bardis, L., Mavrakos, S. Drift Forces on One– and Two– body Structures in Regular Waves, International Conference on Behavior of Offshore Structures (BOSS), 1982

8. Abramowitz, M. and Stegun, I.A. "Handbook of Mathematical", Dover Publications, 1972

Calculation of Potential Flow Around a Given Profile by the Method of Surface Vorticity

A. Kukner

Faculty of Naval Architecture and Ocean Engineering, Technical University of Istanbul, 80626, Istanbul, Turkey

ABSTRACT

In this study, the method of surface vorticity and its applications in computation of two dimensional potential flow around an arbitrary shape are described. A computer program has been developed to calculate the pressure distribution around an arbitrary shape with a different angle of attack.
In the numerical scheme, the velocity field, stream lines, pressure distribution, lift and drag around a given profile with a different angle of attack are obtained. Some typical results of aerofoil calculations are presented. From these results, it can be seen that in practice the solution of potential flow problem around a given profile by the method of surface vorticity would be sufficiently accurate. This method also provides a solid foundation for the extention of the method to real flow where flow separation may occur. This study shows the use of micro computers to obtain rapid solution and graphical representation of flow field, velocity vectors and pressure field around an aerofoil. These graphical representations will give us a physical understanding of the flow profile at a design stage and a better choice of experimental methods to study potential flows around arbitrary bodies.

The developed computer program can be widely used for the application of flow around multi objects, aerofoils, cascade design.

INTRODUCTION

The surface panel methods have remarkable advantages in the field of hydrodynamics and aerodynamics for the design and analysis of two and three dimensional multi objects. There are many kinds of surface panel methods that use different type of surface panels, singularity distribution and boundary conditions. Most of the panel methods are based on the source method which is developed by Hess and Smith[2]. One of the source panel methods is the surface vorticity method which is similar to the source panel method defined by Hess and Smith[2] and was firstly introduced by Martensen[1].In this method, the surface of the body is replaced by an infinitely thin vorticity sheet of strength $\gamma(s)$. The unknown value of vorticity strength $\gamma(s)$ is determined by solving the boundary value problem at control point on each panel. This method is one of the most powerful methods for calculating potential flow. Lewis and Porthouse[3] attempted to simulate the Navier-Stokes equation by means of the surface vorticity method. Lam[5] gave a detailed description of the surface vorticity method to calculate two dimensional potential flow around arbitrary shapes.

Indeed, the problem of calculating potential flow around two dimensional bodies became more attractive subject in the recent years, since its results have practical application in aerofoil design.

In this paper, Martensen's surface vorticity method is

applied to calculate potential flow around a given profile. The velocity field, pressure distribution, lift and drag coefficients are obtained around a given profile by using this method. A special emphasis was made on Kutta condition such as equal pressure on upper and lower surfaces at trailing edge is always satisfied by considering the total circulation at the trailing edge. That is, $\gamma_1(s)+\gamma_N(s)=0$.

FORMULATION OF THE PROBLEM-METHOD OF SURFACE VORTICITY

The concept of vorticity has a significant importance in the study of flow patterns. The surface vorticity method described here is the same as Martensen's sur-face vorticity method[1] and offers the mathematical model for the numerical simulation of a potential flow directly. In this method, the concerned body surface is replaced by an infinitely thin vorticity sheet of $\gamma(s)$ as shown in Fig 1.

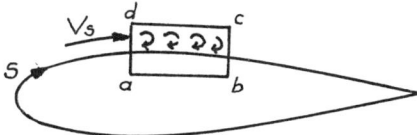

a. Replacement of a body surface by a vortex sheet

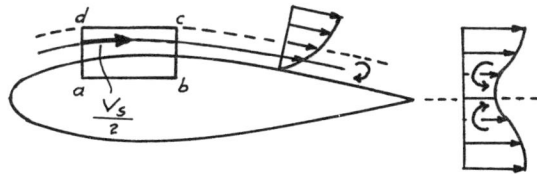

b. Flow situation around an aerofoil due to vorticity

Fig.1

Surface velocity V_s should be equal to the strength of sheet vorticity by considering circulation around abcd.

$$V_s = \gamma(s) \tag{1}$$

The circulation around abcd due to the vortex sheet of length ds is given by

$$\oint_{abcd} \vec{V} \, \vec{ds} = V_s \, ds \tag{2}$$

As it is seen from Fig.1.b at a given location s, the surface vorticity V_s and the strength $\gamma(s)$ remain constant since the modelling of surface vorticity is the replacement of the thin boundary layer in which the surface velocity V_s decreases from a certain value to zero on the body, by a vorticity sheet of strength $\gamma(s)$. Therefore surface vorticity sheet is convected down stream with a velocity $V_s/2$.

If $d\gamma$ is defined as the net vorticity per unit length produced at a given location s in time dt, the vorticity production equation will be obtained as [3]:

$$\frac{d\gamma}{dt} = \frac{d}{ds}\left(\frac{V_s^2}{2}\right) = -\frac{dP}{ds} \tag{3}$$

This equation states that if pressure decreases, vorticity is spontaneously produced or vice versa. In other words, in real flow, effect of viscosity can develop vorticity. It is actually developed in the region of steep velocity gradient(or surface pressure gradient) such as those in the boundary layer near solid surfaces. Then, the action of viscosity diffuse this new created vorticity away from the surface into the main stream where it is going to be a part of boundary layer.

In practice, vorticity production equation cannot be used since the pressure gradient value is not known on the boundary. Thus, for the numerical purposes, let us consider the discretisation of the continuous

distribution of vorticity strength in a vortex   sheet.
In fact, vortex sheet are made amenable   to   numerical
calculation by breaking them   up   into   simple   panels
whose movements are traced as a means of   representing
the motion of the sheets.

Let N be the total number of panels   around   the   body
surface as shown in Fig.2. If the body   is   placed   in
a uniform flow of speed U_∞ which makes an angle α with
the x axis, the convected   velocity   of   vortex   sheet
should be equal to velocity induced from vortex   sheet
on the surface of the body and uniform stream velocity
U_∞.

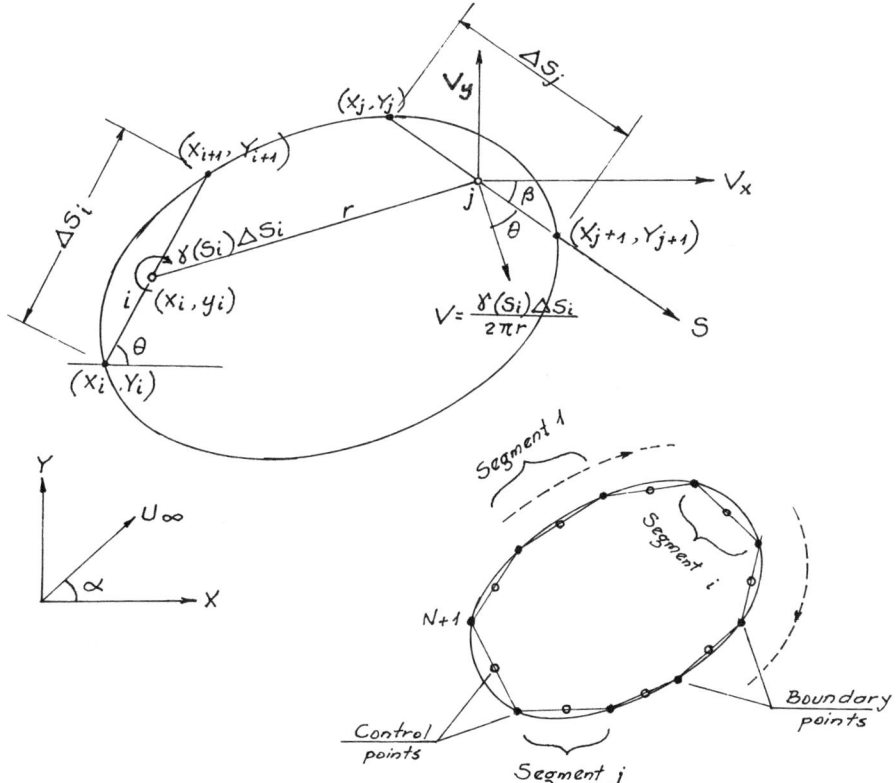

Fig.2 Representing the surface of a body by vortex
      sheets

Now, let us investigate the   effect   of   induction   of

panel i by vorticity element j in clockwise direction. The velocity, V, induced in irrotational flow at a distance r from a line vortex of circulation Γ is

$$V=\frac{\Gamma}{2\pi r}\qquad(4)$$

where the circulation Γ of the panel i is

$$\Gamma=\oint \gamma \; ds = \gamma(s_i)\; ds_i\qquad(5)$$

and $r=\sqrt{(x_i-x_j)^2+(y_i-y_j)^2}$ is the distance between the panels i and j.

The components of velocity, V, in the x and y directions are

$$V_x=V \; Sin\theta \quad , \quad V_y=-V \; Cos\theta\qquad(6)$$

where

$$Sin\theta=\frac{y_j-y_i}{r}\quad , \quad Cos\theta=\frac{x_j-x_i}{r}$$

Since $Cos\beta=\dfrac{dx_j}{ds_j}$ and $Sin\beta=-\dfrac{dy_j}{ds_j}$, the velocity V_{s1}, due to V_x and V_y in the direction of the body surface, s, is given by

$$V_{s1}=V_x \; Cos\beta-V_y \; Sin\beta =V_x \; \frac{dx_j}{ds_j} + V_y \; \frac{dy_j}{ds_j}$$

or,

$$V_{s1}=\frac{\gamma(S_i)}{2\pi}\; dS_i \; [\frac{(y_j-y_i)(dx_j/ds_j)-(x_j-x_i)(dy_j/ds_j)}{(x_j-x_i)^2 + (y_j-y_i)^2}]\qquad(7)$$

Letting,

$$K(S_i,S_j)=\frac{1}{2\pi}\; [\frac{(y_j-y_i)(dx_j/ds_j)-(x_j-x_i)(dy_j/ds_j)}{(x_i-x_j)^2+ (y_j-y_i)^2}]\qquad(8)$$

then equation (7) becomes

$$V_{s1}= \gamma(S_i)\; dS_i \; K(S_i,S_j)\qquad(9)$$

where $K(S_i,S_j)$ is called influence or coupling coefficient which represents the surface velocity induced at the control point on the S_j panel due to a unit vortex strength at the S_i panel. The total induced velocity at panel S_j due to the other vortices is obtained by the integration of V_{s1}.

$$\oint \gamma(S_i) K(S_i, S_j) \, dS_j \qquad (10)$$

On the otherhand, the uniform stream velocity where the body is placed has a component on the body surface s as:

$$V_{s2} = U_\infty \cos\alpha \frac{dx_j}{ds_j} + U_\infty \sin\alpha \frac{dy_j}{ds_j} \qquad (11)$$

The resultant velocity due to induction of vortex sheet on the body surface and uniform stream velocity U_∞ equals to the drifting velocity $V_s/2$. Thus,

$$\oint \gamma(S_i) K(S_i, S_j) dS_j + U_\infty \cos\alpha \frac{dx_j}{ds_j} + U_\infty \sin\alpha \frac{dy_j}{ds_j} = \frac{\gamma(S_j)}{2} \qquad (12)$$

Rearranging, it yields

$$\oint \gamma(S_i) K(S_i, S_j) dS_i - \frac{\gamma(S_j)}{2} = -U_\infty (\frac{dx_j}{ds_j} \cos\alpha + \frac{dy_j}{ds_j} \sin\alpha) \qquad (13)$$

This integral equation which is originally obtained by Martensen, computes the vorticity strength $\gamma(s)$ around an aerofoil directly at each time step.

As we mentioned earlier, the body is assumed to be replaced by N panels of length ΔS_i and strength $\gamma(S_i)\Delta S_i$ located at points which are the mid points of the panels and known as control points or pivotal points. Therefore, equation (13) can be written in terms of these N discrete elements of length ΔS_i and of strength $\gamma(S_i)\Delta(S_i)$. All vortices such as induction, uniform flow and drifting velocities are assumed to be at the control points. Thus equation (13) is reduced to a set of linear equations as follows:

$$\sum_{i=1}^{N} \gamma(S_i) K(S_i, S_j) \Delta S_i = -U_\infty (\frac{\Delta x_j}{\Delta S_j} \cos\alpha + \frac{\Delta y_j}{\Delta S_j} \sin\alpha) \qquad (14)$$

where the infuence coefficients $K(S_i, S_j)$ become

$$K(S_i, S_j) = \frac{1}{2\pi} \left[\frac{-(x_j - x_i)\frac{\Delta y_j}{\Delta S_j} + (y_j - y_i)\frac{\Delta x_j}{\Delta S_j}}{(x_j - x_i)^2 + (y_j - y_i)^2} \right] \quad \text{for } i \neq j$$

(15)

and

$$K(S_i, S_j)\Delta S_j = K(S_j, S_j)\Delta S_j - \frac{1}{2} \quad \text{for } i = j \qquad (16)$$

in which $K(S_j, S_j)$ represent the self induced influence coefficients which are obtained by the derivation of $K(S_i, S_j)$ with respect to L'hospital Rule twice.

$$K(S_j, S_j) = \frac{1}{4\pi} \left[\frac{(dx_j/ds_j)(d^2 y_j/ds_j^2) - (dy_j/ds_j)(d^2 x_j/ds_j^2)}{(dx_j/ds_j)^2 + (dy_j/ds_j)^2} \right]$$

(17)

or,
$$K(S_j, S_j) = \frac{1}{4\pi r_c} \qquad (18)$$

where r_c is the radius of curvature is shown in .Fig 3. and given by

$$r_c = \frac{(dx_j/ds_j)^2 + (dy_j/ds_j)^2}{(dx_j/ds_j)(d^2 y_j/ds_j^2) - (dy_j/ds_j)(d^2 x_j/ds_j^2)} \qquad (19)$$

Thus, equation (16) can be rewritten as:

$$K(S_i, S_j)\Delta S_i = \frac{\Delta S_j}{4\pi r_c} - \frac{1}{2} = \frac{\Delta\theta_j}{4\pi} - \frac{1}{2} \qquad (20)$$

in which $\Delta\theta_j$ gives the angle between two adjacent points as shown in Fig.3.

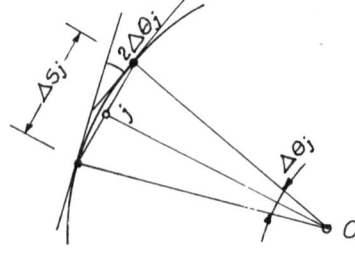

Fig.3 Radius of curvature r_c and the angle $\Delta\theta_j$

As it is seen from Fig.3 the angle $\Delta\theta_j$ is obtained as:

$$\Delta\theta_j = \frac{1}{2} \left| \tan^{-1}\frac{\Delta y_{j-1}}{\Delta x_{j-1}} - \tan^{-1}\frac{\Delta y_{j+1}}{\Delta x_{j+1}} \right| \tag{21}$$

Equation (14) can also be expressed in matrix form as follows:

$$[K(S_i, S_j) \; \Delta S_i][\gamma(S_i)] = [A] \tag{22}$$

where [A] is a column matrix of size N.

$$[A_j] = -U_\infty \begin{bmatrix} \dfrac{\Delta x_1}{\Delta s_1} \; \text{Cos}\alpha \; + \; \dfrac{\Delta y_1}{\Delta s_1} \; \text{Sin}\alpha \\[2ex] \dfrac{\Delta x_2}{\Delta s_2} \; \text{Cos}\alpha \; + \; \dfrac{\Delta y_2}{\Delta s_2} \; \text{Sin}\alpha \\[2ex] \vdots \qquad\qquad \vdots \\[2ex] \dfrac{\Delta x_j}{\Delta s_j} \; \text{Cos}\alpha \; + \; \dfrac{\Delta y_j}{\Delta s_j} \; \text{Sin}\alpha \end{bmatrix}$$

From equation (22), the unknown values of vorticity strengths $\gamma(S_i)$ at each panel are computed by one of known matrix inversion methods.

CALCULATION OF PRESSURE COEFFICIENT, LIFT AND DRAG COEFFICIENTS, AND VELOCITY FIELDS

Since U_∞ is a potential flow, the pressure field can be computed from Bernoulli equation as follows:

$$P_\infty + \frac{1}{2} \rho \, U_\infty^2 = P + \frac{1}{2} \rho \, V_s^2 \tag{23}$$

Sometimes it is more convenient to use the dimensionless pressure coefficient C_p, which is defined as the excess pressure over the free stream value divided by the free-stream dynamic pressure. Thus,

$$C_p = \frac{P-P_\infty}{\frac{1}{2}\,\rho\,U_\infty^2} = 1 - \frac{U_s^2}{U_\infty^2} \tag{24}$$

where P_∞ states the pressure at infinity.

In numerical results, pressure coefficient C_p is ploted by an arrow with the length of $|C_p|$ and perpendicular to the body surface with aslope $-1/(\Delta y/\Delta x)$.

The lift coefficient C_L is defined by the following equation.

$$C_L = \frac{L}{\frac{1}{2}\,\rho\,U_\infty^2\,A} \tag{25}$$

where L is the lift force and A is the projection area of the body on a panel normal to the flow direction. In case of an aerofoil, the projection area A is taken as the chord length of aerofoil,1. Thus the lift coefficient becomes.

$$C_L = \frac{L}{\frac{1}{2}\,\rho\,U_\infty^2\,1} \tag{26}$$

The lift force L is obtained from the pressure forces in the direction perpendicular to the main flow velocity. Simarly, the drag coefficient C_D is given by the following expression:

$$C_D = \frac{D}{\frac{1}{2}\,\rho\,U_\infty^2\,A} \tag{27}$$

where D represents the drag force which is obtained by resolving the pressure forces around the aerofoil tangential to the main flow velocity.

The velocity V at any point (x,y) in the flow field is computed by the summation of the velocity components due to vortex located at point (x_i,y_i) on the body surface and uniform velocity U_∞. Thus, the components of velocity V in the directions of x and y become

$$V_x = U_\infty \cos\alpha + \sum_{i=1}^{N} \frac{(y-y_i)}{(x-x_i)^2 + (y-y_i)^2} \frac{\gamma(S_i)\Delta S_i}{2\pi}$$ (28)

$$V_y = U_\infty \sin\alpha + \sum_{i=1}^{N} \frac{-(x-x_i)}{(x-x_i)^2 + (y-y_i)^2} \frac{\gamma(S_i)\Delta S_i}{2\pi}$$

and the magnitude of the velocity, V is,

$$V = \sqrt{V_x^2 + V_y^2}$$ (29)

at a direction $\phi = \tan^{-1} \frac{V_y}{V_x}$

APPLICATION OF THE METHOD

As a numerical application of the method NACA 0012 symmetrical aerofoil was selected to compute two dimensional potential flow around it. Some typical results of aerofoil calculations are presented in Fig.5 to Fig.19. Numerical experiments show that the number of panels affect the pressure distribution values around the leading and trailing edges of the profile. For this purpose, different number of panels have been chosen and the effect of panel number has been investigated. The results are shown in Figures 5 and 6. As can be seen from the figures, the pressure distribution around the profile becomes more regular as the number of panels increases. More panels were taken, especially, around the leading and trailing edges to provide accuracy. However the number of panels do not make much more difference on the values of lift and drag coefficients. Numerical results show that in practice having 32 panels for simple bodies like circle, cylinder and ellipse, and 40 to 46 panels for shapes with sharp corners will be sufficient.

Special emphasis was put on the Kutta condition to see

its effect on the pressure distribution around the profile. As it is well known, the Kutta condition is a physical condition where the velocity magnitudes on the upper and the lower panels at the trailing edge of the profile are equal. Therefore, summation of the strength of the vorticity on the first and the last panels at the trailing edge must be equal to zero, that is, $\gamma_1(s) + \gamma_N(s) = 0$.

The numerical results of the potential flow calculations around the aerofoil with different angle of attack with and without using the Kutta condition are illustrated in Figures 5 to 11.

CONCLUSION

Calculation of two dimensional potential flow around complicated bodies are described by using surface vorticity method. A computer program for PC's was developed to compute the velocity field, pressure distribution, lift and drag coefficients around any given complicated object. As a numerical application NACA 0012 profile which is placed in a uniform flow has been chosen and pressure fields under different angle of attack with graphical output were obtained. The graphical outputs of the program will help us to understand the flow pattern around any profile during a design stage.

Numerical results showed that the developed computer program can be used for the aerofoil and cascade design, and potential flow calculation around multi bodies. The advantage of this method would be to allow more precise representations of a complicated body configuration to compute the potential flow around it.

ACKNOWLEDGEMENTS

The author would like to thank graduate student H.Ibrahim Keser for his assistance to run the computer programs.

REFERENCES

1. Martensen, E. Die Berechnung der Druckverteilung an Gitterprofilen in ebener Potentialströmung mit einer Fredholmschen Integralgleichung, Zweiter Art. Arch. Rat. Mech. and Analysis, 3, pp 235-237, 1959.

2. Hess, J.L. and Smith, A.M.O. Calculation of Potential Flow About Arbitrary Three-Dimensional Lifting Bodies,Final Technical Report,MDC J5679-01 McDonnell Douglas,Long Beach,California,Oct.,1972.

3. Lewis, R.I. and Porthouse, D.T.C. Recent Advances in the Theoretical Stimulation of Real Fluid Flows, Nort-East Cost Institution of Engineers and Shipbuilders, pp 88-104, 1983.

4. Lam, K. Potential Flow Calculation by Surface Vorticity Method and Computer Graphics, Computers and graphics, an International Journal of Applications in Computer Graphics, Vol.11, No.1, pp. 35-47, 1987.

5. Lewis, R.I. Surface Vorticity Modelling of Seperated Flows from Two-dimensional Bluff Bodies of Arbitrary Shape, Journal of Mechanical Engineering Science, Vol.23, No.1, February 1981.

6. Johnson, F.T. A General Panel Method for the Analysis and Design of Arbitrary Configurations in Incompressible Flows, NASA CR-3079,May 1980.

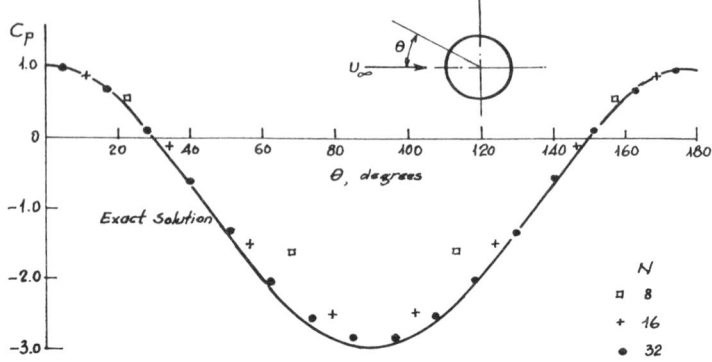

Fig. 4   Comparision of numerical calculation of the pressure distribution around a cirrular cylinder with different panel numbers.

PRESSURE   DISTRIBUTION PLOT

ANGLE OF ATTACK = 0°   LIFTING COEFFICIENT = 0.03
NUMBER OF PANEL= 46   DRAG COEFFICIENT = 0.00
LENGTH OF CHORD= 1
ANGLE OF INCLINATION = 0°
———————————> CP = 1

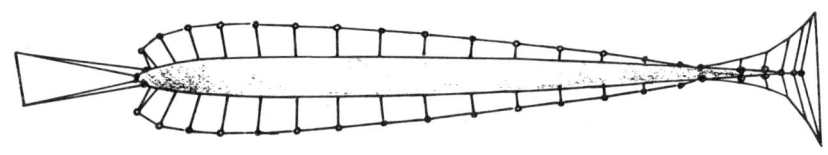

Fig.5. Plot of pressure coefficient around a NACA0012 aerofoil at an angle of attack θ° (Withouth Kutta Condition)

PRESSURE   DISTRIBUTION PLOT

ANGLE OF ATTACK = 8°   LIFTING COEFFICIENT = 0.55
NUMBER OF PANEL = 100   DRAG   COEFFICIENT = 0.88
LENGTH OF CHORD = 1
ANGLE OF INCLINATION = 0°
———————————> CP = 1

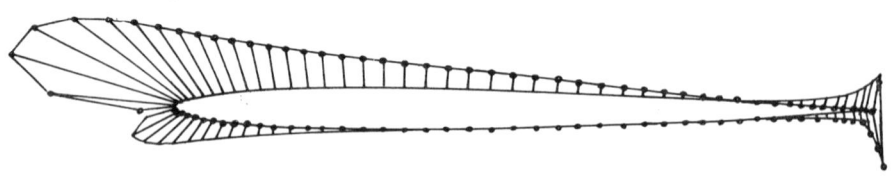

Fig, 6. Plot of pressure coefficient around a NACA 0012 aerofoil at an angle of attack 8° (Withouth Kutta Condition)

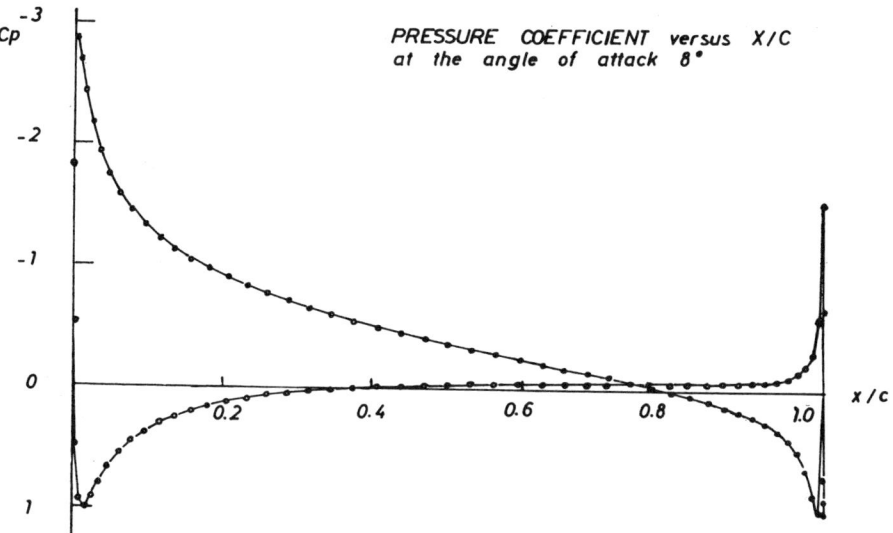

Fig. 7. Pressure Coefficient around a NACA 0012 aerofoil at an angle of attack 8° (Withouth Kutta Condition)

PRESSURE DISTRIBUTION PLOT

$C_p = 1$
Angle of Attack $= 0°$
Length of Chord $= 1$

Lifting Coefficient $= 0.00$
Drag Coefficient $= 0.01$

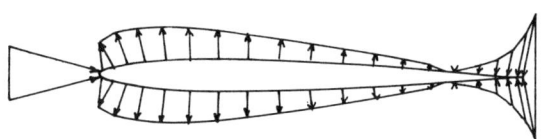

Fig. 8. Plot of pressure coefficient around a NACA 0012 aerofoil at an angle of attack 0° (With Kutta Condition)

PRESSURE DISTRIBUTION PLOT

$C_p = 1$
Angle of Attack $= 8°$
Length of Chord $= 1$

Lifting Coefficient $= 0.98$
Drag Coefficient $= 0.00$

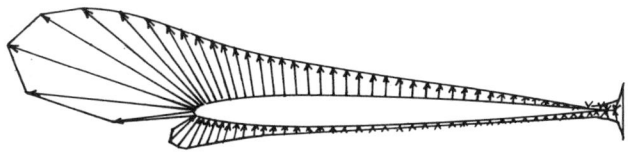

Fig. 9. Plot of pressure coefficient around a NACA 0012 aerofoil at an angle of attack 8° (With Kutta Condition)

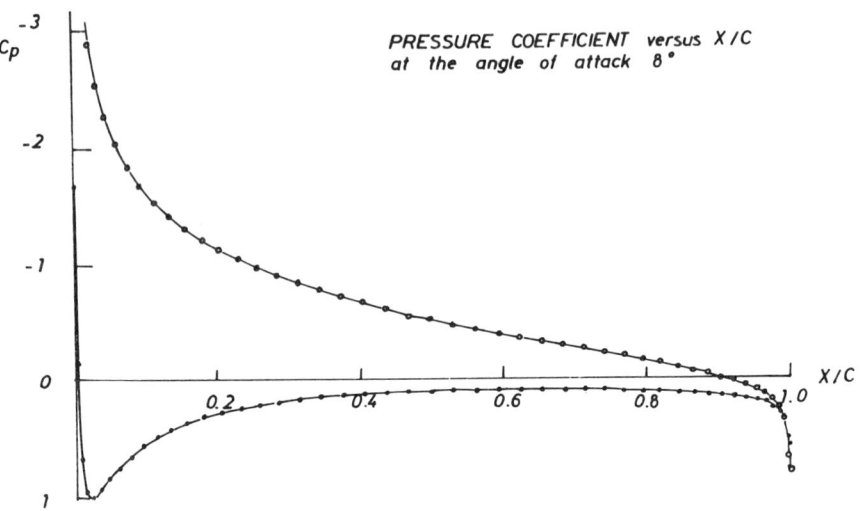

Fig. 10. Pressure Coefficient around a NACA 0012 aerofoil
at an angle of attack 8°

PRESSURE DISTRIBUTION PLOT

$\longrightarrow Cp = 1$
Angle of Attack = 0°
Length of Chord = 1

$C_L = -0.78$
$C_D = 0.01$

$C_L = 0.00$
$C_D = 0.00$

$C_L = 0.78$
$C_D = 0.01$

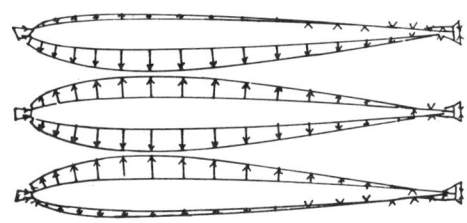

Fig. 11. Plot of pressure coefficient around cascades NACA 0012
aerofoils at an angle of attack 0°

PRESSURE DISTRIBUTION PLOT
AROUND THE LOWER AEROFOIL OF CASCADES

$\longrightarrow Cp = 1$
Angle of Attack = 0°
Length of Chord = 1

$C_L = 0.78$
$C_D = 0.01$

Fig. 12. Plot of pressure coefficient around cascades of NACA 0012
aerofoils at an angle of attack 0°

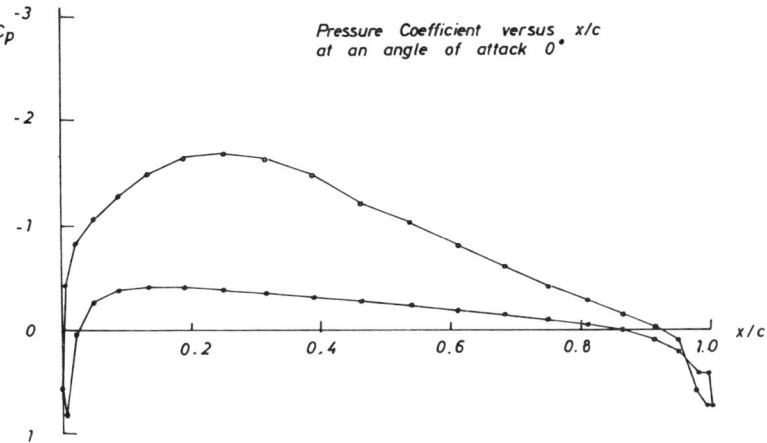

Fig. 13 Pressure Coefficient around the lower aerofoil of cascades
at an angle of attack 0.°

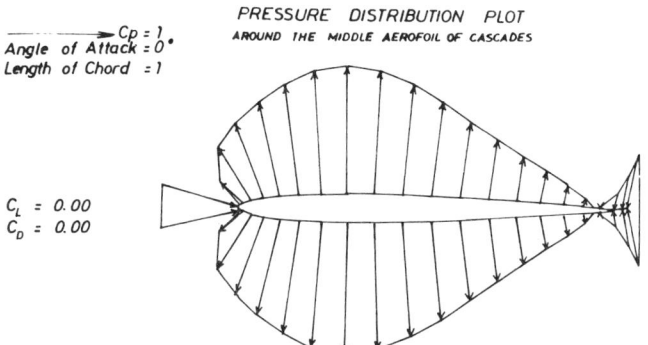

Fig. 14. Plot of pressure coefficient around cascades of NACA 0012
aerofoils at an angle of attack 0°

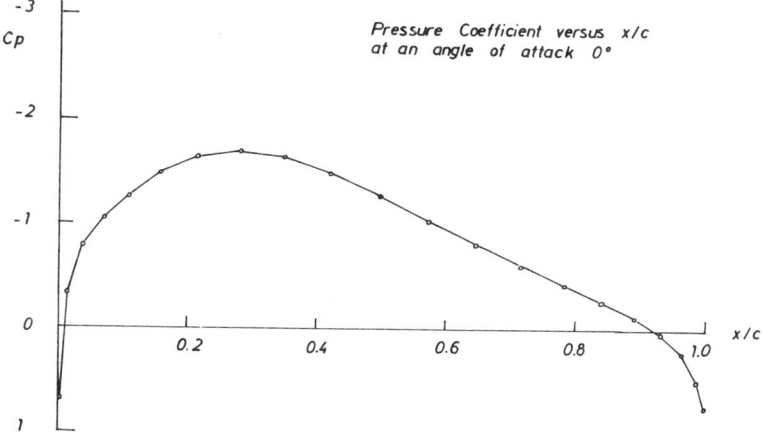

Fig. 15. Pressure Coefficient around the middle aerofoil of cascades
at an angle of attack 0°

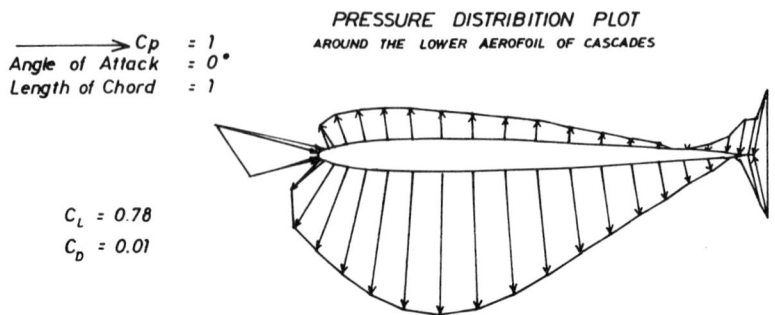

PRESSURE DISTRIBITION PLOT
AROUND THE LOWER AEROFOIL OF CASCADES

$\longrightarrow Cp$ $= 1$
Angle of Attack $= 0°$
Length of Chord $= 1$

$C_L = 0.78$
$C_D = 0.01$

Fig. 16. Plot of pressure coefficient around cascades of NACA 0012
aerofoils at an angle of attack 0°

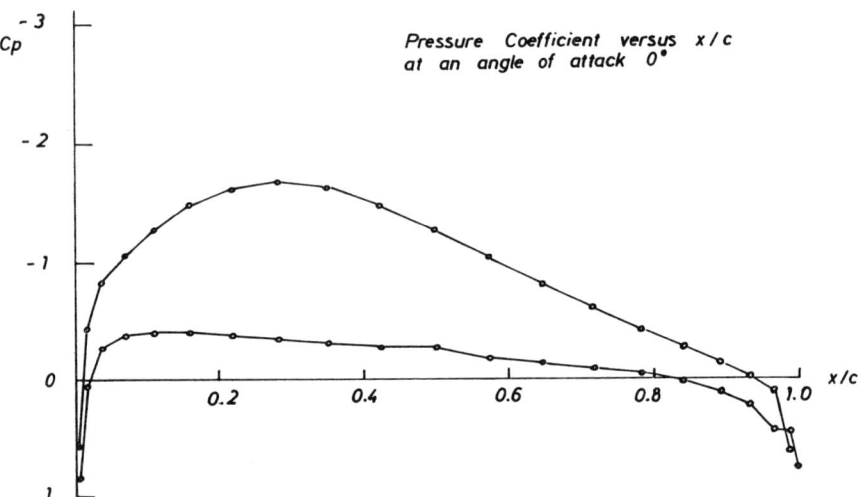

Cp

-3

-2

-1

0

1

Pressure Coefficient versus x/c
at an angle of attack 0°

0.2   0.4   0.6   0.8   1.0   x/c

Fig. 17. Pressure Coefficient around the upper aerofoil of cascades
at an angle of attack 0°.

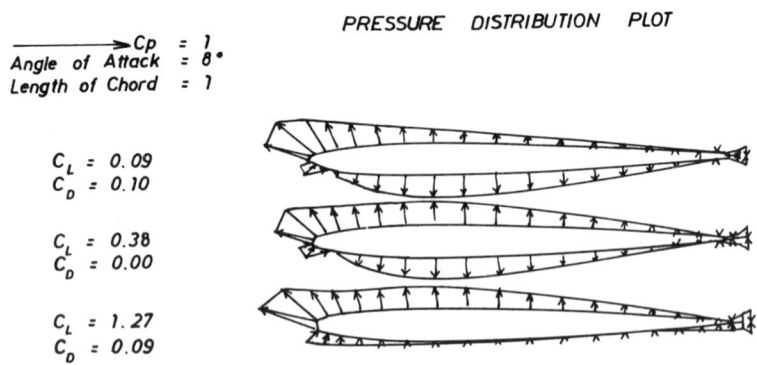

PRESSURE DISTRIBUTION PLOT

$\longrightarrow Cp$ $= 1$
Angle of Attack $= 8°$
Length of Chord $= 1$

$C_L = 0.09$
$C_D = 0.10$

$C_L = 0.38$
$C_D = 0.00$

$C_L = 1.27$
$C_D = 0.09$

Fig. 18. Plot of pressure coefficient around cascades of NACA 0012
aerofoils at an angle of attack 8°

SECTION 3: SHIP MOTIONS AND SEAKEEPING

The Use of High Definition Graphics in Seakeeping and Ship Manoeuvring

P.A. Wilson

Department of Ship Science,
University of Southampton,
SO9 5NH, U.K.

Introduction

Traditional methods of presenting data that are generated by computer programs that predict the seakeeping adequacies of surface ships have often been limited by the capabilities of two dimensional graphical displays and their complimentary hard copy output devices. The typical display is often, say, of the heave motion at the centre of gravity of the ship, as a function of the wave frequency or wave length for a given combination of ship speed and ship heading to the wave system. This in itself, provides useful information, but is sadly lacking for comparative purposes. To achieve a better visualization of the achieved performance of a ship in a random seaway requires much greater data to be generated than has previously been displayed.

Allied to this problem in seakeeping is the presentation of the manoeuvring capabilities of a ship in open water, or restricted water of finite depth and width. The modern display mechanisms for such simulators are excellent, but exceedingly expensive. This paper will seek to show that it is possible to present the manoeuvring information in a concise but high definition standard.

Equally important information that needs careful thought in its display is the time simulations of the dynamic rigid motions of a ship in a random seaway. This paper will explore the techniques that are presently used for these simulations, and will seek to provide guidance in the choice of the number of frequencies that are needed in the representations. This will also give guidance in budgetary terms. The simulations will be the traditional long crested seas as well as the much more useful engineering tool of short crested seaways.

Mathematical Models and Techniques

Seakeeping

Advances in the prediction of the behaviour of surface ships have been significant over the past twenty years. The techniques have often been classified into those due to surface wave effects, seakeeping, and those in calm water, ship manoeuvring. The traditional tools of seakeeping have been the determination of the rigid body motion, often in the frequency regime as well as the vibrational node analysis. These tools and techniques have very good correlation between theory, model experiments and full scale trials and performance, especially in respect to the vertical

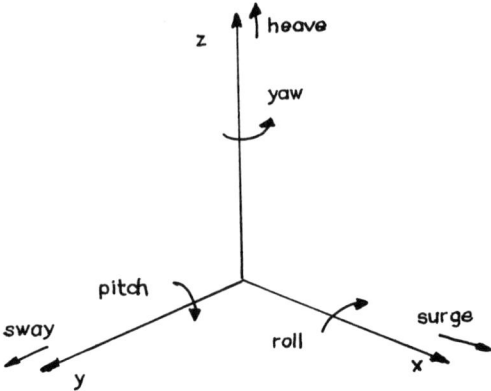

Figure 1

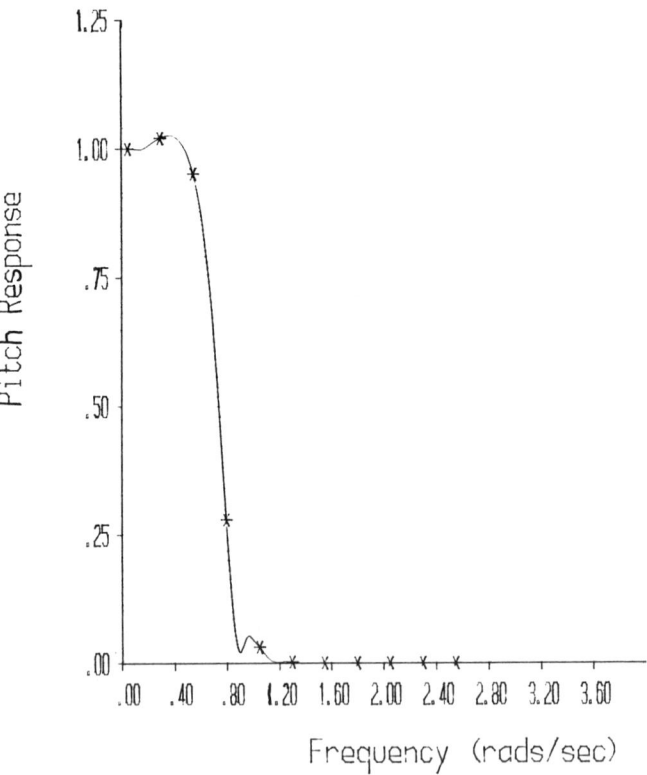

Figure 2a

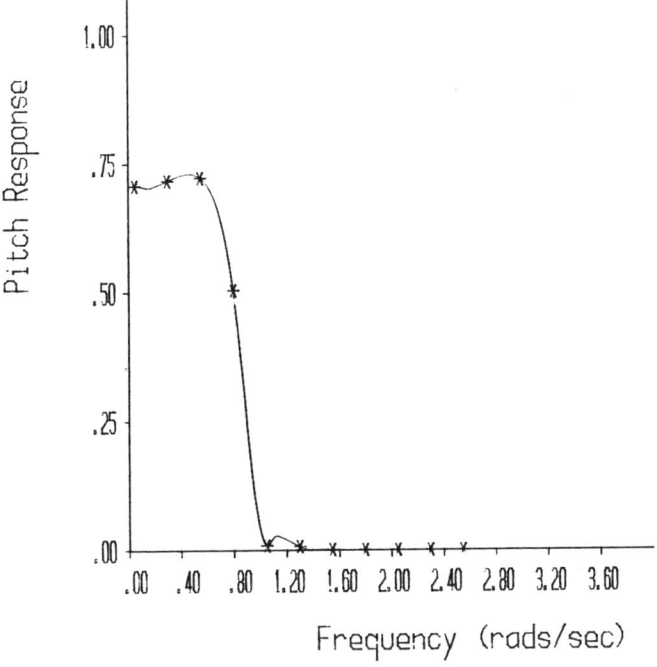

Figure 2b

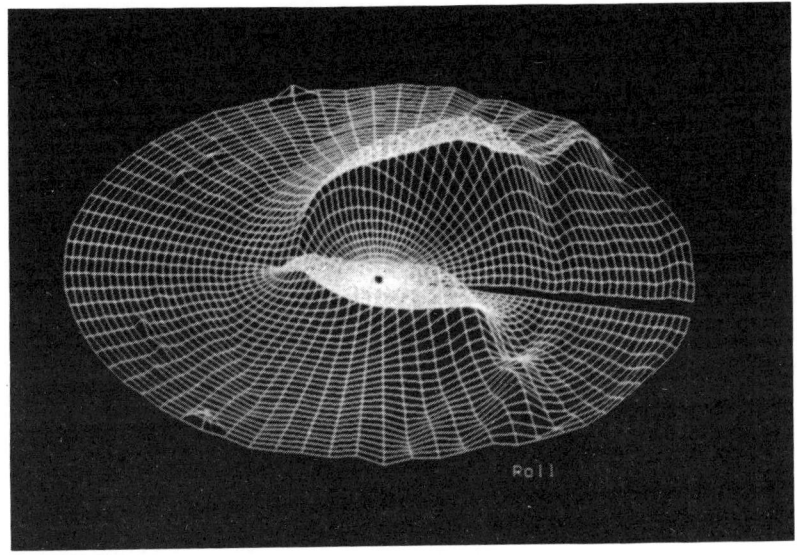

Figure 3

plane responses of heave and pitch. The roles, in ship design, of the dynamic responses have not yet unfortunately reached the status of the traditional hydrostatic calculations of stability. The actual use of seakeeping results generated from computer programs of the theories, is still at a very rudimentary level, in my humble opinion. The conservative method of presenting the data is indicated in figure 2a, which shows the pitch response of a three thousand tonne vessel moving at 18 knots, at a heading angle of 180 degrees. This indicates that at low wave frequencies, large wavelengths, that the ship pitches with the slope of the wave; whilst at large wave frequencies, short wave lengths, the ship hardly pitches. Figure 2b shows the pitch response for the same ship, but at different wave heading angle. The characteristics are completely different. Obviously a multitude of such graphs could be produced. An alternate method of presentation of such data is in three dimensional form. Figure 3 shows the same ship roll response for a large number of wave headings for a range of wave frequencies, on a high definition graphics device. This display allows the viewpoint of the observer to be changed, but still retaining the total information of the ship heave response.   There is still the problem of how to store this information for the use in reports, papers or presentations. The results from the lateral plane responses can equally be dealt with in the same manner, but of course there are still the difficulties with the theories and the numerical evaluation not being accurate enough for most heading angles.

The ultimate aim is of course to present data that theoretically models the ship motion in random seaways, such as shown in Figure 4, including all the surface flow effects. As a first step along these lines the results of the frequency domain need transforming to the time domain. There are at least two major approaches to this problem. The first is still under considerable development, by for instance [1,2]. This involves the problem being formulated by splitting the velocity potential into three terms; the first is associated with the steadily moving ship, the second with the wave system and finally the third with the motion of the body.

$$\Phi_T = Ux + \Phi_o (x,y,z,t) + \Phi (x,y,z,t) \qquad (1)$$

The boundary conditions that the velocity potential have to satisfy are those on the free surface, together with the kinematic condition on the hull; the far field condition which requires radiated waves travelling away from the body, and finally that the initial conditions of motion need prescribing.

The commonest method of solution is to use a Greens function type of solution in the time domain. The velocity potential then can be written as

$$\Phi (P,t) = \frac{1}{4\pi} \iiint_{\tau \ S} \left[\Phi (Q,\tau) \frac{\partial G}{\partial n} - G(P,Q,t-\tau) \frac{\partial \Phi}{\partial t} (Q,\tau) \right] dS d\tau \qquad (2)$$

where P = (x,y,z) is the field point
and   Q = (ξ,η,ζ) is the source point.

and

$$G(P,Q,t-\tau) = \left[\frac{1}{r} - \frac{1}{r'}\right] \delta(t-\tau) + H(t-\tau) \; \bar{G} \; (P,Q,t-\tau)$$

where

$$\bar{G}(P,Q,t-\tau) = 2 \int_0^\infty \sqrt{gk} \; \sin \left[\sqrt{gk} \; (t-\tau) \right] e^{k(z+\zeta)} \; J_0 \; (kr) \; dk \qquad (3)$$

This is a very long winded calculation that is proving very challenging at the present time.

An alternative approach is to use the Cummins intro-differential equation approach, which is stated succinctly for the case of heave motion as

$$m\ddot{z} = A(\omega) \; \ddot{\zeta} + B(\omega)\dot{\zeta} + C\zeta \qquad (4)$$

or
$$m\ddot{z} = F \qquad (4a)$$

where $A(\omega)$ is the frequency added mass coefficient.
$B(\omega)$ is the frequency added damping coefficient.
$C(\omega)$ is the stiffness coefficient.
ζ is the wave elevation.

As such this cannot be used to calculate the time histories of a vessel in irregular waves since the coefficients are frequency dependent. This is so because the local wave elevation is not known a priori, and hence the $A(\omega)$ and $B(\omega)$ cannot be calculated. Cummins and Ogilvie replaced F above by the following equation:

$$F(t) = a_0\ddot{\zeta} + b_0\dot{\zeta} + c_0\zeta + \int_0^t K(t-\tau) \; \dot{\zeta} d\tau \qquad (5)$$

where $K(t-\tau)$ is termed a memory effect function, which in reality is the effect of all previous wave motions from the start up of the motion. This allows the definition of the added mass and damping coefficients to be defined in terms of the memory function $K(t-\tau)$

$$A(\omega) = a_0 - \frac{1}{\omega} \int_0^t K(\tau) \sin(\omega\tau) \, d\tau \tag{6}$$

$$B(\omega) = b_0 + \int_0^t K(\tau) \cos(\omega\tau) \, d\tau \tag{7}$$

$$C(\omega) = c_0$$

where a_0, b_0 and c_0 are all constants.

This allows the use of Fourier transforms to compute the motion of the ship in an irregular seaway by a suitable specification of the wave profile.

The basic problem then resolves about the simulation of the sea surface in the time domain. This can be summarised as the simulation of a random homogeneous process with zero mean defined by a spectral density function $S(\omega)$. This is given by the equation

$$f(x) = \sum_{m=1}^{N} A_m \cos(\omega_m x + \epsilon_m) \tag{8}$$

where ϵ_m are angles that are uniformly distributed in the range o to 2π, and

$$A_m = \sqrt{(2S(\omega_m)\Delta\omega)} \tag{9}$$

$$\omega_m = m \, \Delta\omega \tag{10}$$

$$\Delta\omega = \omega_u/N \tag{11}$$

ω_u is the upper cut off frequency, and N is the number of wave frequency components.

This could be the time realization of the free surface

$$\eta(x,z,t) = \sum_{i=1}^{N} \sum_{j=1}^{M} A_{ij} \cos (k_i (x \cos\vartheta_j + y \sin\vartheta_j) - \omega_i t + \epsilon_{ij}) \tag{12}$$

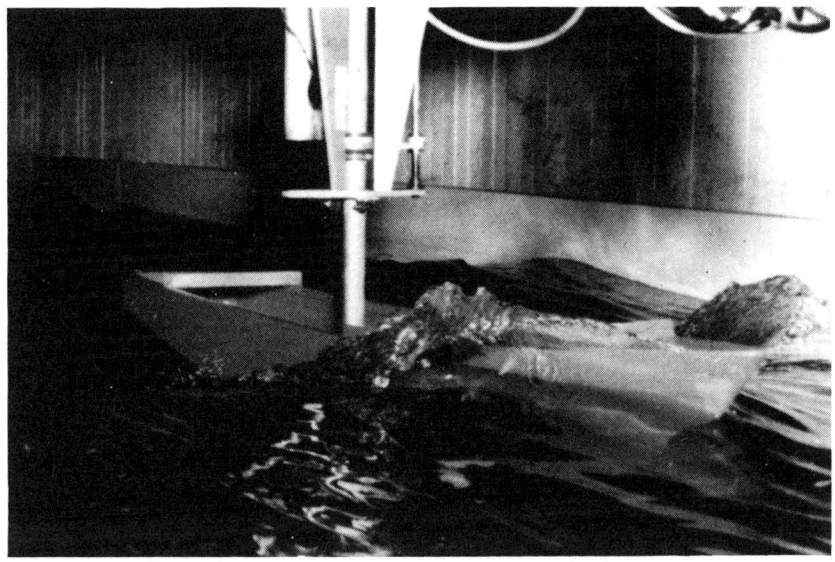

Figure 4

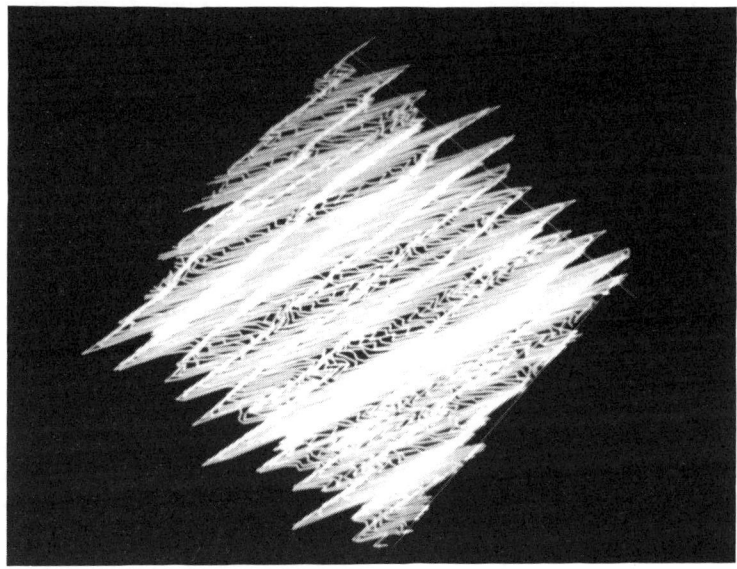

Figure 5

where $\qquad A_{ij} = \sqrt{2S(\omega_i, \vartheta_j)\, \Delta\omega\Delta\vartheta}$ $\qquad\qquad$ (13)

$S(\omega_i, \vartheta_j)$ is the wave spectral density function allowing for the wave spreading function.

$k_i = \omega_i^2/g$, are the random phase angles uniformly distributed in the range 0 to 2π.

Thus the calculation using equation (8), is the realization of

$$\eta(t) = \sum_{i=1}^{N} A_i \cos(-\omega_i t + \epsilon_i) \qquad\qquad (14)$$

where x = 0 and z = 0 in equation [12]

The obvious numerical method that is employed to evaluate equation [14] is the Fast Fourier Transform (FFT). These are obviously freely available for most values of N, but the swiftest transforms are obtained by powers of 4. A typical realization is given in Figure 5.

The problem of the realization of equation [12] is that it is two dimensional in nature. Two and three dimensional FFT's exist. Alternatively if an algorithm is only available that is one dimensional then the time element is not insignificant. Thus for example a spatial distribution at N_t time instances requires N_t 2-D FFT's to be calculated. The alternative method is to permform N_t . N. M single FFT's. The results of such a calculation is displayed in Figure 6.

The next stage is therefore to place the ship in this digitally generated seaway and then perform the calculations according to Cummins method. Thus the motion at many points in the ship, together with their phases in relation to the wave system will need calculating.

As before, with the simple frequency response described above, the major problem is making a permanent record. The slides that accompany this paper show the fine detail that can be obtained, but to plot this on a two dimensional graph produces a rash of lines that often prove indistinct.

Manoeuvring

The use of the high definition graphics devices for ship manouervring have long been recognised. Their role is often complicated by the use of special imaging hardware and software. The main drawback of these systems is the cost. If it is accepted that an acceptable standard of display can be useful to the design engineer or ship designer then the sort of computer program detailed in [3,4,5] is of a great deal of use.

If the force vector F acting in the system given in Figure 1 has components (x,y,z) and H is the vector moment on the body, then relative to this axis system the equations of surge, sway and yaw motion are:

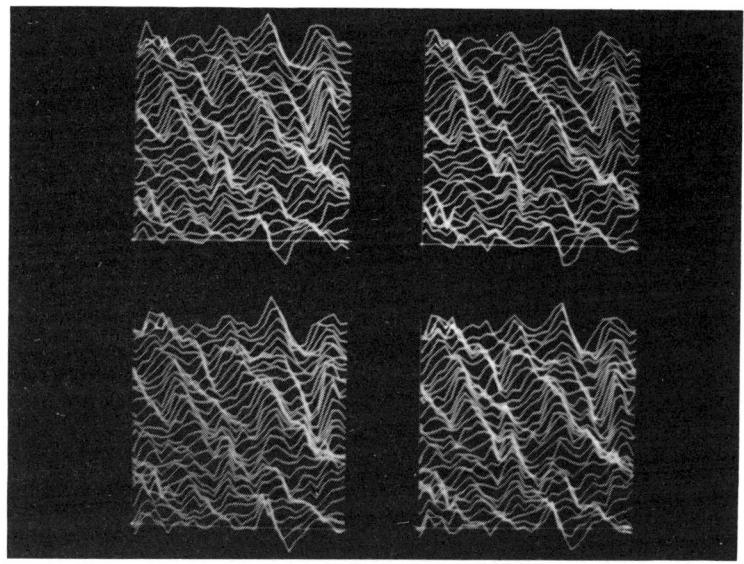

Figure 6

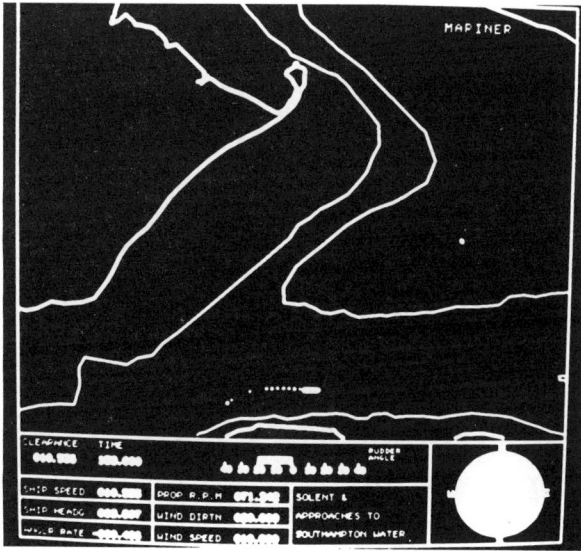

Figure 7

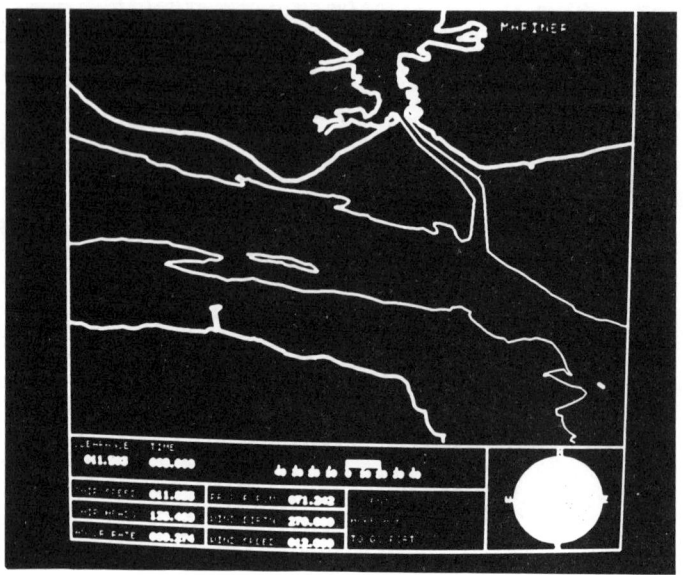

Figure 9

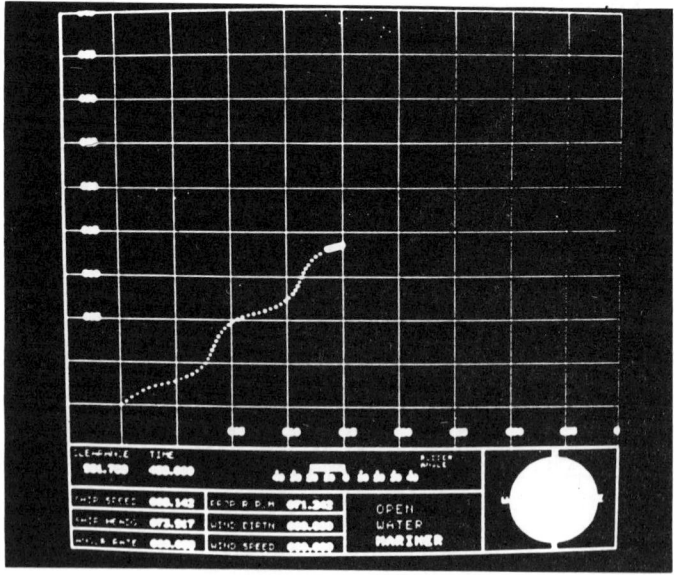

Figure 8

$$X = m(\dot{u} - rv) \tag{15}$$

$$Y = m(\dot{v} + ru) \tag{16}$$

$$M = I_z \dot{r} \tag{17}$$

where u, v are the velocity components, $\dot{u}$, $\dot{v}$, $\dot{r}$ are the acceleration components, r is the yaw rate of turn, m is the ship mass I_z is the virtual mass inertia. The traditional methods of determining the manoeuvring components is to follow the derivative approach [6]. The determination of the various derivatives is outlined in [7,8]. The equations are essentially a series of statistical determinal parameters that best fit the experimental and full scale data. They in no way are meant to have any physical significance, however, they do allow the ship engineer and designer the flexibility of some major ship parameters, such as ship beam, draught, and also depth variation in water. This is obviously a necessity since these critical ship manoeuvres will mainly take place in shallow water and probably, also, a width limited channel. A major problem that accompanies such an approach is the determination of the ship resistance. This obviously affects the surge equation. Allied to this problem is that of propeller information in all four quadrants of motion. Unfortunately only a limited amount of propeller data is available in all four quadrants.

Once all these practical and numerical problems have been overcome, the results of the ship manoeuvres need display. The high quality display of the approaches to Southampton are shown in Figure 7. The ship can be placed anywhere in the frame of reference, together with its initial condition being specified. If the simulation is of an accident then all the environmental parameters can be input. It is possible to have a close up of the ships path displayed as shown in Figure 8.

Alternatively, the program can be used as a ship trials simulator. In the example shown in Figure 9, an open water zig-zag path is displayed. A grid has been superimposed to help determine the Namoto indices.

Conclusion

The use of high definition graphic displays has been explored for both seakeeping and ship manoeuvring. In the case of ship manoeuvring it must be said that the results are adequate for the ship designer, but are probably unsatisfactory for a ship operator wishing to train crew.

In the case of seakeeping, the displays show the capabilities of the calculation. These will prove extremely useful to the ship designer.

References

1.   Beck, R.F., Liapis, S.J. (1987) Transient Motions of Floating Bodies at Zero Foward Speed. J.S.R., Vol. 31, No. 3, pp.164-176.

2.   Beck, R.F., Magee, A.R. (1990) Time Domain Analysis for Predicting Ship Motion. IUTAM, Brunel University.

3.   Wilson, P.A., Lewis, G.D.W. (1986) Predicting Ship Manoeuvring Characteristics for Preliminary Design, CADMO86.

4.   Wilson, P.A. (1988) The Prediction of Manoeuvring Characteristics of Surface Ships in the Design Spiral.   Technology Common to Aero and Marine Engineering, Chapter 17.

5.   Wilson, P.A., Squires, M.A., Seakins, A.P. (1990) Enhanced Preliminary Design Ship Manoeuvring Simulator Techniques. Symposium on Modelling and Control of Marine Craft, Exeter University, April 1990.

6.   Abkowitz, . (1970) A Series of Lecture Notes on Ship Manoeuvring.

7.   Clarke, D. (1983) The Application of Manoeuvring Criteria in Hulll Design using Linear Theory. T.R.I.N.A., Vol. 125.

8.   Inoue, S., Hirano, M., Kijana, K. (1981) Hydrodynamic Derivatives in Ship Manoeuvring. I.S.P., Vol. 28, No. 321.

Modelling of the Combined Yaw and Sway for Ship Manœuvring Models

M.M.A. Pourzanjani

School of Engineering, University of Exeter, Exeter, EX4 4QF, U.K.

ABSTRACT

In almost all ship manœuvring mathematical models, the assumption of linearity and simple addable terms is made. This assumption is conceptually wrong since the fluid forces involved in ship manœuvring are highly non-linear, and strictly speaking superposition cannot be applied. This paper suggests a semi-empirical method where the combined yaw and sway forces and motions can be estimated. The method is an extension to the slender body theory where full scale trials data from a number of ships is also used in development of the algorithm. The equations in their simplest form can be used for simulation of ship manœuvring. In their most comprehensive form they can be used for ship design. Simulation results for a couple of ships will be included in the paper.

INTRODUCTION

Ship manœuvring mathematical models have been in use within the maritime community for decades. In the early days, very simplified models were produced mainly to study the steering characteristics and design of primitive autopilots for marine craft (e.g. see Sperry[13]). These models have since been developed much further and are being used in areas such as Port design, Ship design, auto-pilot design, training and research as discussed by Pourzanjani[9].

One particular application which is of interest to this study is the employment of ship manœuvring mathematical models in ship simulators. Points which should be taken into account when developing a model for this purpose are: Simplicity, Accuracy, Applicability.

In recent years microcomputers have been targeted as the potential machines for ship simulators and for this reason simplicity of the developed algorithm would be essential. Hence methods such as CFD which involve high computational time would not be suitable for real time simulation.

Ship simulators are being used for various purposes, and for some applications such as Port design exercises it would be necessary to have a model with a high degree of accuracy.

The developed model should be <u>applicable</u> to simulation of a wide range of ship size and type. It would not be a cost effective exercise to produce a model which can predict the behaviour of a specific ship only.

Various approaches have been adopted when describing the ship dynamics behaviour, a couple of which will be discussed here briefly, i.e. Hydrodynamic derivative type modelling, direct force model approach. All of these have their origin in the Newton's second law (i.e $\sum F's = Mass \times Acc$) of motion. For a 3 degrees of freedom model and a body fixed coordinate system the equations of motion can be written as:

$$\sum X = m_1 \dot{u} - m_2 v \dot{\psi}$$
$$\sum Y = m_2 \dot{v} + m_1 u \dot{\psi}$$
$$\sum N = I_z \ddot{\psi} \tag{1}$$

The main difference between the two approaches mentioned above is the way the left hand side of the equations (1) is represented as explained in the following sections.

HYDRODYNAMIC DERIVATIVE TYPE MODELLING

The most widely used, accepted and , perhaps well developed model, is the traditional hydrodynamic derivative type modelling approach (Crane et al[4]) where the sum of forces and moments acting on the body (the left hand side of the equations 1) are expanded using Taylor series expansion as a function of various variables. Using the symmetry of the hull and some other considerations, some of the terms can be omitted. Further higher order terms which are assumed to have little influence on the forces are also neglected. The left hand side of the first of the equations (1) can be written as:

$$\sum X = \frac{\partial X}{\partial \dot{u}} \dot{u} + \frac{\partial X}{\partial u} u + \frac{\partial X}{\partial v} v + \frac{\partial X}{\partial \dot{v}} \dot{v} + \frac{\partial X}{\partial \dot{\psi}} \dot{\psi} + \frac{\partial X}{\partial \ddot{\psi}} \ddot{\psi} \tag{2}$$

This approach has produced very close match between the predictions and the full-scale trials. The major shortcomings of this approach can be summarised as follows:

1- There seems to be a disparity between various research workers as to which terms to be included in the mathematical model.

2- The hydrodynamic derivative terms don't bear any relation to the actual physics of the problem.

3- determination of the coefficients require scale model testing which may take up to a few months and is very expensive.

SLENDER-BODY APPROXIMATION

A different approach has also been adopted in recent years, where slender-body approximation is employed to estimate the forces acting on the marine hull forms, as discussed by Pourzanjani[8], Gadd[5] and Oltman[7] and what follows is a summary of this method.

A slender body of revolution fully submerged in steady unbounded irrotational ideal-fluid flow, with its longitudinal axis in the horizontal plane inclined at an angle α to the main-stream is considered. For this body, the total force acting on it is assumed to be the superposition of the forces generated by the two components one parallel and one at right angle to the major axis of the body.

Potential-flow lateral force and lift

For a slender body, the flow in a transverse plane is assumed to be locally two-dimensional. The transverse momentum of the displaced fluid due to the presence of the body can be shown to be

$$U \sin \alpha \, \rho \, S \, dx \qquad (3)$$

the rate of change of lateral momentum between the planes at x and $(x + dx)$ is

$$U^2 \sin \alpha \, \cos \alpha \, \rho \frac{dS}{dx} dx \qquad (4)$$

hence, the lateral force up to the point having the maximum cross-sectional area $S1$ is

$$Y_b = 0.5 \, \rho \, U^2 \, \sin 2\alpha \, S1 \qquad (5)$$

A force Y_s equal in magnitude to that on the leading part of the body would develop on the after body, but with the opposite sign. Therefore, the resultant of these two forces is a pure couple on the body:

$$Y_p = Y_b - Y_s = 0$$
$$N_p = 0.5 \, \rho \, U^2 \, \sin 2\alpha \, V \qquad (6)$$

where V is the volume of the body. These equations apply only to bodies which are pointed at both ends. For bodies which do not satisfy this condition, Y_s is smaller, its value depending on the difference between the maximum cross-sectional area and the area at the after-end:

$$Y_s = 0.5 \, \rho \, U^2 \, \sin 2\alpha \, (S1 - S3) \qquad (7)$$

if we consider the boundary layer growth along the length then equation (7) may be written as:

$$Y_s' = (1 - K)Y_s \qquad (8)$$

where K is an empirical correction factor which represent the ratio of the effective cross-sectional area at the stern to $S1$. When the body is pointed at the stern, $Y_s = Y_b$ and $Y_s' = (1 - K)Y_b$ so that

$$Y_p = Y_b - Y_s' = K \, Y_b \qquad (9)$$

The value of K must depend on the geometry of the body (especially its after-part) and on the Reynolds number of the flow. Adopting this representation of the effective potential-flow lateral force, the corresponding lift force and lift coefficient are given by

$$Lift_p = Y_p \, \cos \alpha = 0.5 \, K \, \rho \, U^2 \, S1 \, \sin 2\alpha \, \cos \alpha \qquad (10)$$

where $S1 = \pi D^2/4$, $b = S1/L\,D$ and D is the maximum diameter of the body.

Viscous cross-flow lateral force and lift

The cross-flow is assumed to be locally two-dimensional and is represented by the flow of the real fluid past a two dimensional body whose cross-section is the same as the local cross-section of the slender body. Since the free-stream speed of the cross-flow is $U \sin \alpha$, the local lateral force due to viscous cross-flow on strip dx of the body may be written as

$$dY_c = 0.5 \, \rho \, U^2 \, C_{nl} \, \sin^2 \alpha \, d \, dx \qquad (11)$$

where C_{nl} is the local cross-flow drag coefficient based on the local draft d. Integrating along the length, the total cross-flow force may be written as:

$$Y_c = 0.5 \, \rho \, U^2 \, C_n \, sin^2 \alpha \, S2 \qquad (12)$$

where $S2 = \int dd x$ is the projected lateral area of the body and C_n should be interpreted as an effective average value of the local coefficient C_{nl}. The lift force due to the viscous cross-flow is then

$$Lift_c = Y_c \cos \alpha = .0.5 \, \rho \, U^2 \, C_n \, \sin^2 \alpha \cos \alpha \, S2 \qquad (13)$$

Even for a body of revolution whose cross-sections remains circular all along the body, C_{nl} is expected to vary with x because of the variation of the local cross-flow Reynolds number and because of the coupling, neglected in this simple approach, between the axial and transverse components of flow in the boundary layer. Moreover, C_n must also depend on α, for the same reasons. For ship hulls, there is the further complication arising from the dependence of C_{nl} on the local shape of the hull cross-section and it is clearly impossible to predict C_n theoretically. From what is known about drag coefficients of two-dimensional bodies with shapes roughly similar to hull cross-sections, C_n is expected to be of the order of magnitude of unity. Because of these uncertainties in the value of C_n and because for a typical hull $S2$ is only slightly smaller than LD, it is convenient to set $S2/LD = 1$ in equations (13), thus absorbing this purely geometrical factor in the definition of the effective C_n.

Total lateral force and and total lift coefficints

These are simply the sums of the corresponding potential and viscous cross-flow contributions and are given by

$$Y = Y_p + Y_c = \rho U^2 \, \sin \alpha (S1 \, K \, \cos \alpha + 0.5 \, L \, D \, C_n \sin \alpha) \qquad (14)$$

HULL DRAG FORCES

Hull drag-forces are assumed to be made up of the contributions associated with the transverse and axial components of the flow, respectively. Thus, we neglect the effects on the drag force of any coupling between the two flows.

Transverse-flow drag

This is given by

$$Drag_y = Y \, \sin \alpha \qquad (15)$$

Axial-flow drag

At zero incidence, this is the straightforward hull resistance to forward motion, which can be estimated by the model scale techniques. At finite incidence, this part of the drag force is dependent on the component of the velocity which is parallel to the major axis of the body i.e. $U \cos \alpha$, so

$$Drag_a = 0.5 \, \rho U^2 \cos \alpha \, S1 \, Cd0 \qquad (16)$$

where $Cd0$ is the hull drag coefficient at zero incidence. Since we are neglecting any wave-making drag, the total drag force is the sum of the two drag components:

$$Drag = Drag_y + Drag_a \qquad (17)$$

YAWING MOMENT

Potential-flow

The yawing moment generated by the potential transverse flow may be calculated by considering the potential-flow lateral-force on an elementry strip of thickness dx of the body and taking moment about the midships:

$$dN_p = (\frac{L}{2} - x)dY_p = J(\frac{L}{2} - x)\frac{dS}{dx}\, dx \tag{18}$$

where $J = 0.5\rho U^2 \sin 2\alpha$ is used to simplify notation. Integrating from $x = 0$ to $x = L$ we the total potential-flow yawing moment is obtained about the midships:

$$N_p = J\,(V - 0.5\,L\,S3) \tag{19}$$

where $S3$ is the transverse area at the stern and V is the volume of the body. As it stands, this expression takes no account of how the boundary layer affects the potential flow outside it. In previous sections the correction factor K was introduced, which allowed the effective stern area to be expressed as $KS1$. To determine an approximation to the effective volume, the boundary layer effects on the forebody are assumed to be negligible, but that the effective cross-sectional area of the afterbody increases at a constant rate from the actual $S1$ amidships to the effective $KS1$ at the stern. Then the effective volume can be shown to be

$$V_e = V + \frac{L(KS1 - S3)}{4} \tag{20}$$

where $S3$ is the actual stern area. The potential-flow yawing moment corrected for boundary layer displacement effects is obtained by rewriting equation (19) as

$$N_p = J(V_e - 0.5\,L\,K\,S1) = J\left(V - \frac{L(S3 + KS1)}{4}\right) \tag{21}$$

Viscous cross-flow

On the assumption of symmetry of the hull about YZ plane, the cross-flow lateral force would act amidships and the yawing moment would be zero. The boundary-layer growth in the longitudinal direction, however, distorts the symmetry about YZ plane; also marine hull forms are known to be unsymmetrical about the YZ plane, which tends to produce a yawing moment. This effect has been ignored in this simple approach.

RESISTANCE TO YAWING

Hull forms have a tendency to resist the yawing motion. In its simplest form this resistance may be assumed to be proportional to the yaw rate squared,

$$N_v = K_r\, r|r| \tag{22}$$

where K_r is assumed to be constant and its numerical value for any particular ship may be calculated from the results of the steady-state turning-circle trial which is available for most ships. It is obvious that the values of K_r obtained in this manner will represent the steady state of the same trial adequately, but errors will be introduced during the transient periods or steady state of other manœuvres. The following method which is based on a theoretical reasoning by considering the moments due to the rotation and cross-flow effects of a flat plate are suggested.

The fluid resistance to yawing motion of the hull may be estimated by employing the strip as in previous sections. Consider an elementary strip of thickness δx, which is at a distance x from the C.G. The force and the resulting moment due to the pure yawing of the vessel experienced by this strip may be expressed as

$$dF = 0.5\,C_f\,\rho\,v^2\,D_l\,dx = 0.5\,C_f\,\rho\,r^2\,x^2\,D_l\,dx$$
$$dM = x\,dF = 0.5\,C_f\,\rho\,r^2\,x^3\,D_l\,dx \tag{23}$$

where C_f is the force coefficient and D_l is the local draft. Integrating along the length to obtain the moment,

$$M = 0.5\,\rho\,r^2 \int C_f\,D_l\,x^3\,dx \tag{24}$$

Non-dimensionalising the moment and approximating the integral,

$$M' = \frac{M}{0.5\rho\,r^2\,L^3\,D_m}$$
$$M' = \int C_f\left(\frac{D_l}{D_m}\right)\left(\frac{x}{L}\right)^3 d\left(\frac{x}{L}\right)$$
$$\simeq 0.25\,C_f \int_b^a c\,d(f) \tag{25}$$

$$\text{where}\quad a = \left(\frac{L_2}{L}\right)^4, \qquad b = \left(\frac{L_1}{L}\right)^4, \qquad c = \frac{D_l}{D_m}, \qquad f = \left(\frac{x}{L}\right)^4$$

where D_m is the maximum draft, L_1 is the distance between the C.G. and the astern and L_2 is the distance between the C.G. and the stem, so that $L = L_1 + L_2$. Ships when manœuvring do not necessarily pivot about their C.G., and indeed most of the time this point is somewhere on the longitudinal axis and away from the C.G.

To examine the effects of variation of the pivot point on the fluid moment resistive to yawing motion, the situation is simplified by considering a two-dimensional rectangular flat plate (i.e $L_1 = L_2$ and $D_l = D_m$), rotating about a point at a distance P from the C.G. and on the longitudinal axis of the body. Neglecting the end effects and potential-flow contributions, the moment due to cross-flow effects alone may be described as follows,

$$dF = 0.5\,\rho\,r^2\,x^2\,D\,dx$$
$$dM = 0.5\,\rho\,r^2\,x^2\,(x - P)\,D\,dx$$
$$M = 0.5\rho\,r^2\,D \int_{P-l}^{P+l} x^2\,(x - P)\,dx$$
$$M = 0.5\rho r^2 D\left(\frac{L^4}{32} + \frac{L^2 P^2}{4} - \frac{P^4}{6}\right) \tag{26}$$

From equations 22 and 26 it follows that,

$$K_r = \rho\,D\,L^4\,a$$
$$K_r = \rho\,D\,L^4\left(\frac{1}{64} + \frac{P^2}{8L^2} + \frac{P^4}{12L^4}\right) \tag{27}$$

Figure 1 shows the effects of variation of pivot point on the fluid resistive moment to yawing.

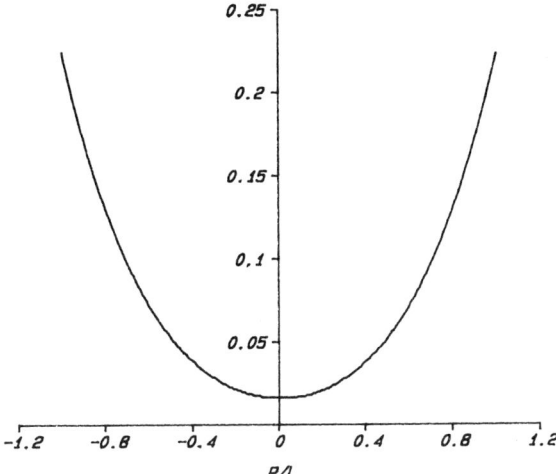

Figure 1. Variation of the pivot point with K_r

COMBINED SWAY AND YAW

In developing the algorithm for hull hydrodynamics, all the attention was focussed on the drifting forces. It was later assumed that superposition theorem may be applied to add the effects of the yaw and drift for a vessel engaged in a manoeuvre.

A different approach which may be adopted, is to examine the forces acting on a hull form due to the drift and yaw at the same time.

Angle of Incidence

Angle of incidence is an important parameter, particularly in the Force mathematical model, since hull hydrodynamic forces and rudder forces are dependent on its accuracy. It can be calculated using the relationship

$$\alpha = \tan^{-1}\left(\frac{v}{u}\right) \tag{28}$$

This relationship gives the incidence at the centre of gravity, and unless the ship is drifting with some incidence and zero yaw rate, the local fluid velocities and hence the local incidence angle along the length would vary, being smaller at the bow and greater at the stern when moving ahead and turning. One particular problem is to estimate the flow angle at the rudder postion.

To overcome this problem an effective drift angle for the rudder which is dependent on the manoeuvring regime can be assumed, or having only a multiplier to the drift angle to get the incidence angle at the rudder position.

$$\alpha_r = M_r \quad \alpha \tag{29}$$

The concepts of effective drift angle or a simple multiplier for "α_r" are very poor representations of this variable since it is also dependent on the yaw rate and speed. A more accurate way of calculating "α_r" is developed here.

Figure 2 represents a ship engaged in a turning circle manœuvre. It can be seen that the radius of turning circle changes along the length, being larger at the stern, which causes a larger drift angle.

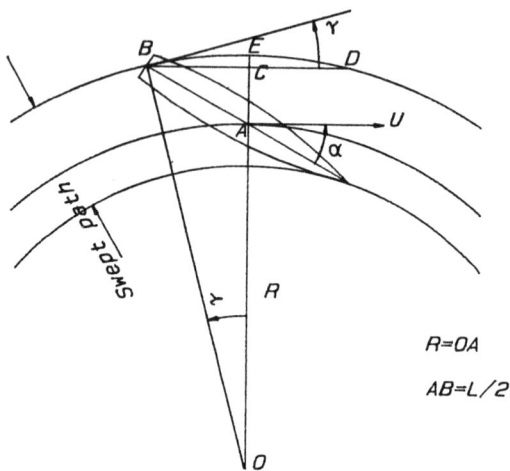

Figure 2. A vessel engaged in a turning circle manœuvre

The geometric incidence angle at the rudder "α_{rg}" is calculated from the geometry of figure 2 (flow angle at point B in figure 2). The angular velocity is constant along the length but the local fluid velocities and radius of turn vary. If we assume that the rudder position is half the length abaft of the C.G. and the difference between the incidence at C.G. and rudder is denoted by "γ" then,

$$\gamma = \tan^{-1}(\frac{BC}{OC})$$
$$BC = (\frac{L}{2})\cos\alpha, \qquad OC = (\frac{U}{r}) + (\frac{L}{2})\sin\alpha$$
$$\alpha_{rg} = \alpha + \gamma \tag{30}$$

The equations 30 describe the geometric angle of incidence at the rudder position for different speeds and yaw rates. These equations, however, tend to overestimate the true angle of incidence at the rudder due to the fact that the hull and propeller distort the idealised flow as shown in figure 2. It is assumed that there is a linear relationship between the geometric and true angle of incidence at the rudder:

$$\alpha_r = \alpha_{rg} \ \ K_r \tag{31}$$

The exact value of K_r for any particular ship may be determined from the open water characteristics of the rudder and the force curves of the hull fitted with the rudder.

Full-scale steady-state turning-circle data from four different ships have been used with equations 30 and 31 to calculate α_r and results are presented in table 1.

Table 1 α_r calculated for four different ships

Ship Type	C_b	R	Rud_a	α	α_r	Ref.
VLCC	0.825	246	35P	30	40.3	2
B. Carrier	0.804	235	35S	29	40.3	12
Mariner	0.6	424	35P	14	24.0	6
Container	0.565	340	35P	12	21.5	1

Values of α_r for the two ships with higher block coefficients (VLCC and B. Carrier) show that the flow angle at the rudder is greater than the rudder angle which indicates that the rudder has an adverse effect on the turn. This shows that the hull forces and moments dominate the turning characteristics and indicate the directional instability of these ships. This may be the reason for ships built in more recent years to have a maximum rudder angle of as much as 50° either side as opposed to 35° on most conventional ships.

Hull Forces

From equation (5) the potential-flow side-force from the bow portion of the hull may be written as:

$$Y_b = \int_0^{L/2} 0.5\,\rho\,U^2 \sin 2\alpha \frac{dS}{dx} dx \tag{32}$$

Refering to the previous section on the angle of incidence and the fact that the angle of icidence is not constant along the ship's lenght, equation 32 may be written as:

$$Y_b = \int_0^{L/2} 0.5\,\rho\,U^2 \sin 2\alpha_l \frac{dS}{dx} dx \tag{33}$$

where α_l is the local incidence angle, and as mentioned above varies along the length during a change in the manœuvring regime. From the geometry of a vessel engaged in a manœuvre with a combination of yaw and drift, α_l can be expressed in terms of α (incidence at the C.G) as follows:

$$\alpha_l = \alpha + \gamma$$
$$\gamma = \tan^{-1} \frac{(x-2)\cos\alpha}{\frac{U}{r} + (x - \frac{L}{2})\sin\alpha} \tag{34}$$

Equation 33 may now be expressed as:

$$Y_b = \int_0^{L/2} 0.5\,\rho\,U^2 \sin 2(\alpha + \gamma) \frac{dS}{dx} dx \tag{35}$$

If we substitute γ from equations 34 into equation 35 then although the integral may still be determined, the resulting algorithm is so long and complicated that we lose the required simplicity of which was one of the main goals of this approach. Therefore simplification to this algorithm is made by appealing to the full-scale data.

Turning circle data for the USS Compass Island which are presented as a matrix of different speeds and rudder angles are used in figure 3 to represent the variation along

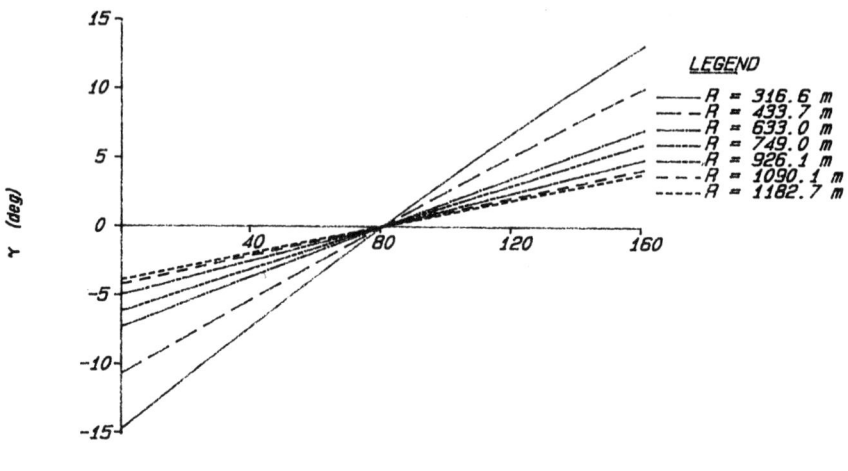

Figure 3. Variation of γ with length, Compass Island

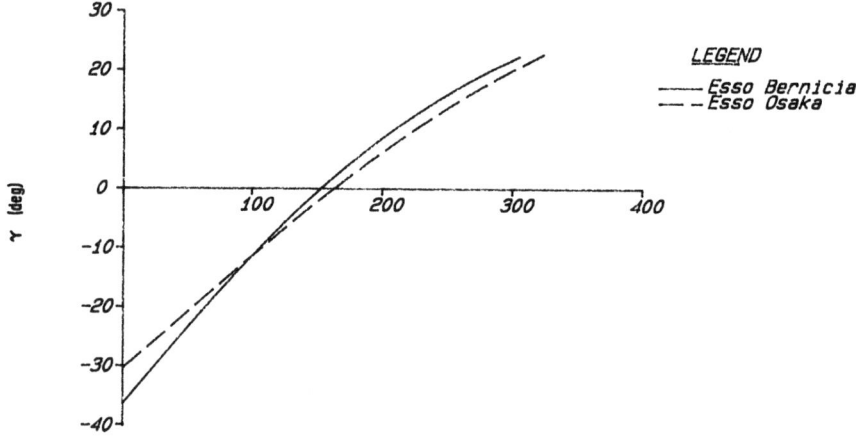

Length (m)
Figure 4. Variation of γ with lenght, Esso Osaka and Esso Bernicia

the length of the increment to the angle of incidence γ. Figure 4 is similar to figure 3, but is for Esso Osaka (Crane[2]) and Esso Bernicia (Clarke et al.[3]).

In these figures, it is noticed that the variation of γ along the length can approximately be represented by a straight line. This linear relationship is more applicable to the bow region and also to the turning circles with larger radius of turn (in other words where the viscous effects are not the dominant features). If we allow this assumption to be made, then from figures 3 and 4 we can express γ as

$$\gamma = \frac{2Kk}{R}\left(x - \frac{L}{2}\right) \tag{36},$$

where Kk varies as a function of the radius of turn. To examine this variation full-scale data from different ships are used to plot Kk against normalised radius of turn

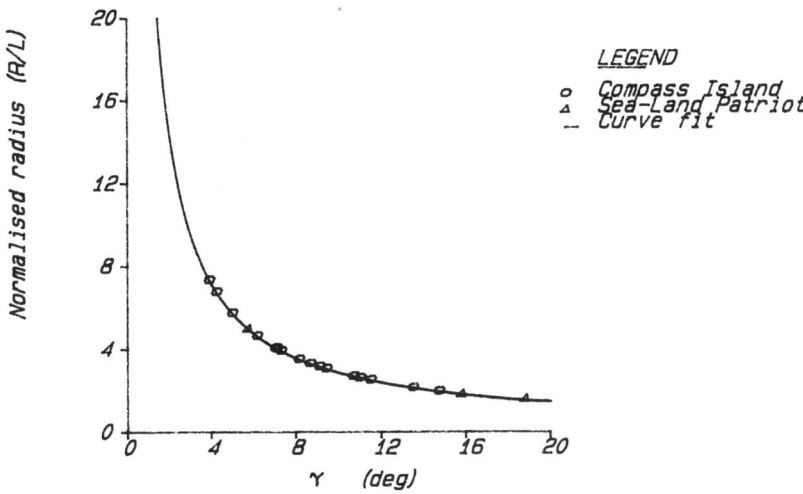

Figure 5. Variation of γ with normalised radius of turn

$[= R_s/L_{pp}]$ (figure 5).

The rate of change of the transverse cross-sectional area with length varies from ship to ship and may be found by refering to the lines plan of any particular ship. For the purpose of our simplified calculations if we assume that this variation is linear, i.e $dS/dx = 2S_1/L$, equation 35 may be written as

$$Y_b = 0.5\,\rho\,U^2\,\frac{2S_1}{L}\sin(2\alpha - 2\beta)\int_0^{L/2}\cos\left(\frac{4Kk}{R}x\right)dx+$$

$$0.5\,\rho\,U^2\,\frac{2S_1}{L}\cos(2\alpha - 2\beta)\int_0^{L/2}\sin\left(\frac{4Kk}{R}x\right)dx$$

$$Y_b = 0.5\,\rho\,U^2\,S_1\left(\sin(2\alpha - 2\beta)\frac{\sin(2\beta)}{2\beta} + \cos(2\alpha - 2\beta)\frac{1 - \cos 2\beta}{2\beta}\right) \qquad (37)$$

where Kk is calculated from figure 5 and $\beta = Kk\,L\,R^{-1}$. Similar approach is adopted by considering the variation of the fluid velocities and local angle of incidence to estimate the potential-flow effects in the stern region and also to estimate the cross-flow effects.

Experiments may be carried out to validate and examine the accuracy of the formulae suggested above. Rotating-arm facilities may be used where the angle of incidence and radius of turn can be set for any trials. This facility, however, is very expensive to employ and is not available in most towing-tank establishments. One other possiblity is to use a wind-tunnel to measure the forces. Since the fluid flow is in a straight line in the tunnel, the scale model should be transformed so that the curvature of the flow is simulated with a curved body as discussed by Pourzanjani[11].

SIMULATION RESULTS

The model as described above has been incorporated in the ship manœuvring suit at Exeter University and a number of trials both in off-line mode and real-time simulations have been carried out. Figures 6 and 7 are the comparison between the model prediction and the full-scale results of turning circle trials for a Mariner Type vessel. There appears to be a good agreement between the simulation results and the full-scale trials. similar results have been achieved for other ships of different type and sizes. More work will be required on the verification and validation of the method as suggested in this paper. This could take the form a systematic scale model and full scale tests.

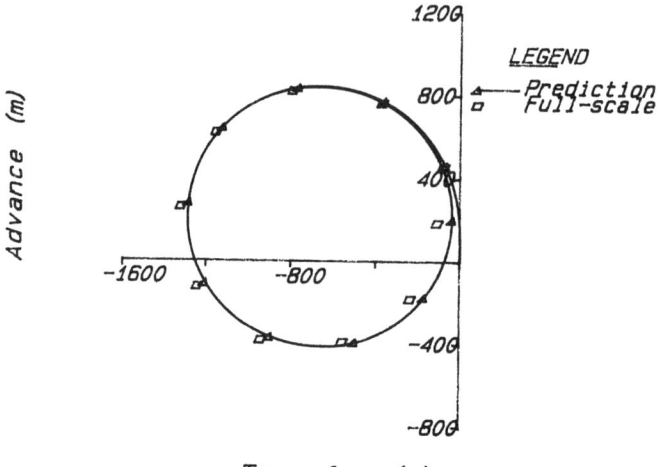

Figure 6. Turn 10° Port from 15 knots; position diagram

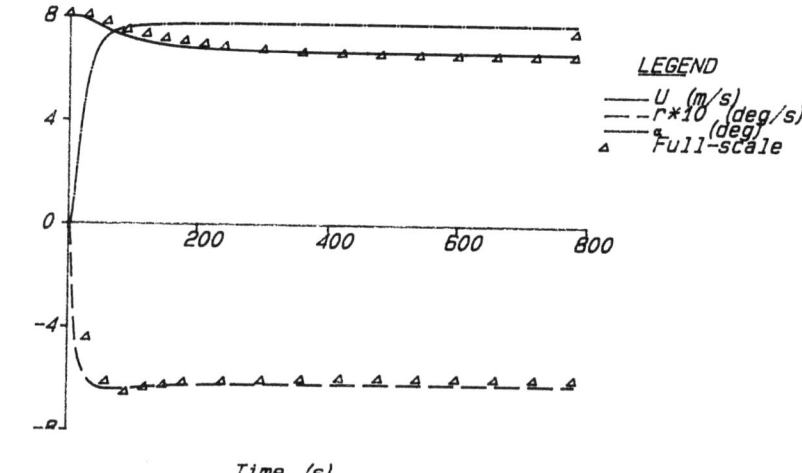

Figure 7. Turn 10° Port from 15 knots; U, r & α

References

1- Baumler, R. J., Watanabe, T. and Huzimura, H., "Sea-Land's D9 Containership: Design, Construction, and Performance", Transactions of SNAME, Vol. 91, 1983, pp. 225-256.

2- Crane, C. L. Jr., "Manœuvring trials of the 278000-DWT Esso Osaka in shallow and deep waters", EXON International Company, Report No. EII. 4TM.79, Jan. 1979.

3- Clark, D., Patterson, D. R. and Wooderson, R. K., "Manœuvring trials with the 193000-tonnes d.w tanker Esso Bernicia", BSRA report NS 295, 1970.

4- Crane, C. L., Eda, H. and Landsburg, A.,"Principles of Naval Architecture", Chapter IX, SNAME, 1989.

5- Gadd, G. E., "A Calculation method for Forces on Ships at Small Angles of Yaw", paper issued for written discussion, RINA, London 1984.

6- Morse, R. V. and Price, D., "Manœuvring characteristics of the mariner type ship (USS Compass Island) in calm seas", Sperry Polaris Management, Sperry Gyroscope Company, New York, Dec. 1961.

7- Oltman, P. and Sharma, S. D., "Simulation of Combined Engine and Rudder Maneuvers Using an Improved Model of Hull-Propeller-Rudder Interactions", Fifteenth symposium on Naval Hydrodynamics, Hamburg, 1984.

8- Pourzanjani, M. M., Zienkiewicz, H., and Flower, J. O.,"A Hybrid Method of estimating Hydrodynamically Generated Forces for use in Ship Manœuvring Simulations", International Shipbuilding Progress, Vol. 34, No 399, 1987.

9- Pourzanjani, M. M. "Requirements and Applications of Ship Dynamics Simulators", Proceedings of the 1990 UKSC conference on computer simulation, Brighton, September 1990.

10- Pourzanjani, M. M. "Formulation of the force math., model of ship manœuvring", International Shipbuilding Progrress, Vol 37, No. 409, April 1990.

11- Pourzanjani, M. M. "An alternative for rotating arm facilities: Curved model technique", Internal report, School of Engineering, Exeter University, 1990.

12- Samsung Shipbuilding & Heavy Industries Co., Ltd., "Sea trial data of the vessel Iron Pacific", Port Kembla, Australia, 1986.

13- Sperry, E.,"Automatic Steering", Transactions of SNAME, 1922.

SECTION 4: SHIP STRUCTURES

An Optimum Structural Design System Using Reanalysis Method

S.W. Park(*), J.K. Paik(**), I.S. Nho(*) and H.S. Lee(*)

(*) Structure and Welding Technology Dept., Korea Research Institute of Ship and Ocean Engineering, Daejeon, 302-343, Korea
(**) Dept. of Naval Architecture, Pusan National University, Pusan, 609-735, Korea

Abstract

In usual, tremendous amount of computational efforts is needed in the optimization process for such large and complicated structures as ship structures because the structural analysis should be repeatedly carried out with changing the design variables. In this study, a reanalysis technique, Modified Reduced Basis Method, is incorporated into the structural analysis program in order to reduce the number unknowns of the structural stiffness equation in the optimization process. For the validity of the present program, optimum designs for a simple 1×2 grillage and an actual container ship's midhold are performed. The present solutions are compared with other numerical results in the view point of the accuracy and computing time. It is observed that the present system gives accurate design results and also the computing time can be greatly reduced.

1 Introduction

Nowadays, for the minimum weight or minimum cost design of ship structures, it is general to apply the optimum design technique directly to the numerical structural analysis such as finite element method, paralleled with the design procedure according to classification society rules[1-6]. However, the iterative process of structural analysis is inevitable due to the continuous change of design variables in order to achieve an optimum design. In the design of ship structures which are composed of various types of structural members, the computing time of the repeated structural analysis is extremely long and impractical in the view point of the design efficiency. Here, a reanalysis technique is applied to the optimum design process in order to improve the efficiency of the structural analysis.

The efficiency and usefulness of the reanalysis techniques were very rapidly increased in 1960's with the development of computers, and the researches for this area are still active[7-10]. Generally, the reanalysis techniques can be applied to both categories of structural analysis method; displacement and flexibility method. They are also applied to dynamic structural analysis to calculate the eigenvalues of the system.

In this paper, we formulate the optimum structural design problem by using the Reduced Basis Method which can produce more accurate solutions than other reanalysis techniques even though the variations of design variables are large. An integrated optimum structural analysis/design program is developed, and verifies its efficiency and accuracy by obtaining the minimum weight design of an actual ship structure.

2   Formulation of Minimum Structural Weight Design

In this paper, the structural weight is chosen as an abjective function to achieve a minimum weight design. The dimensions of structural members, for example, thicknesses or cross sectional areas, are design variables within prescribed constraints. The displacements and stresses of each structural member will be calculated by the finite element analysis, and should not exceed the given criteria for safety. Then, this problem can be expressed as equation (1).

$$
\begin{aligned}
&Find & &d_i & &(1)\\
&Minimize & &W = W(d_i)\\
&Subject\ to & &r_k - \bar{r}_k \leq 0 & &k = 1,...,l\\
& & &\sigma_j - \bar{\sigma}_j \leq 0 & &j = 1,...,m\\
& & &d_i - \bar{d}_i \leq 0 & &i = 1,...,n
\end{aligned}
$$

where

d_i : design variables
$\bar{d}_i$: allowable values of design variables
W : objective fuction (structural weight)
r_k : nodal displacements
$\bar{r}_k$: allowable values of nodal displacements
σ_j : element stresses
$\bar{\sigma}_j$: allowable values of element stresses
l : total number of degrees of freedom
m : total number of elements
n : total number of design variables

The design variables and element stresses described in equation (1) are related with the element type of the finite element structural analysis program, and are summarized in Table 1.

Table 1   Design Variables and Stresses for Each Finite Element Type

Types of Element		Design Variable	Types of Stress
Truss Element		Sectional Area	Axial Stress
Beam Element	'H' Shape	Thickness of Upper/Lower Flange Plate, Thickness of Web Plate, Breadth of Flange Plate, Depth of Web Plate	Axial Stress, Transverse & Vertical Shear Stress, Transverse & Vertical Bending Stress, Effective Combined Stress
	Arbitary Shape	Axial Area, Transverse & Vertical Shear Area, Transverse & Vertical bending Moment of Inertia, Torsonal Monment of Inertia	
Membrane/Plate Element		Plate Thickness	Biaxial Bending Stress Shear Stress Combined Stress

The mathematical optimization techniques, such as direct search methods and tangent gradient methods, need a large number of search steps. In every search step, the structural analysis should be performed to obtain the stresses and displacements in equation (1) as constraints. Since this procedure requires the extreme calculating time, the reanalysis technique is considered to be very effective to reduce the computing time. In the following section, more detail procedure for the application of reanalysis technique will be described.

3   Reanalysis Technique by the Reduced Basis Method

From the general finite element formulation, the equilibrium equations (2) and (3) can be written for the initial design state and modified design state, respectively.

$$[K]\{r\} = \{R\} \tag{2}$$

$$[K^*]\{r^*\} = \{R\} \tag{3}$$

where

$[K]$: stiffness matrix for initial design
$[K^*]$: stiffness matrix for modified design
$\{r\}$: nodal displacement vector for initial design
$\{r^*\}$: nodal displacement vector for modified design
$\{R\}$: design load vector

It is assumed that the design loads are not changed during the modification. Generally, most of the calculating time in the finite element structural anlaysis are expensed in the calculation of the nodal displacement vectors in equations (2) and (3). The Reduced Basis Method(RBM) can calculate the nodal displacement vector more effectively for the modified structure in equation (3).

$$\{r^*\} \cong [\Psi]\{c\} = c_1\{\tilde{r}_1\} + c_2\{\tilde{r}_2\} + ... + c_q\{\tilde{r}_q\} \tag{4}$$

where

$[\Psi] = [\{\tilde{r}_1\}, \{\tilde{r}_2\}, ..., \{\tilde{r}_q\}]$: matrix of reduced basis vectors, $\{\tilde{r}_i\}$
$\{c\} = [c_1, c_2, ..., c_q]^T$: vector of unknown coefficients

The number of q in equation (4), which is close to the number of design variables (n), is much smaller than the total degrees of freedom of the structure $(q \ll l)$. Substituting equation (4) into equation (3) gives

$$[K^*][\Psi]\{c\} = \{R\} \tag{5}$$

$$[K_r]\{c\} = \{R_r\} \tag{6}$$

where

$[K_r] = [\Psi]^T[K^*][\Psi]$: reduced basis stiffness matrix
$\{R_r\} = [\Psi]^T\{R\}$: reduced basis load vector

We can obtain the matrix equation (6) of reduced degrees of freedom from l to q. Consequently, this method makes us more effectively calculate the nodal displacement vector of the modified structure from equations (4) and (6) than from equation (3) which has very large number of degrees of freedom. Because equation (6) has reduced degrees of freedom, and equation (4) is only requiring the simple summation algorithm, we may expect the total calculation time be reduced.

Due to the assumption included in equation (4), the RBM gives an approximate solution, the accuracy of which is absolutely dependent on the selection of the basis vectors in equation (4). The basis vectors must be linearly independent and can approximate very well the actual structural behaviors. In 1974, Noor and Lowder[13] proved that the Modified Reduced Basis Method (MRBM) was excellent in both the solution's accuracy and the algorithm's efficiency compared with other approximate methods.

Therefore, this MRBM is adopted in our system also. The first basis vector $\{\tilde{r}_1\}$ of MRBM is set to the initial nodal displacement vector. Next n basis vectors are chosen as the first-order sensitivity vectors of the nodal displacement with respect to the design variable. Then, the basis vectors can be written as follows.

$$\{\tilde{r}_1\} = \{r\}_{d_i = d_i^0} \tag{7}$$
$$\{\tilde{r}_{i+1}\} = \frac{\partial\{r\}}{\partial d_i}, (i = 1, 2, ..., n)$$

where

$\{\tilde{r}_1\}$: displacement vector for initial design value
$\{\tilde{r}_{i+1}\}$: first-order sensitivity vectors for ith design variable

The n's first-order sensitivity vectors obtained above include unavoidable errors because the change of the element stiffness does not linearly relate with the change of the design variables. Also, numerical errors result from the size of infinitesimal increments of design variables. Those errors are increased with the increase of the difference between the modified design and initial design, and this affects directly to the accuracy of equation (4). In this paper, in order to reduce these errors, the following convergence criterion of the reanalysis is used.

$$\varepsilon \le \frac{d_i^* - d_i^0}{d_i^0} \tag{8}$$

where

ε : convergence criterion of reanalysis
d_i^* : design point of modified structure

When any design variable exceeds the given criterion during the optimization process, the initial design point will be reset and all of the basis vectors at that point will be recalculated. The value of ε used here is 0.5. The suggested infinitesimal incremental value of each design varibable is 0.1 % of the corresponding initial value. This value is expected to be smaller than the minimum incremental step size in the optimization process.

4   Development of Optimum Structural Design Program

As shown in Fig. 1, the program is composed of four modules: optimization module (Optimizer), structural analysis module (Analyzer), File A (storing input/output files), and File B (storing the sensitivity files).

The optimization module in our system uses six algorithms : Hooke & Jeeves' Direct Search Method, Nelder & Mead's Simplex Search Method, Powell's Conjugate Direction Method, Flexible Tolerance Method, Improved Direct Search and Feasible Direction Method, and Random Search Method. The structural analysis module gives Optimizer the values of element stresses from the finite element structural analysis. The Analyzer stores the basis vectors, initial stiffness matrix and other system information in File B at every necessary stage. Once input data are prepared for the system, the optimum point of the design variables and structural analysis results at the point can be obtained automatically.

Fig. 2 shows the flow diagram of the structural analysis module. The module can be divided into two parts. One is complete structural analysis process to calculate the basis vectors, and the other part is the reanalysis process based on MRBM. The substructural analysis technique is used partly in the module for the purpose of increasing the computational efficiency. Then, the stiffness matrix of system can be divided into two parts.

$$[K] = [K_o] + [K_s] \tag{9}$$

where

$[K_o]$: stiffness matrix of the members independent with design variables
$[K_s]$: stiffness matrix of the members dependent with design variables

In equation (9), the stiffness matrix $[K_o]$ is calculated only once at the initial design stage, and added to $[K_s]$. Stiffness matrix $[K_s]$ should be calculated at every analysis stage.

5   Numerical Calculations and Discussions

Two examples are selected for the numerical calculations. The first example is to design a 1×2 grillage structure. This simple example will show the validity of the system. The practical aspect of the system will be shown by the second example, which is to design an actual container ship.

5.1   Minimum Weight Design of a 1 x 2 Grillage

As shown in Fig. 3, this design problem is defined as the minimum weight design of a grillage composed of one girder in the longitudinal direction and two stiffeners in the transverse direction. The end conditions of the girder are fixed and those of the stiffeners are simply supported. The sectional areas of the girder and the stiffeners are design variables. The only constraint condition is that the maximum bending stress of each structural member should not exceed the allowable stress. Then, equation (1) can be rewritten as follows.

$$
\begin{aligned}
&Find \qquad\quad d_1 \ and \ d_2 \\
&Minimize \quad Weight = c_1 \times d_1 + c_2 \times d_2 \\
&Subject \ to \qquad\quad \sigma_j \leq \sigma_a
\end{aligned}
\tag{10}
$$

where

d_1 : cross sectional area of girder
d_2 : cross sectional area of stiffener
c_1 : length of girder × no. of girder = $120in \times 1$
c_2 : length of stiffener × no. of stiffener = $100in \times 2$
σ_j : maximum bending stresses of each member
σ_a : allowable stresses of each member

The second moment of inertia and section modulus are assumed to be a function of the cross sectional area[14].

$$
I = 1.007[A/1.480]^{2.65} \quad : \text{sectional moment of inertia}
\tag{11}
$$
$$
Z = [A/1.480]^{1.82} \quad : \text{section modulus}
$$

In Table 2, three optimization algorithms, Improved Direct Search and Feasible Direction Method, Hook & Jeeves' Direct Search Method, and Nelder & Mead's Simplex Search Method, are used with and without the reanalysis technique. As shown in Table 2, the reanalysis technique used in this paper is effective regardless of the optimization algorithms. We obtain the very accurate solutions in spite of using the reanalysis technique. The application of the reanalysis technique is not significant to reduce the computing time because the total degrees of freedom of this model is very small. The reanalysis convergence criterion of 0.5 is a proper size.

Table 2   Optimum Design Results of 1x2 Grillage

Optimizer		Direct Search & Feasilbe Direction Method			N - M Method		H - J Method
Reanalysis Convergence Criterion(ϵ)		Full Analysis	0.5	1.0	Full Analysis	0.5	0.5
Initial	d_1	50.0	50.0	50.0	50.0	30.0	30.0
Point	d_2	50.0	50.0	50.0	50.0	10.0	10.0
Optimum	d_1	22.49	22.49	22.49	22.50	22.50	22.45
Point	d_2	6.157	6.157	6.115	6.407	6.299	6.187
Weight		3,930	3,930	3,930	3,980	3,960	3,914
Constraint Value(Kpsi)	g_1	11.44	11.44	11.44	11.56	11.54	11.41
	g_2	0.06	0.06	0.06	0.35	0.29	0.01
	g_3	20.0	20.0	20.0	20.0	20.0	20.0
	g_4	-0.23	-0.23	-0.23	0.8	0.6	0.0
No. of Iter.		90	90	90	103	65	57

5.2   Minimum Weight Design of Container Ship's Midhold

We have applied our program for the minimum weight design of container ship's one midhold under the maximum operating load conditions as shown in Fig. 4. First of all, the optimum design for the longitudinal members is carried out by using the Rule Scantling Program for Container Ship[12]. The optimization of the transverse strength members composed of side web frame, bilge web, and bottom floor is performed using the present system.

As shown in Fig. 5(a), a quarter of the hold is modeled in the finite element analysis by considering the symmetry of the hold. To minimize the total nodes of the model, we use the idealized membrane and beam elements as shown in Fig. 5(b) which is equivalent to the stiffened plates of ship structures[11]. Generally, three types of the design load conditions are considered : full load condition with sagging wave position, ballast condition with hogging wave position, and full load condition with rolling heeded position. Among them, the last two conditions, as shown in Fig. 6, are applied in this calculation since the first one has relatively little effect on the transverse strength of the ship structure. In order to describe the heeled condition exactly, we assume that the boundary conditons are same as that of the ballast load condition. This assumption is practically needed because it is impossible to apply the different boundary conditions simultaneously in our program even though the ballast load condition and heeled load conditions must be considered simultaneously. However, considering that our main interest of the heeled condition is transverse strength of double bottom and web frame and the change of boundary conditions will give little effect on this part, we can see that the above idealization is sufficiently realistic. The constraints of the problem is that the shear stresses and combined stresses of the all strength members should not exceed the shear buckling stresses and yielding stresses, respectively. Therefore, we can formulate this structural optimization problem as equation (12).

$$Find \qquad d_i \ (i = 1, 5) \tag{12}$$
$$Minimize \quad Weight = W(d_i)$$
$$Subject\ to \qquad \tau_j \leq \tau_{cr}$$
$$\sigma_{cj} \leq \sigma_c$$

where

d_1 : plate thickness of bottom floor
d_2 : plate thickness of bilge part
d_3, d_4, d_5 : plate thicknesses of web frame
$Weight$: total weight of transverse members
τ_j : shear stresses of transverse members
τ_{cr} : shear buckling stress $(= CE(t/b)^2$)
σ_{cj} : combined stresses of transvers members
σ_c : combined yielding stress
t : plate thickness
C : coefficient of shear buckling of plates
E : Young's modulus of elasticity
b : stiffener spacing in web plates

In Table 3, computing time and solution's accuracy are compared for the two cases: with and without reanalysis technique. The final optimum points of the two cases are equal, and the computing time by the reanalysis technique is about 1/3 of the time by complete structural analysis.

Table 3   Optimum Design Results of Transverse Strength Member

		Initial Design	Optimum Reanalysis Design	Optimum Complete Analysis
Design Variable	d_1	15.0 mm	10.0 mm	10.0 mm
	d_2	15.0	10.0	10.0
	d_3	11.0	9.5	9.5
	d_4	11.0	7.5	7.5
	d_5	11.0	7.5	7.5
Total Weight (m^3/section)		0.948115	0.67058	0.67058
C.P.U. (PRIME 6350)			2,346 sec	6,100 sec

Since the initial design values are calculted from the Rule Scantling Program which have too much safety margin, the optimum results show that the weight of transverse members in the hold can be reduced to 70 % of the initial design. Even though this optimum values are obtained not considering the corrosion effects, the possibility to reduce the total weights of ship structures by this approach is proved.

6   Conclusion

An optimum structural design program is developed to include various mathematical optimization algorithms and general structural analysis codes by the finite element method. The efficiency and usefulness of the structural reanalysis technique are verified in this paper by linking to the optimization process. The results of the optimum design of actual container ship show that the computing time due to the iterative process of the optimization has been reduced to 1/3 of the computing time without the reanalysis technique.

7   Acknowledgement

The work described in this paper is a part of the project, "Development of Computer Program for Controlling Ship Structural Weight and Construction Cost (I),(II)" [11,12] carried out under the sponsorship of the Ministry of Science and Technology, KOREA.

We would like to thank Dr. J.G. Shin for his comments and suggestions to this paper.

References

1. Shin, J.G.,"Minimum Weight Design of Midship Section Structure Using Optimization Technique," J. of the Society of Naval Architects of Korea, Vol.17, No.4, 1980.

2. Na, S.S., Min, K.S., Urm, H.S., and Shin, D.H., "Minimum Weight Design of Ship Transverse Members Using Finite Element Method," J. of the Society of Naval Architects of Korea, Vol.22, No.3, 1985.

3. Yim, S.J. and Yang, Y.S.,"Computer Aided Optimal Grillage Design by Multiple Objective Programming," J. of the Society of Naval Architects of Korea, Vol.25, No.1, 1988.

4. Chowdhury, M. and Caldwell, J.B., "An Automated Design Scheme: Weight Minimi zation of Single Hulled Ship Compartment," International J. for Numerical Methods in Engineering, Vol.20, p1763-1790, 1984.

5. Hughes, O.F., Ship Structural Design : A Rationally Based, Computer Aided, Optimization Approach, John Willey & Sons, Inc., 1983.

6. Haftka, R.T. and Prasad, B., "Programs for Analysis and Resizing of Complex Structures," Computers & Structures, Vol.10, 1979.

7. Paik, J.K.,"Reliability-Based Optimum Design for Tubular Frame Structures," J. of Ocean Engineering and Technology of Korea, Vol.2, No.1, 1988.

8. Arora, J.S., "Survey of Structural Reanalysis Techniques," J. of the Structural Div., ASCE, Vol.102, No. ST4, 1976.

9. Kirsh, U., Optimum Structural Design Concepts, Methods and Application, McGraw Hill Co., 1983.

10. Abu Kassim, A.M. and Topping, B.H.V., "Static Reanlysis: A Review," J. of Structural Engineering, Vol.113, No.5, 1987.

11. Lee, H.S., Nho, I.S., Paik, J.K., and Park, S.W., "Development of Computer Program for Controlling Ship Structural Weight and Construction Cost (I)," Report of Korea Research Institute of Ship and Ocean Engineering, UCN072-895.D, 1987.

12. Lee, H.S., Nho, I.S., Paik, J.K., and Park, S.W., "Development of Computer Program for Controlling Ship Structural Weight and Construction Cost (II)," Report of Korea Research Institute of Ship and Ocean Engineering, UCN136-1215.D, 1988.

13. Noor, A.K. and Lowder, H.E., "Approximate Techniques of Structural Reanalysis," Computers & Strctures, Vol. 4, 1974.

14. Moses, F. and Onoda, S., "Minimum Weight Design of Structures with Apprication to Elastic Grillage," International J. for Numerical Methods in Engineering, Vol. 1, 1969.

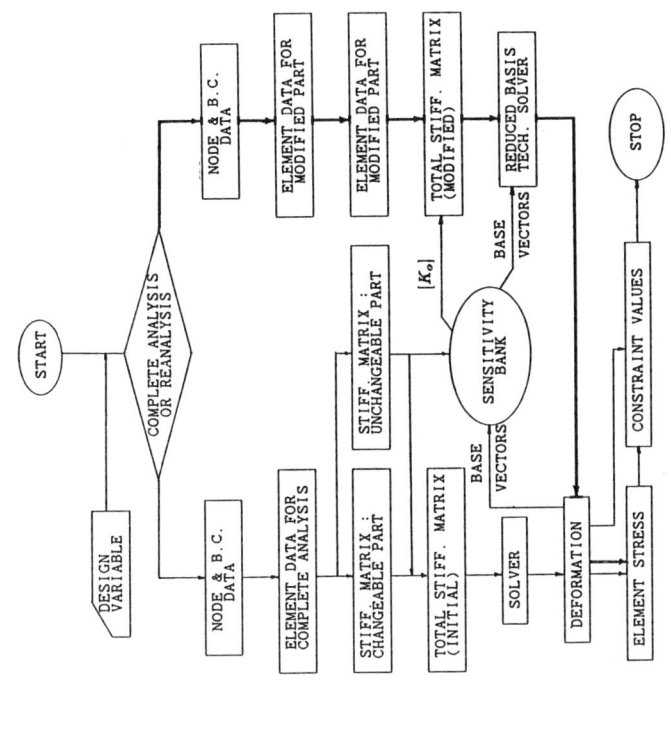

Fig. 2   Flowchart of structural analysis module

Fig. 1   Diagram of optimum structural design system

1. Input for initial design point, formulations of constraint and objective function, and selection of optimizer
2. Output for optimum design results
3. Data/results of structural analysis for modified design
4. Data/results of structural analysis for initial design
5. Design variables (d_i)
6. Constraint and objective function values $(g_j(d_i), f(d_i))$
7. Information for new initial design point
8. Base vectors (Ψ)
9. Initial point (d_i^0) and infinitesimal increaments of design variables $(\triangle d_i)$
10. Base vectors (Ψ) and initial stiffness matrix (K_o)

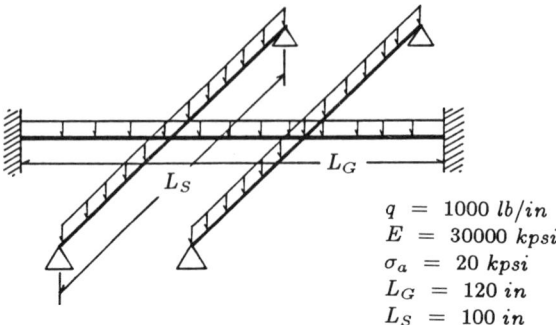

$$
\begin{aligned}
q &= 1000 \ lb/in \\
E &= 30000 \ kpsi \\
\sigma_a &= 20 \ kpsi \\
L_G &= 120 \ in \\
L_S &= 100 \ in
\end{aligned}
$$

Fig. 3   1×2 grillage : configuration and design condition

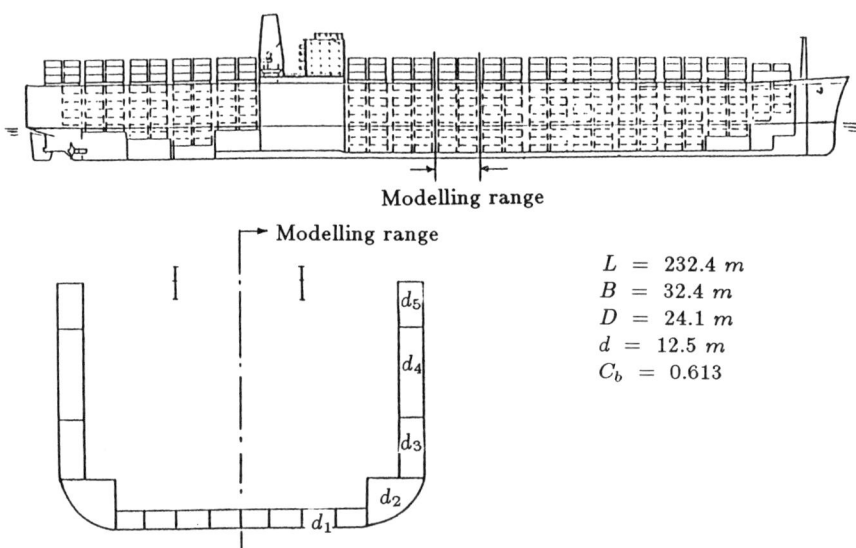

Modelling range

Modelling range

$$
\begin{aligned}
L &= 232.4 \ m \\
B &= 32.4 \ m \\
D &= 24.1 \ m \\
d &= 12.5 \ m \\
C_b &= 0.613
\end{aligned}
$$

Fig. 4   Modelling range in transverse strength analysis of container ship's midhold

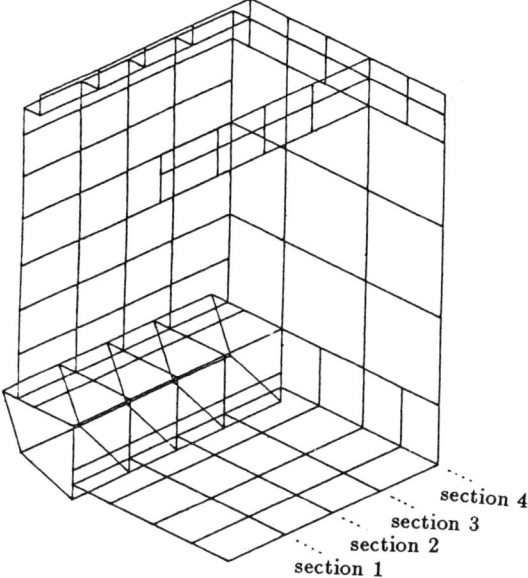

section 4
section 3
section 2
section 1

Fig. 5(a)   FEM model of transverse strength analysis

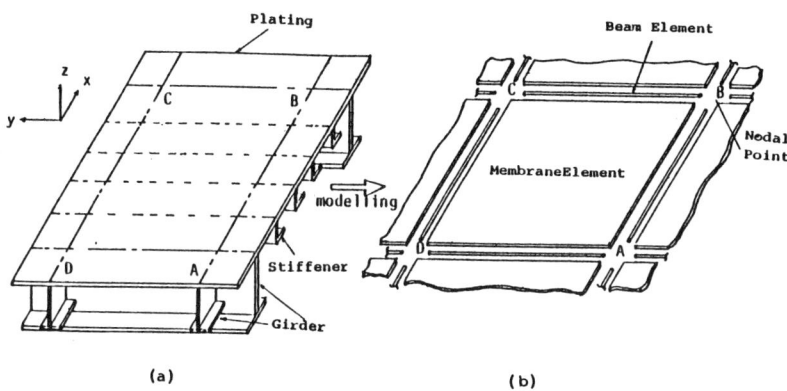

(a)

(b)

Fig. 5(b)   Idealization of structural elements of stiffened panels

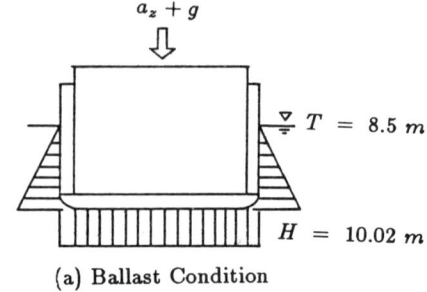

(a) Ballast Condition

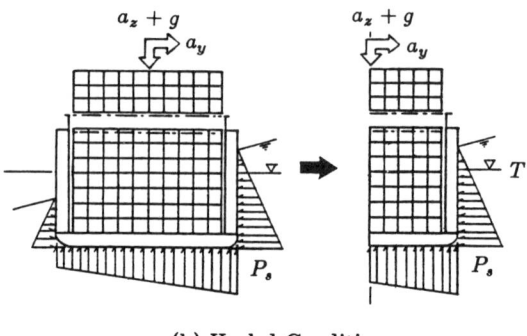

(b) Heeled Condition

Fig. 6   Loading condition of transverse strength analysis

An F.E. Approach to Yacht Structural Analysis

R.A. Shenoi

*Department of Ship Science,
University of Southampton, Southampton,
SO9 5NH, U.K.*

1. INTRODUCTION

A great deal of interest has been generated in research on aspects of yacht design due, mainly, to events such as the America's Cup and Whitbread round-the-world race. The principal focus of much of this work has been on the hydrodynamic aspect, with little attention being devoted to the structural and engineering context. However, there is growing recognition that structural design and layout of the load bearing members can have a significant impact on yacht performance.

In the past, structural design of yachts and small craft relied mainly on "rules of thumb" gathered from experience. The experience was derived from a rather hazardous learning curve where trial-and-error was the primary tool. The "bottom line" in structural design was too often drawn only when failure in a member had taken place. Recently, for example, there have been a series of structural failures in hulls, decks and masts of (International Offshore Rule or) I.O.R. one tonners, 3/4 tonners and maxi yachts.

A contributory factor, in this respect, has been the search for materials that yield lighter and optimised structural layouts. These are mainly polymer-based fibre reinforced plastic materials. Early usage, in this respect, centred around chopped strand mat type of cloth which could be modelled (in a planar sense) by use of isotropic, simple beam theory. However with the advent of different types of cloth weaves, giving more "directional" properties, simple beam theory models and empirical rules by themselves have become insufficient for proper response prediction.

The cumulative impact of high stakes, competition, changing technology and large financial backing is forcing design teams to seek better methods to carry out detailed structural analysis of yachts. Two such projects were carried out at Southampton University (1,2). The overall purpose of these projects was to study the applicability of numerical, F.E. techniques to conduct a global examination of yacht hull/deck structure involving both quasi-static response analysis and modal/ eigenfrequency analysis.

The objectives of this paper are to briefly outline the problem background concerning yacht structure analysis and the manner in which F.E. techniques were applied in one

specific case.
2. DESIGN PARTICULARS

The basic dimensions have been obtained by analysing many 3/4 tonners and cruiser-racers of that size. Figure 1 shows the nature of the body plan used and gives details of the principal particulars.

Following the general philosophy, the sail plan, mast and rigging system has been designed to be representative of a yacht of this length. The mast and sail arrangements are shown in Figure 2.

The structural scantlings were based on the use of E-glass and epoxy resin. A first estimate of laminate details was obtained using classification society rules (3). Table 1 summarises typical modulus values for the hull and deck laminates. A general arrangement of the structural details is shown in Figure 3.

The F.E. modelling was done using an educational version of the commercially available suite - ANSYS (4) - which was mounted on the University's VAX 11/750 cluster of computers. Because of symmetry along the yacht centre-line and also due to space limitations, only half the yacht was modelled. Figure 4 shows the global model used for the linear static analysis.

3. DEVELOPMENT OF FORCE AND CONSTRAINT SYSTEMS

3.1 Constraint System

In most F.E. models, the constraints are usually in the form of several nodes being restrained in translation and/or rotation, thus producing a reaction force at those selected nodes. However, a yacht floating in water is in a "free-free" state. There are two ways of modelling this in a linear static analysis context.

The first is to consider the water supporting the yacht as a spring - see Figure 5. There are two problems associated with such representation. Firstly, the hydrostatic force is not a linear function and so is not governed by the $F = K.d$ relationship. Secondly, even if the relationship is assumed for small displacements, the constant K is difficult to obtain. Hence, this method was not used.

The second approach is to "restrict" the motion at two specific points on the model. These points are where the yacht hull cuts the waterline in its fundamental mode of vibration, i.e. the nodes. If the yacht is not in total dynamic equilibrium, there will be reaction forces at these points - which is unrealistic. This was avoided by running the analysis several times with the hydrostatic and keel forces being incremented until the reaction forces were negligible and equilibrium was achieved.

The nodal points are obtained, using the iterative procedure, by modelling the yacht as a (complex) 3D beam. The mass of the hull was assumed to be negligible in comparison with that of the keel. The latter was modelled as a generalised mass element (4).

The force system acting on the yacht is unsymmetric whereas only half the yacht is modelled. Because of this, two analyses have had to be run - symmetric and antisymmetric. The boundary conditions for these runs are shown in Figure 6.

3.2 Force System

The force system adopted in the present course of studies has been restricted to a quasi-static mode principally because of resource restrictions. Hydrodynamic forces caused by wave, viscous and friction drags were beyond the scope here.

3.2.1 Deck Forces

These were derived by creating a separate F.E. model of the mast and rigging system of the yacht. The mast and spreaders were modelled using 3D beam elements while the shrouds, backstays, runners and forestay were modelled using "tension only" cable elements. The principal purpose of this analysis was to derive the reaction forces at the base of the mast and shrouds.

The forces input into this model are schematically illustrated in Figure 7. The wind loads on the sails were derived by dividing them into various segments radiating out from the clew - see Figure 8. In order to take into consideration the camber in a sail, each segment is assumed to be an arc of a circle. So, to obtain the forces on the edges of the sail, the force in the middle of the arc is found. This is then resolved at the end of each segment, as shown in Figure 8. The total forces at the clew are found by summing the forces resolved at the "clew" end of each segment.

The output of the mast analysis, using the above sail forces as input, comprises tensions in forestay, shroud, backstay and main sheet and the compression in the mast. This is then divided into the symmetric and antisymmetric load cases as shown in Table 2. The application points of these forces, in the yacht model, are at nodes as close to their real positions as possible.

3.2.2 Hydrostatic Force Distribution

The total hydrostatic force must equal the total downward force when the yacht is in dynamic sailing equilibrium. Here, it is insufficient to assume static equilibrium as "weight equals buoyancy". In the dynamic case, there will be a downward force due to sail and mast forces and an upward component due to the hydrodynamic keel force. Thus the total buoyancy (or quasi-hydrostatic) force will be the sum of the weight of the structure (and other components) and the vertical force component from sail mast and keel forces. In order to reach this equilibrium, the analysis had to be run several times until the reactions at the rigid body restraint nodes nearly equalled zero. The hydrostatic force is then distributed over the length of the bottom of the vessel in accordance with the sectional area curve.

Two assumptions were made in the distribution of this process. Firstly, the forces were assumed to act equally on all submerged nodes (at a given cross-section). Secondly, it was assumed that the forces act vertically rather than prependicular to the hull surface. The effects of these assumptions are negligible when compared with the magnitude of the deck force system. As in the case of the deck forces, the hydrostatic force too was resolved into the symmetric and antisymmetric cases.

3.2.3 Keel Forces

This system comprises the weight of the keel itself and the hydrodynamic side force component on the keel. The former is given in Figure 1. The latter is taken to be the force required to keep the yacht from assuming too high a leeway angle. Once again, the forces are divided into the symmetric and antisymmetric cases. These are applied as close to the hydrodynamic centre of pressure of the keel as possible.

4. LINEAR STATIC ANALYSIS

Thus was used to obtain deflection and stresses of the yacht when the load system was applied. The model, as shown in Figure 4, included both longitudinal and transverse stiffeners. The F.E. code was written with symmetric and antisymmetric force systems forming two load steps. The overall analysis statistics are listed in Table 3.

A typical displacement plot of the deformed structure is shown in Figure 9. Here, an upward scaling factor of ten has been applied in order to make the displacements visible. The figure indicates that major deflection occurs at the mast area and the bow. This is caused by the mast compression and forestay load. The deflection caused by the shrouds is visible but small in magnitude.

The principal stress in the symmetric load case and the Von Mises equivalent stress in the antisymmetric load case are shown in Figures 10 and 11 respectively. An initial overall comparison between the two shows that stresses pertaining to the symmetric condition are much higher than those in the asymmetric condition.

Considering the symmetric case, the maximum stress in the logitudinal direction, of magnitude 143 MN/sq.m, occurs at the bottom girder and is primarily caused by the mast compressive load. The shell stiffeners near the mast are compressive in nature with a magnitude of 52 MN/sq.m due to windward shroud load.

Transverse stresses in the symmetric case are less in both tension and compression - 99 MN/sq.m and 77 MN/sq.m respectively. These are at their highest near the keel bolt region.

Shear stresses are low in comparison with the longitudinal and transverse stresses. The maximum tensile shear stress occurs at the windward shroud and forestay attachment points. The forestay causes the shear to be distributed along the bottom girder; thus large shear stresses are created near fore-end at the hull bottom. Fairly large compressive shear stresses, of magnitude 20 MN/sq.m, occur near the keel bolts; these are caused by the side force acting on it.

In the antisymmetric analysis, the highest tensile stress of 39 MN/sq.m occurs at the backstay and mainsheet attachment points. The maximum compressive stress, of magnitude 37 MN/sq.m occurs on the side of the hull at the coach roof front.

The largest (tensile) shear stress in the antisymmetric case is of the order 6.3 MN/sq.m and occurs at the backstay and windward shroud. The largest compressive shear stress, of magnitude 7.9 MN/sq.m occurs at in the foredeck near the coach roof.

A comparison with similar work (5), albeit on a 12m yacht, indicates that the results obtained here are plausible. For example, the maximum longitudinal stress in both cases occurs at the mast base in the yacht bottom. The stress in the 12m case was 100 MN/sq.m (in comparison with the 143 MN/sq.m in this case). Again, the stresses around the mast are similar in both cases - 100 MN/sq.m for the 12m boat and 90 MN/sq.m in this case. However, there are differences with regard to transverse and shear stresses magnitudes and locations.

5. EIGENFREQUENCY ANALYSIS

The objective here is to determine the natural frequencies and corresponding mode shapes of the structure. This involves the solution of the undamped and free harmonic equation:

$$[M]\ddot{x} + [K]x = 0$$

where [M] and [K] are the mass and stiffness matrices respectively.

The structure that has been modelled for this run included all transverse and longitudinal stiffeners. The keel and rudder were removed so as to negate any spurious deflections taking place. Lumped mass elements were used instead. Table 4 lists the analysis statistics.

A plot of the fundamental mode shape is shown in Figure 12. The frequency for this mode is approximately 4 Hz. Mode shapes of frequencies less than 2 Hz and greater than 10 Hz appear to be unrealistic and unsustainable.

In order to investigate the effects of stiffening, a separate run was conducted without stiffeners. It was noticed that the inclusion of longitudinal stiffening had a major effect on the mode shapes. The deflection of the cockpit was prevented almost completely. The whole deflection of the deck, at 2.57 Hz, followed a smooth curve resembling a free-free beam mode shape.

In the completely stiffened structure, natural frequencies are higher than in the unstiffened case. The deflection of the whole structure is negligible, thus showing the effect of stiffeners to increase overall stiffness of the structure. However, there was a small localised deflection at the keel mass region.

6. CONCLUDING REMARKS

Overall, this paper has briefly outlined the finite element modelling of yacht structure. A simplified, preliminary approach to load determination has been given. Linear static analysis of the structure indicated reasonable correlation with similar studies on another yacht. The results are realistic; high stresses are found in areas where they would be expected to occur. Furthermore the stresses are also of the correct order of magnitude.

However, there is a need to verify and validate these results. The first task is to have a more comprehensive load calculation routine. This should include both hydro- and aerodynamic effects. Ideally, this ought to be validated by experimentation. The response of the structure too ought to be monitored in a practical sense. After this has been done, and answers compared favourably with F.E. analyses then there will be a more global acceptance of such techniques for yacht structural analysis.

7. ACKNOWLEDGEMENTS

The computation and work outlined in this paper was carried out under the author's supervision by Messrs. Yates and Barr, students in the Department of Ship Science.

8. REFERENCES

1. O.G.K.Yates : "A Study of Yacht Structures", Ship Science Report, Southampton University, 1989.

2. D.Barr : "Yacht Structural Analysis with the Finite Element Method", Ship Science Report, Southampton University, 1990.

3. "Rules and Regulations for the Classification of Yachts and Small Craft", Lloyd's Register of Shipping, London, 1983.

4. G.J.Desalvo and J.A.Swanson : "ANSYS Finite Element Package Users Manual", Version 4.3, Swanson Analysis Systems Inc., Houston (PA), 1986.

5. D.Boote, V.Ruggiaro, N.Sironi, A.Vallicelli and B.Finzi : "Stress Analysis for Light Alloy 12m Yacht Structures", Proceedings of the Seventh Chesapeake Sailing Yacht Symposium, S.N.A.M.E., 1985.

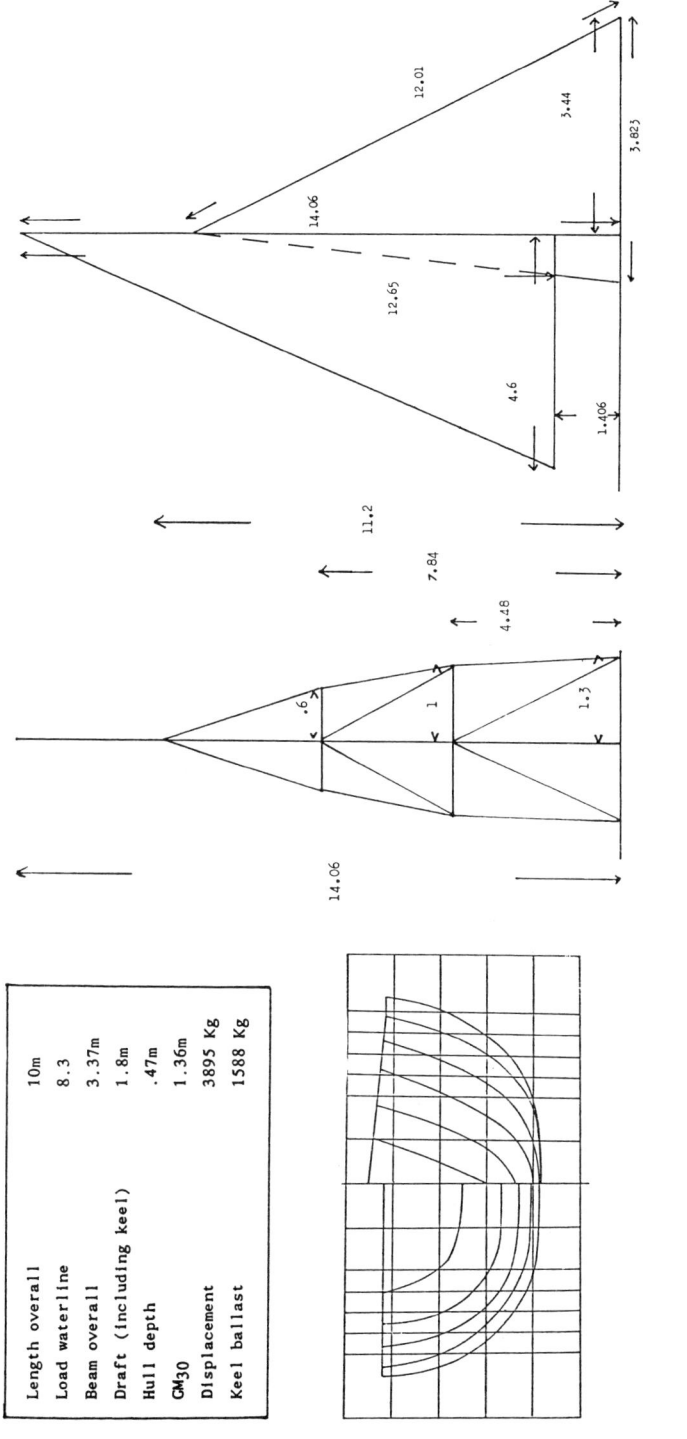

Length overall	10m
Load waterline	8.3
Beam overall	3.37m
Draft (including keel)	1.8m
Hull depth	.47m
GM30	1.36m
Displacement	3895 Kg
Keel ballast	1588 Kg

Figure 1: Yacht Particulars and Body Plan

Figure 2: Mast System (distances in metres)

<u>Hull Frames and Girders</u>

Frame	Calculated Modulus/cm^3	Depth /mm	Laminate thickness /mm	Face Area /cm^2	Section Modulus /cm^3
Transverse:					
Side frame	8	40	5	2	11
Centre floors	13.57	60	5	2	19
Web:					
Side frame	20.57	60	5	2.3	20
Centre floors	50.58	80	5	4.7	50
Bottom					
Longitudinals	20.725	60	5	2	19

<u>Deck and Coach Roof Beams and Girders</u>

Frame	Calculated Modulus/cm^3	Depth /mm	Laminate thickness /mm	Face Area /cm^2	Section Modulus /cm^3
Transverse Beams	5.3	20	5	2	5.3
Girders					
(1)	47.8	80	5	4.25	48
(2)	35.09	60	5	4.60	35
(3)	31.56	80	5	2.2	32
Coach roof					
Transverse Beams	5.3	20	5	2	5.3

Table 1: Hull and Deck Scantlings

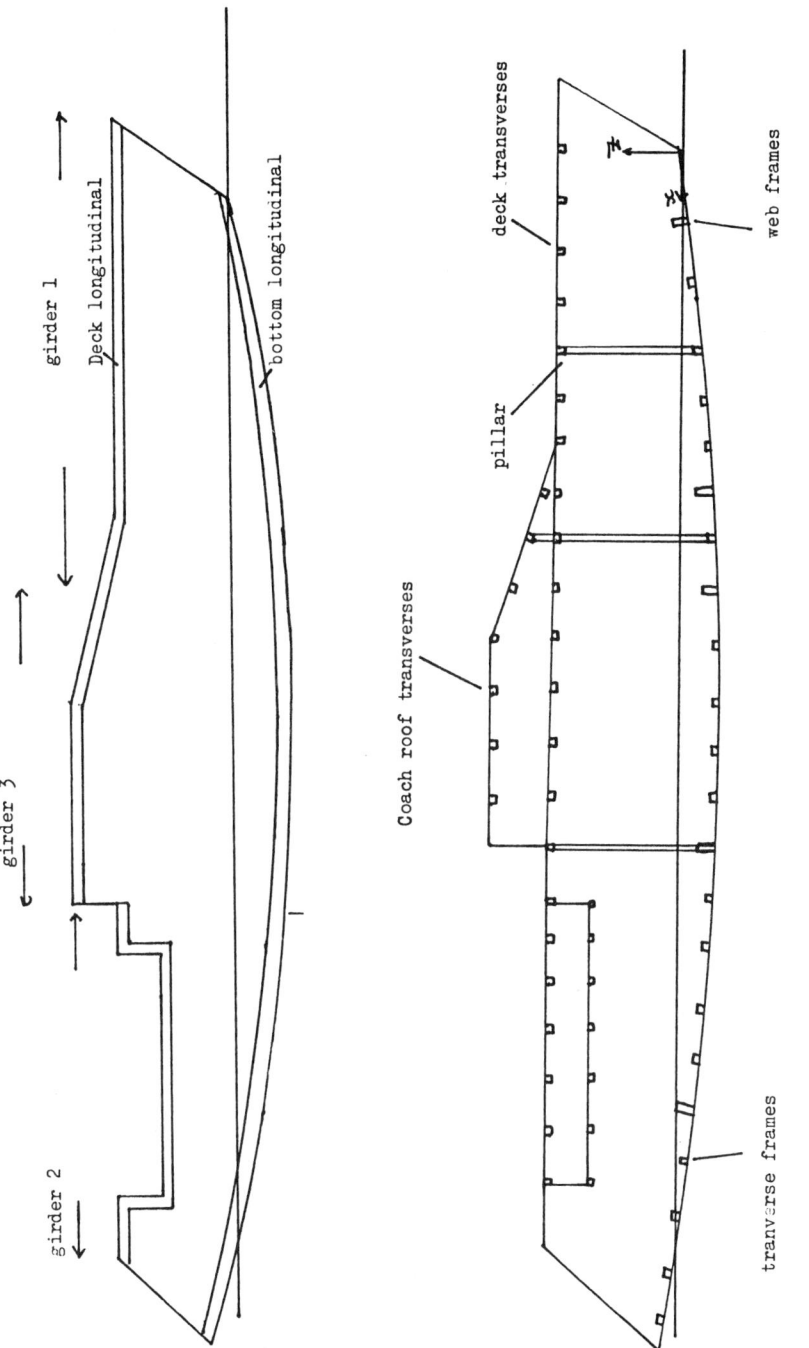

Figure 3:   Structural Arrangement

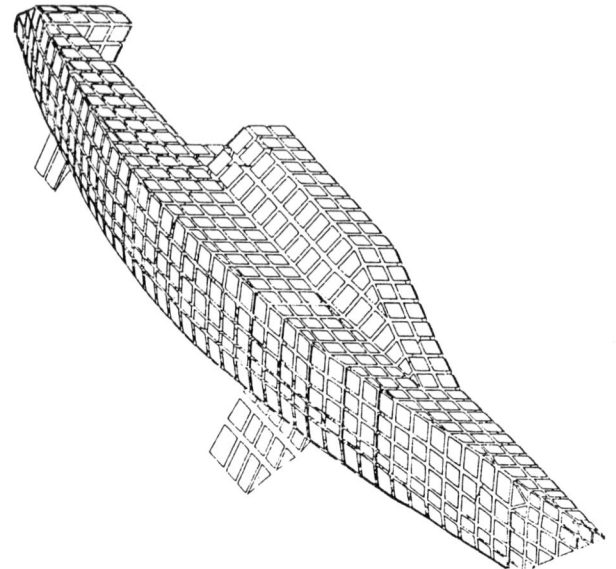

Figure 4: F.E. Model for Static Analysis

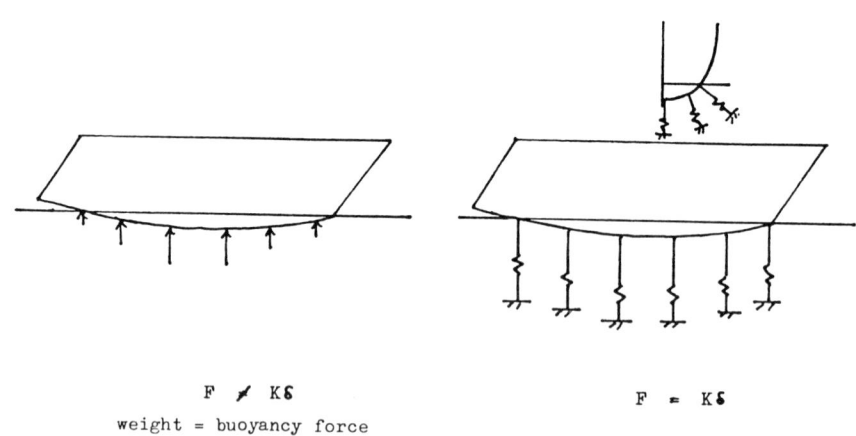

$$F \neq K\delta$$
weight = buoyancy force

$$F = K\delta$$

Figure 5: Hydrostatic Representation: One Possibility

Symmetric Boundary Conditions

$$UY = 0$$
$$ROTX = 0$$
$$ROTY = 0$$

on the centre line

Degrees of freedom

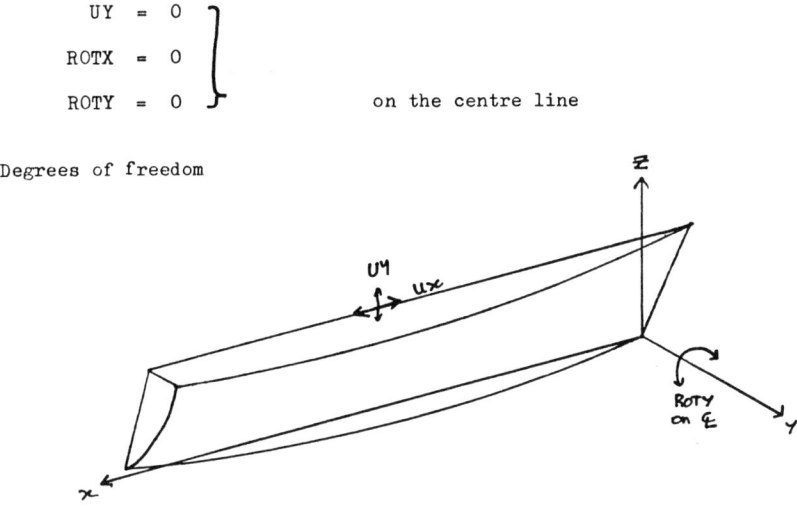

Antisymmetric Boundary Conditions

$$UX = 0$$
$$UY = 0$$
$$ROTY = 0$$

on the centre line

Degrees of freedom

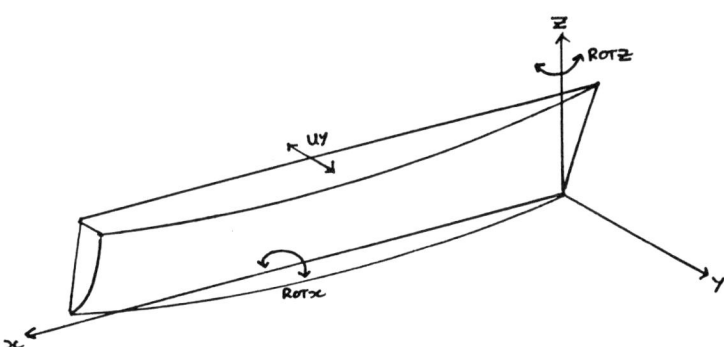

Figure 6:   Boundary Conditions

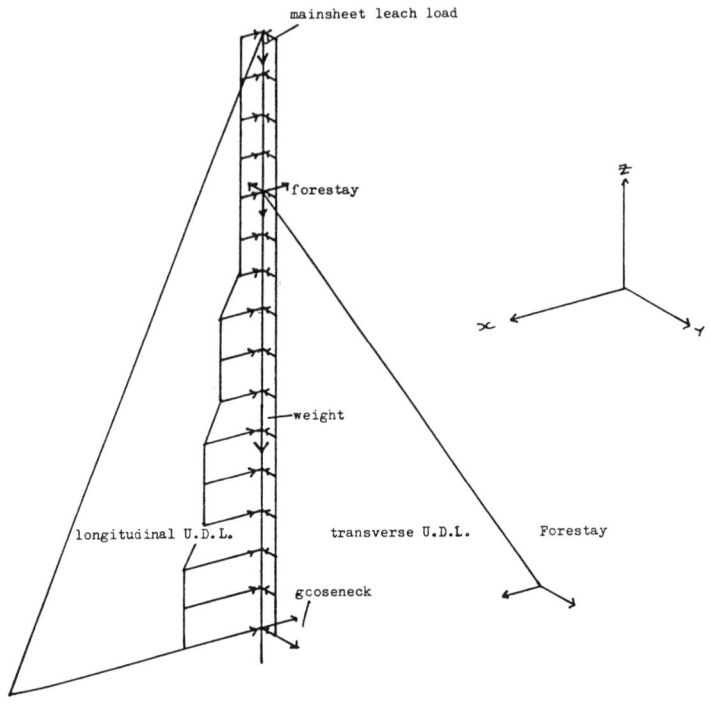

Figure 7:   Mast Forces

		Fx/N	Fy/N	Fz/N
Forestay	Symm	1617.9	0	4884.4
	Antisymm	0	-228	0
Shroud	Symm	752.9	-2722.6	18786.5
	Antisymm	752.9	-2722.6	18786.5
Mast compression	Symm	-446.9	0	-27149.5
	Antisymm	0	-489.4	0
Main sheet	Symm	-438	0	1302.2
	Antisymm	0	-519.8	0
Backstay	Symm	-184.6	-31.3	395
	Antisymm	-184.6	-31.3	395

Table 2:   Deck Forces

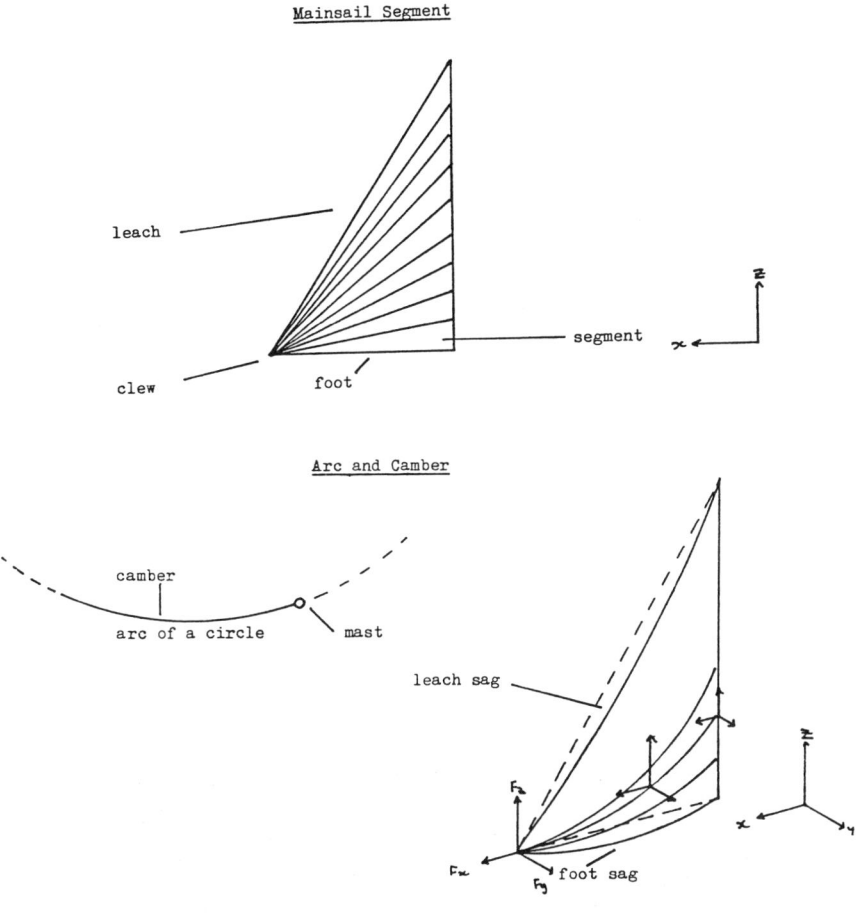

Figure 8: Sail Load Modelling

Degrees of Freedom	4671
Maximum wave front	180
Number of elements	851
Number of nodes	890

Table 3: Linear Static Analysis Statistics

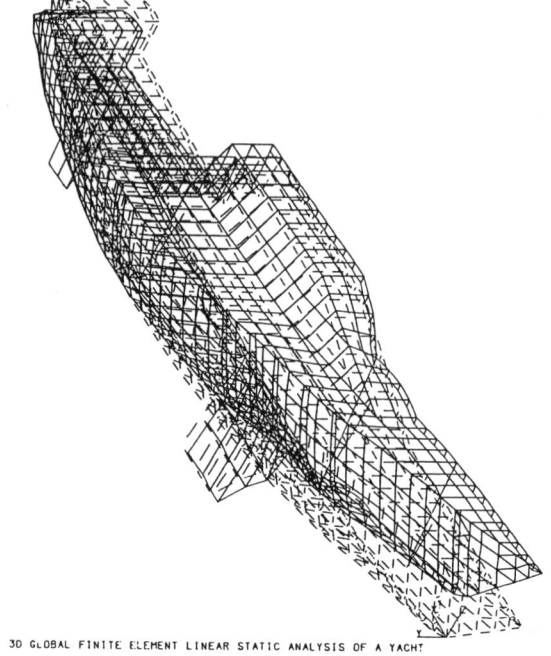

30 GLOBAL FINITE ELEMENT LINEAR STATIC ANALYSIS OF A YACHT

Figure 9:   Deflection Pattern of Yacht

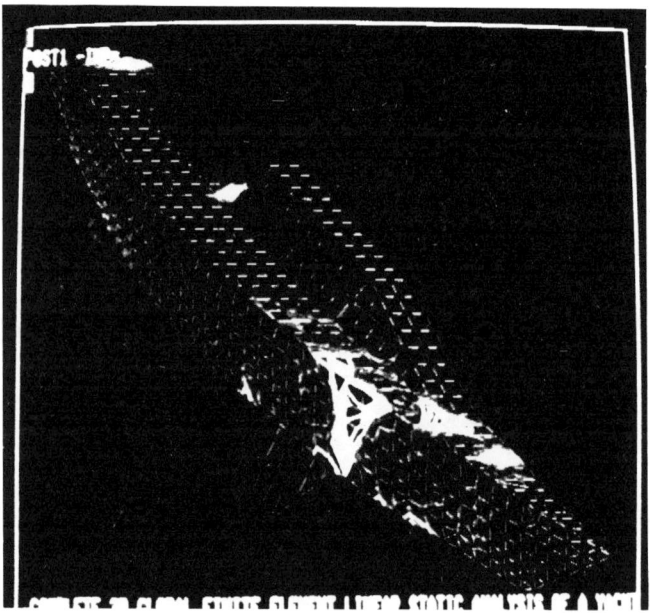

Figure 10:   Principal Stress - Symmetric Case

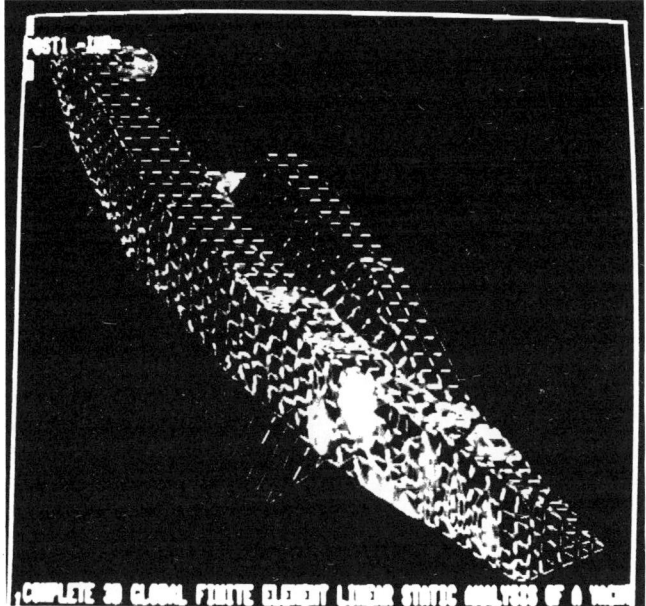

Figure 11:   Von Mises Equivalent Stress - Antisymmetric Case

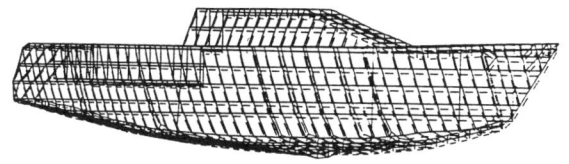

Figure 12:   Fundamental Mode Shape

Master degrees of freedom	40
Total degrees of freedom	4157
Solution time	5½ hours
Number of elements	770
Number of nodes	820
Maximum wave front	181

Table 4: Vibration Analysis Statistics

SECTION 5: SHIP OPERATIONS AND NAVIGATION

A Review of Mathematical Models used in Ship Manoeuvres

J. Chudley, M.J. Dove and N.J. Tapp
Department of Marine Science and Technology,
Polytechnic South West,
Plymouth, United Kingdom

ABSTRACT

The Ship Control Group at Polytechnic South West is developing, in conjunction with a UK marine electronics company and consultants, an Integrated Navigation System that will automatically steer the vessel along a predetermined track, avoiding any obstructions and taking the necessary precautions to avoid collision with another vessel. Central to the research programme is the development of simple, yet accurate, mathematical models of own ship and target ships. The paper describes the research into different types of model and concludes that a non-linear modular type could be used for 'own ship' and a simple linear model for the 'target ship'.

INTRODUCTION

Recent accidents at sea, together with a series of oil crises that have increased the price of fuel oil, have made ship owners and operators more safety and economy conscious; this in turn has made the requirements on ship steering more demanding, particularly in confined waters, where extensive manoeuvring is needed. It is therefore important to be able to predict the path of the ship precisely. It has been suggested [1] that 85% of all marine collisions and groundings are due to human error and of these 90% occur in coastal waters [2]. On this evidence alone there is a case for research and development into automating the control and guidance systems which are installed in ships. Mariners on a sea passage are likely to experience periods of relatively uneventful sailing, interspersed with periods requiring careful attention and substantial decision-making – such as traversing a busy seaway or entering port. The potential hazards of such a regime are twofold: on the quiet stretches, a false sense of security can lead to impending danger being overlooked until it is too late; conversely, information input at busy times may overload the decision-making process (i.e. the Officer of the Watch), leading to ill-judged actions or dangerous delays in manoeuvring. Both of these problems could be obviated by an electronic monitoring system, which would analyse sensory inputs such as ARPA and navigational information, and give reasoned and pertinent advice to the mariner on the bridge. If such a system was available to advise the Officer of the Watch, perhaps disasters such as those involving the Exxon Valdiz and the Marchioness could be avoided.

Development of a production system for fully automated ship control probably lies well into the future. The technology exists, but there are other considerations governing the instrumentation installed on a vessel, such as cost and legislation which may pose constraints in the immediate future but may subsequently be relaxed. In connection with development towards automatic navigation, research at Polytechnic South West (formerly Plymouth Polytechnic) is underway, to maintain the vessel not only on course but also on track. To undertake this, more than just positional information is required by the autopilot. That is, in particular, velocity feedback in the two dimensions of surge and sway and rate of turn are necessary in order to stabilize the system. While such measurement devices are available they are rarely found in commercial shipping due to financial constraints. To overcome the problem of providing the appropriate measurements, the use of Kalman filtering techniques may be adopted. Research in this area has been underway for a number of years and Dove [3] has shown the concept of Kalman filtering as applied to marine navigation; combining state estimates from measurements with those from a mathematical model. This has further been investigated by Miller [4].

The overall aim of the work is to investigate, design and develop an integrated navigation and collision avoidance system to provide advice to the master of the vessel. This is a wide spectrum to cover and involves two full time research staff. A schematic diagram of the system is shown in figure 1.

The system will;

  i)   interface to the ship's navigational aids,
  ii)   perform the mathematical model computations,
  iii)   perform the filter computations,
  iv)   display an electronic chart showing ship status, desired track

and information on target vessels,

v)   interface to the radar,

vi)   run heuristics for collision avoidance (ACAS) [5],

vii)   make modifications to the mathematical model if necessary,

viii)   present track information.

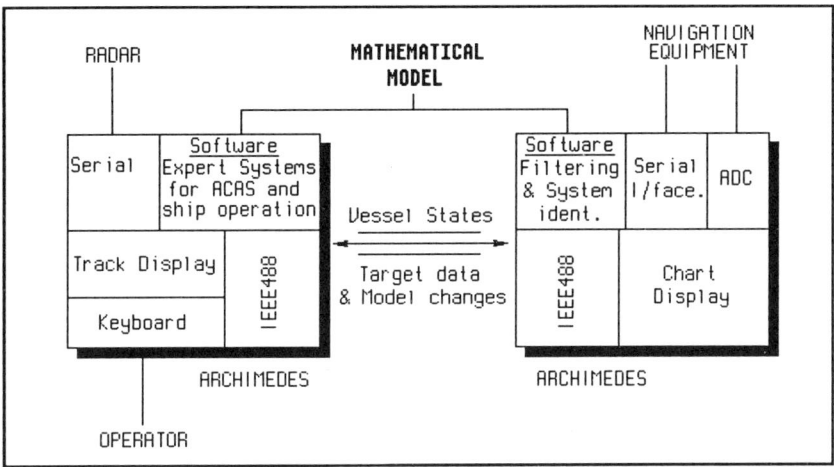

*Figure 1. Prototype System Schematic showing that
the central element is the mathematical model.*

The project comprises of a team approach through the formation of the
'Ship Control Group' involving a wide range of disciplines including control
and guidance, navigation, naval architecture, computing, artificial
intelligence, mathematical modelling and signal processing. The prototype
system will be fitted on board one of the Polytechnic's research vessels
with the aim of having it operational within a period of two years.

MATHEMATICAL MODELLING AT POLYTECHNIC SOUTH WEST

Central to the overall system being developed is the mathematical model.
Mathematical modelling has been undertaken by the Ship Control Group for
a number of years, culminating in the modular manouevring model
developed for use in a marine simulator by N.J. Tapp [6].

Between different research establishments there is little commonality of
ship manouevring mathematical models and hydrodynamicists have over the
years developed models of various forms and fidelity. The major reason
for this is the complexity of the flow phenomena around the hull, propeller
and rudder particularly on the subject of generation and losses of vorticity
and surface waves [7]. The mathematical model designed for the system

must be capable of representing a wide range of ship types and configurations, machinery and propulsion/steering devices. The many different types of mathematical model can generally be placed under one of four headings, namely;

    i)   Input-Output relationship model,
    ii)   An holistic model,
    iii)   A force mathematical model,
    iv)   A modular manouevring model.

Recently a study has been undertaken as to the type and complexity of the required model for the proposed integrated navigation system. The two types looked at in detail are the holistic model and modular manouevring model.

The holistic model

This type of model is highly formal and systematic. It treats the hull-water interface as a black box and models the system as a complete entity. It is based on the premise that a manoeuvre is a small perturbation from an equilibrium state of steady forward motion at a nominal service speed. It has been used successfully for the simulation of ship manoeuvres by the application of rudder control by Strom-Tejsen [8] and in a modified form has been applied to engine manoeuvres by Crane [9] and Eda [10], despite the fact that such manoeuvres can hardly be described as small perturbations. Dand [11] describes this type of model as;

" A model which performs satisfactorily when taken as a whole, but does not allow individual elements to be changed readily as the design is changed."

The modular manoeuvring model

Current research on ship manoeuvring modelling tends to favour this type of model. The Mathematical Model Group (MMG) of the Society of Naval Architects of Japan, first published a paper describing a model of this type in 1978 [12]. This was subsequently followed by various papers on the subject [13-14], and a further refined model in 1984 to simulate various ship manoeuvring motions in harbour [15]. Research in Germany, by Oltmann and Sharma [16], is based on the modular concept, as is the modular manoeuvring model developed at British Marine Technology Ltd (BMT) between 1983 and 1984.

A modular manoeuvring model is one in which the individual elements, such as the hull, propeller, rudder, engines, and external influences, of a manoeuvring ship are each represented as separate interactive modules. Each module, whether it relates to hydrodynamic or control forces or external effects is self-contained. The modules are constructed by reference to the detailed physical analysis of the process being modelled. The system as a whole is then modelled by combining the individual elements and expressing their interaction by other physical expressions.

The equations of motion for a modular manoeuvring model are generally expressed by;

$$m\dot{u} - mrv = X_H + X_P + X_R + X_E$$

$$m\dot{v} - mru = Y_H + Y_P + Y_R + Y_E \tag{1}$$

$$I_z\dot{r} = N_H + N_P + N_R + N_E$$

where the suffixes H, P, R and E denote components of hull, propeller, rudder and external forces.

The model arranged in this way lends itself to a number of applications. For example it allows research on one particular module and the effect that module has on the system model as a whole. This is invaluable when trying to determine the effect of various rudder areas on the manoeuvring performance of a vessel. Previously, a series of captive model tests had to be undertaken to select optimal rudder area. Advances in any particular field of related research can be incorporated into a module and into the system as a whole without having to alter other system modules. Other advantages of this approach are the expansion facilities it allows. In addition to the modules shown in equation set (1) extra modules can be employed, to simulate bow thrusters and stern thrusters for example. Hence the model can be tailored to suit a number of applications and such effects as ship to shore and ship to ship interaction can be investigated. Gradually a very sophisticated model incorporating all of the more specialised attributes can be developed.

MODEL FORMULATION AND SIMULATION

The study involved the simulation of a number of different vessels, however, to show the adaptability of the models the paper will show the simulation results for two completely different vessels;

i)   278000 dwt tonnage tanker,

ii)   2000 tonne converted dredger engaged in the European coastal trade.

The different models used in the study were;

i)   Linear holistic   -   3 degrees of freedom,
                              -      4 degrees of freedom,
ii)   Non-linear holistic   -   3 degrees of freedom,
iii)   Modular   -   3 degrees of freedom.

A description of each model follows along with example plots. A comprehensive range of results can be found in ref [17].

Linear holistic - 3 degrees of freedom
A floating body can move in all six degrees of freedom of motion - translation along three orthogonal axes and rotation about each of the three axes - surge, sway, heave, roll, pitch and yaw. Although work has been carried out on six degree of freedom models [18], it is not usual for a vessel to be represented by all six equations. The equations describing ship motions in the horizontal plane, which typically covers the most practical needs of ship simulators, are a particular case of the general equations of the six degrees of freedom, and are therefore reduced to surge, sway and yaw. The equations are further simplified if the origin of the ship co-ordinate system is selected to coincide with the mass centre of the vessel. The 3 degree of freedom linear holistic model is of the simplest possible form derived from Abkowitz [19]. This type of model performed reasonably well when rudder movements were relatively small but when performing a complete turning circle the results were inaccurate. For exampla, the linear model generally has turning circle of approximately half that of the ship being modelled.

Linear holistic - 4 degrees of freedom
It was decided to expand the equations to include a fourth degree of freedom, namely roll [20]. Results are shown for the 2000 tonne converted dredger in figs .2-3.

The results obtained were not an improvement over the 3 degree of freedom model and it was decided at this stage not to use the 4 degrees in the modular model. The roll equation would be required if the navigation system was ever to be installed in a long thin warship or perhaps, on smaller craft where roll influences a turning manoeuvre.

Non-Linear Holistic Model
The Ship Control Group at Polytechnic South West has used this model in past research. The selection of the important non-linear terms were made by reviewing the work of Strom-Tejsen, Lewison [21], Gill [22-23] and Eda and Crane [24]. The non-linear functions of the control parameters

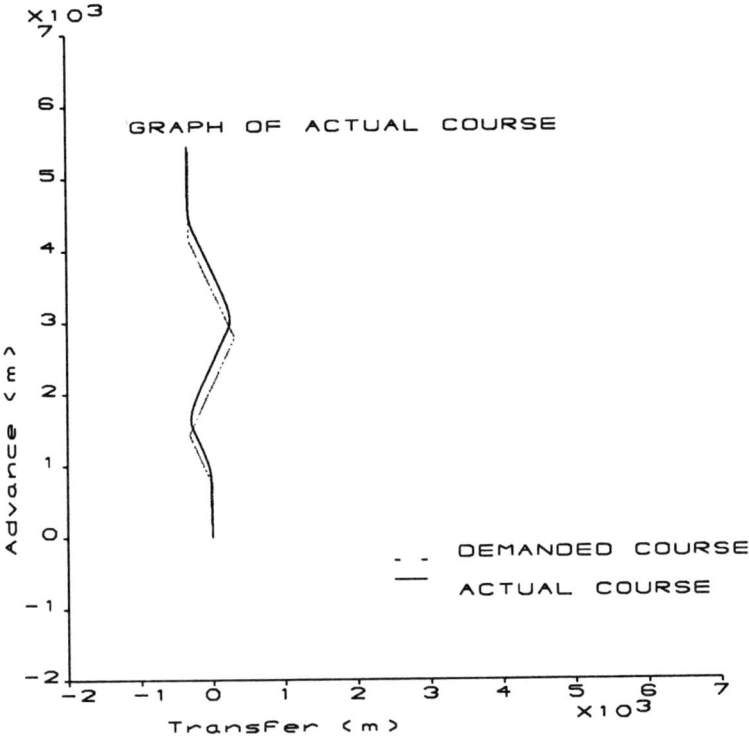

Fig 2. Linear 4 degree holistic model. Approach speed 5 m/s. Course 0 degrees. Deviation 20 degrees.

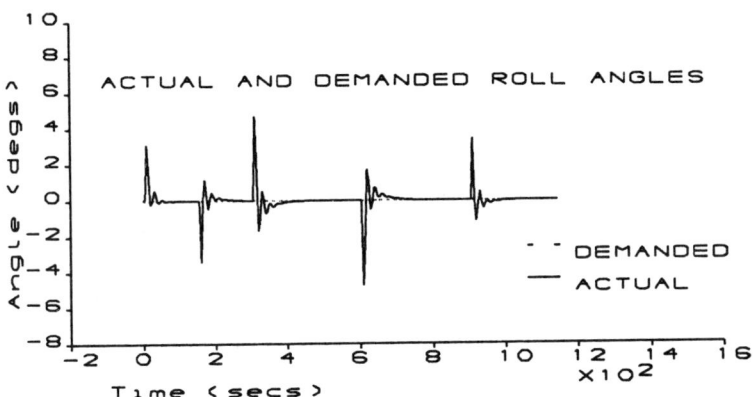

Fig 3. Roll Angle.

(rudder and propeller) were also required in the final non-linear equations of motion.

The complete set of the holistic model non-linear equations of motion as used by the Ship Control Group has been described in various papers [25-26] and has been shown to give accurate representation of the three degrees of ship motion in all manoeuvring situations. A comparative evaluation of the mathematical model was made with full scale measurements taken by Morse and Price for the USS Compass Island [27]. The USS Compass Island was constructed with a Mariner type hull form, and a complete set of hydrodynamic coefficients for this class of vessel, have been measured by Chislet and Strom-Tejsen [28] using the Planar Motion Mechanism Test.

Although this model gives accurate simulations of ship manoeuvring it does not allow rudder, propeller or hull geometry to be changed with ease; a major requirement of the system model is that it should be adaptable.

Modular Model

This model has the general form of equation set (1). Taking this equation set each of the modules can be looked at in turn.

Hull Forces and Moments The hull forces and moments module contains all the hydrodynamic data which is specific to the hull alone. They can be expressed by the following equations;

$$X_H = X_{\dot{u}} \dot{u} + X_{vr} vr + X_{vv} v^2 + \frac{u}{|u|} X_{rr} r^2 + R_H$$

$$Y_H = Y_{\dot{v}} \dot{v} + Y_{\dot{r}} \dot{r} + Y_{uv} v + \frac{u}{|u|} Y_{ur} r + Y_{vvv} v^3 + Y_{rvv} rv^2 \qquad (2)$$

$$N_H = N_{\dot{r}} \dot{r} + N_{\dot{v}} \dot{v} + N_{ur} r + \frac{u}{|u|} N_{uv} v + N_{vvv} v^3 + N_{rvv} rv^2$$

The equations are a further development of previous research work on the holistic type model with the important non-linear terms being similar to enable comparisons of the models to be made. The multiplier $\frac{u}{|u|}$ included in some of the terms is to correct the sign of the derivative during astern motion of the ship.

The term R_H in the surge equation of equation set (2) represents the ships resistance on a straight course and is modelled by the following expression;

$$R_H = X_u u + \frac{u}{|u|} X_{uu} u^2 + X_{uuu} u^3 \qquad (3)$$

Propeller Forces and Moments In order to model the motion of a ship for both ahead and astern motion it is important to determine correctly the propeller forces and moments. To cover all manoeuvring regimes, Tapp adopted the method of modelling the propeller forces and moments published by Oltmann and Mikelis [29]. This method is based on the knowledge of the thrust coefficient;

$$C_T^* = \frac{2 T_P}{\rho A_o (up^2 + cp^2)} \qquad (4)$$

for the whole range of the advance angles ε for the propeller.

Propeller Forces and Moments for a Single Screw Ship

$$X_P = (1 - t_P) T_P$$

$$Y_P = Y_{nn} n^2 \qquad (5)$$

$$N_P = N_{nn} n^2 \qquad \text{where } N_{nn} = Y_{nn} L/2$$

assuming that the screw is located at a distance $L/2$ from the LCG.

Propeller Forces and Moments for a Twin Screw Ship Obviously the modelling of a twin screw ship is a more complex problem than a single screw. It is not the intention of this paper to present these equations as the results that will be shown are based on single screw vessels. Basically, the surge term is a summation of the effect of both propellers as is the yaw term. The sway term is dependent on a number of factors including rotation of propellers and their operating condition i.e. port propeller ahead and starboard propeller astern.

Rudder Forces and Moments In common with the propeller modelling it is important to calculate accurately the rudder control forces and moments in order to model correctly the turning and course keeping performance of the ship. From Hirano et al [30], using Tapp's adopted sign convention, the forces and moments induced on the ship due to rudder action are given by;

$$X_R = (1 - t_R) F_N \sin(\delta)$$

$$Y_R = -(1 + a_H) F_N \cos(\delta) \qquad (6)$$

$$N_R = (1 + a_H) F_N X_R \cos(\delta)$$

where ;

> F_N is the normal force produced by the rudder. a_H and t_R are correction factors to adapt the open-water characteristics of the rudder to behind-hull conditions. a_H can be determined from knowledge of the form factor of the hull, C_B. The value of t_R can be estimated from the reduction in forward speed of the ship when turning.

External Disturbance Forces and Moments A number of modules can be used to describe various external effects which, in keeping with the modular structure, are treated simply as additional forces and moments imposed on the basic hull hydrodynamics. The required complexity and operating conditions of the model determines the external force modules needed. These can include, for example, such effects as wind, tide, thrusters, bank effects, tugs, anchorage, ship to ship interaction, and squat. Tapp's model, for use in a marine simulator, had external forces and moments modules for wind and tide, the wind module being based on research by Isherwood [31].

Results of the Modular Model
Exxon International published a report in 1979 detailing the performance of a modern supertanker [32], describing full-scale trials of the 278000 dwt Esso Osaka. The modular model outlined was verified by using it to simulate the full scale results given in the report.

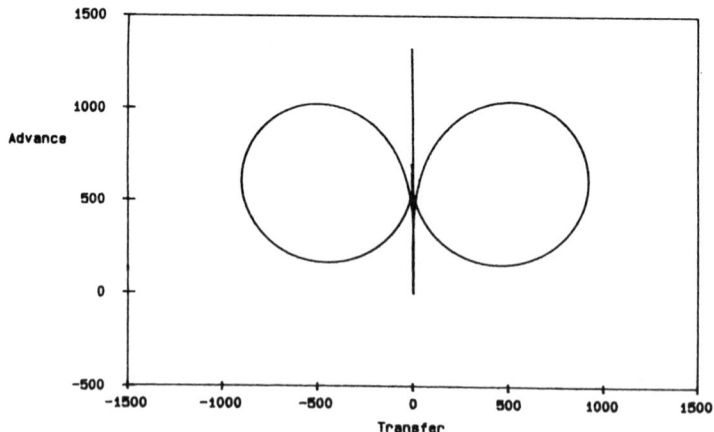

Fig 4. 35 degree turning circles and 0 degree course setting in deep water

The simulated results obtained were comparable with the full scale trials results. Simulations were carried out at varying depths and with wind and tide influences. It was noted that on some simulated turning circles and Kempf manoeuvres, the model tended to respond slowly to rudder alterations, particularly to rudder angle alterations. This aspect of the model performance is being investigated in the next phase of the research programme.

CONCLUSION

The ultimate aim of the research into modelling at Polytechnic South West is to develop an adaptable model that can be implemented into the navigation system onboard any vessel. The requirements of the specification is to be able to input the principal dimensions of the vessel and calculate the hydrodynamic coefficients for the model without sea trials, hence keeping initial cost down. Without doubt the model that is most practical and accurate is the modular type model. Its adaptability is shown by the allowance of a new propeller being fitted with a new pitch; by inputing the new dimensions manually through a keyboard, the propeller module would be re-calculated, altering the model without sea trials.

The majority of the hydrodynamic coefficients can be derived adequately by various analytical means. The surge terms can be found using a program adapted from the work by Holtrop et al [33]. The linear acceleration and velocity components for sway and yaw can be calculated using such formulae as that developed by Clarke [34]. The propeller and rudder modules can be found as outlined earlier. The problem coefficients are the third order hull terms, and investigations are in progress, to discover the relationship between their value and the principal dimensions of the vessel. These terms are at present evaluated from physical model tests or sea trials.

The investigation into the different levels of complexity of the model showed inaccuracies in the simple linear model. The navigation system will, however, require two levels of model for two distinctly separate tasks:

i) The more complex non-linear modular form is required to model the vessel in the navigation system.

ii) A simple linear form represents the target vessel in the ACAS element of the system. It is intended to include a number of models representing different vessel categories; on detecting a target vessel on the ARPA, the system will plan its own manoeuvre taking due account of the estimated manoeuvring capabilities of the target vessel.

A simplified protoype system is fitted onboard the Polytechnics' research vessel and encouraging results are being obtained. The model in use at present will need to be expanded to simulate;

i)   slow speed operations of a vessel in the pilotage phase of a voyage,
ii)   stopping in narrow channels,
iii) ship handling procedures in an emergency,
iv)   anchoring procedures,
v)   manoeuvring in shallow water,
vi)   berthing,
vii) use of bow and stern thrusters.

This section of work is central to the overall research at the Polytechnic. It is intended to improve the model in use on the research vessel, so that the software can be tested under operational conditions. The complete integrated navigation system will ultimately be fitted in the vessel, displaying the required data on VDU's, thus creating a central navigation console.

REFERENCES

1.   Panel on Human Error in Merchant Marine Safety. *Human Error in Merchant Marine Safety*. National Acadamy of Science, Washington D.C., 1976.

2.   COCKROFT A.N. *Collisions at Sea*. Safety at Sea, June 1984.

3.   DOVE M.J. *A Digital Filter/Estimator for the Control of Large Ships in Confined Waters*. PhD Thesis, Plymouth Polytechnic, 1984.

4.   MILLER K.M. *A Navigation and Automatic Collision Avoidance System for Marine Vehicles*. PhD Thesis, Polytechnic South West, 1990.

5.   BLACKWELL G.K., STOCKEL C.T. *Strategic Planning for Intelligent Real-Time Management*. Summer Computer Simulation Conference, Calgary, Canada, July 1990.

6.   TAPP N.J. *A Non-Dimensional Mathematical Model for use in Marine Simulators*. MPhil Thesis, Plymouth Polytechnic, 1989.

7.   ANKUDINOV V.K. *Ship Manoeuvrability Assessment in Ship Design - Simulation Concept*. Proceedings of the International Conference on Ship Manoeuvrability, London, R.I.N.A., Paper No 19, April-May 1987.

8.   STROM-TEJSEN J. *A Digital Computer Technique for the Prediction of Standard Manoeuvres of Surface Ships*. Research and Development Report, No 2130, Dept of U.S. Navy, 1965.

9.   CRANE C.L. *Manoeuvring Safety of Large Tankers : Stopping, Turning and Speed Selection*. Transactions S.N.A.M.E., Vol 81, 1973.

10.   EDA H. *Digital Simulation Analysis of Manoeuvring Performance*. Proceedings of the 10th ONR Symposium of Naval Hydrodynamics, Cambridge, Mass/USA, 1974.

11.   DAND I.W. *On Modular Manoeuvring Models*. Proceedings of the International Conference on Ship Manoeuvrability, London, R.I.N.A., Paper No 8, April-May 1987.

12.   OGAWA A. KASAI H. *On the Mathematical Model of Manoeuvring Motion of Ships*. International Shipbuilding Progress, Vol 25, 1978.

13.   KOSE K. *On a New Mathematical Model of Manoeuvring Motions of a Ship and its Applications*. International Shibuilding Progress, Vol 29, No 336, 1982.

14.   INOUE S. HIRANO M. KIJIMA K. TAKASHINA J. *A Practical Calculation Method of Ship Manoeuvring Motion*. International Shipbuilding Progress, Vol 28, No 325, 1981.

15.   KOSE K. *On a New Mathematical Model of Manoeuvring Motions of Ships*. Proceedings of the 3rd International Conference on Marine Simulation, Rotterdam, 1984.

16.   OLTMANN P. SHARMA S.D. *Simulation of Combined Engine and Rudder Manoeuvres Using an Improved Model of Hull-Propeller Interactions*. Proceedings of the 15th Symposium on Naval Hydrodynamics, ONR, 1984.

17.   CHUDLEY J. *An Investigation Into Integrated Navigation Systems*. MPhil/PhD Transfer Report, Polytechnic South West, To be published.

18.   MATTHEWS R. *A Six Degrees of Freedom Ship Model for Computer Simulation* Proceedings of the 3rd International Conference on Marine Simulation, Rotterdam, 1984.

19.   ABKOWITZ M.A. *Lectures on Ship Hydrodynamics, Steering and Manoeuvrability*. Hydro-Og Aerodynamisk Laboratorium, Lyngby, Denmark, Report No Hy-5, May 1964.

20.   FRANCIS R.A. *An Investigation Into the Validity of a 4 Degree of Freedom Model*. Undergraduate Honours Project, Polytechnic South West, To be published.

21.   LEWISON G.R.G. *The Development of Ship Manoeuvring Equations*. National Physical Laboratory, Report Ship 176, Dec 1973.

22.   GILL A.D. *The Identification of Manoeuvring Equations from Ship Trial Results*. Trans R.I.N.A. Supplementary Papers, Vol 118, July 1976.

23.   GILL A.D. *The Analysis and Synthesis of Ship Manoeuvring*. Transactions R.I.N.A., Vol 122, 1980.

24.   EDA H. CRANE C.L. *Steering Characteristics of Ships in Calm Water and Waves*. Presented at the annual S.N.A.M.E. meeting, New York, Nov 1965.

25.   BURNS R.S. DOVE M.J. BOUNCER T.H. STOCKEL C.T. *A Discrete, Time Varying, Non-linear Mathematical Model for the Simulation of Ship Manoeuvres*. Proceedings of the 1st Intercontinental Maritime Simulation Symposium, Control Data Corporation, Schliersee, W. Germany, June 1985.

26.   BURNS R.S. DOVE M.J. BOUNCER T.H. STOCKEL C.T. *Mathematical Modelling and Computer Simulation of Large Ships During Tight Manoeuvres*. Proceedings of the International Conference on Computer Applications in the Operation and Management of Ships and Cargoes, London, R.I.N.A., Nov 1985.

27.   MORSE R.V. PRICE D. *Manoeuvring Characteristics of the MARINER Type Ship (USS Compass Island) in Calm Seas*. Sperry Polaris Management, Sperry Gyroscope Company, New York, Dec 1981.

28.   CHISLETT M.S. STROM-TEJSEN J. *Planar Motion Mechanism Tests and Full Scale Steering and Manoeuvring Predictions for a MARINER Class Vessel*. International Shipbuilding Progress, Vol 12, May 1965.

29.   MIKELIS N.E. *A Procedure for the Prediction of Ship Manoeuvring Response for Initial Design*. Proceedings of the 5th International Conference on Computer Applications in the Automation of Shipyard Operation and Ship Design, North Holland Publishing Company, Trieste, 1985.

30.   HIRANO M. TAKASHINA J. MORIYA S.
A Practical Method of Ship Manoeuvring Motion and its Application. Proceedings of the International Conference on Ship Manoeuvrability, London, R.I.N.A., Paper No 17, April-May 1987.

31.   ISHERWOOD   M.A.   *Wind   Resistance   of   Merchant   Ships*. Transactions R.I.N.A., Vol 115, Nov 1983.

32.   CRANE C.L. Jnr *Manoeuvring Trials of the 278000 dwt Esso Ossaka in Shallow and Deep Waters*. Exxon International Company Report, No E11.4TM.79, 1979.

33.   HOLTROP J. *A Statistical Re-analysis of Resistance of Propulsion Data* International Shipbuilding Progress, Vol 31, 1984.

34.   CLARKE D et al. *The Application of Manoeuvring Criteria in Hull Design Using Linear Theory*. Transactions R.I.N.A., Vol 125, 1983.

NOMENCLATURE

A_o	Propeller Disc Area
cp	Tangential propeller velocity
I_z	Moment of inertia about z axis
L	Ship length between perpendiculars
m	Mass of ship
n	Propeller revolution rate
N	Hydrodynamic turning moment
N_v, N_r, etc	Yaw hydrodynamic coefficients
r	Yaw rate
T_p	Propeller thrust
t_p	Thrust deduction fraction
u	Forward velocity
up	Axial propeller velocity
v	Lateral velocity
X	Hydrodynamic surge force
X_v, X_r, etc	Surge hydrodynamic coefficients
Y	Hydrodynamic sway force
Y_v, Y_r, etc	Sway hydrodynamic coefficients
δ	Rudder angle
ρ	Density of sea water

In this paper shorthand notation has been adopted;

$$X_u = \frac{\partial X}{\partial u} \quad , \quad \overline{X}_{uu} = \frac{1}{2} X_{uu} = \frac{1}{2} \frac{\partial^2 X}{\partial u^2}$$

Development of Marine Autopilots

M.J. Dove and C.B. Wright
Ship Control Group, Polytechnic South West, Plymouth, United Kingdom

ABSTRACT

The paper commences by surveying the development of marine autopilots in vessels engaged in trading within European coastal waters. In particular, autopilots that are used with gyro compasses. Few ships of any size are built today without an autopilot. Autopilots are well established navigational aids in modern commercial and military shipping: in their basic form they will maintain a ship on course in the open seas. Shipowners recognize this equipment as an investment with a return on capital for many reasons. These include reduced manpower, improved fuel economy, accurate course keeping and less wear on machinery. In fact, the Ship Control Group at Polytechnic South West (formerly called Plymouth Polytechnic) has been undertaking research for many years in related areas such as high precision navigation, weather routing, automatic collision avoidance and control in port approaches. The immediate goal of the group is to produce a system that will help the mariner as fully as possible.

The paper continues by discussing the use of adaptive autopilots; such autopilots can automatically adjust or 'adapt' themselves to environmental changes. It suggests possible improvements leading to reduced operating costs and improvements in course and track keeping.

The paper concludes by discussing the development of the 'cost function'. The autopilot controls are adjusted to optimize this term, thus producing minimal consumption of fuel.

INTRODUCTION

Marine autopilots are well established and are part of the modern, electronic navigational system; different types are available for ships. In their basic form they will maintain a ship on course in the open seas such as the North Sea, i.e course-keeping. If the ship deviates from the selected course, the autopilot will automatically adjust the rudder to bring the ship back to the selected heading. A more advanced autopilot may be required to navigate a ship through a desired path. For example, within the coastal waters of the English Channel a ferry may need to perform a circular manoeuvre, i.e., course-changing. Figure 1 illustrates a typical operator's panel for such an autopilot. This man-machine interface is relatively simple; it needs to be. The navigator's controls are a joy-stick used to select

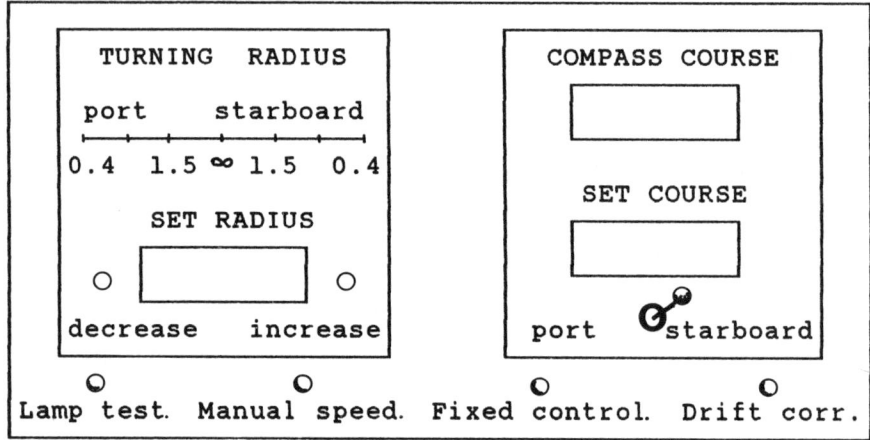

Figure 1. Control panel for an advanced autopilot.

a new course and push buttons used to select the turning radius. If a malfunction develops the helmsman may override the autopilot; this will give the navigator immediate direct access to the steering gear. Finally, the essential input to the autopilot system is the heading derived from the gyro-compass and is displayed on the panel. Refinements such as ship's speed input will also be required to produce a constant radius. Moreover, the echo-sounder may also provide an input since water depth will affect the speed of the vessel if it turns within shallow waters. However, the gyrocompass was and will, in the foreseeable future, remain the heart of the marine autopilot system; autopilots only differ from the algorithm that analyses and controls input information.

Yet, will either technology or legislation allow a marine vessel to sail unmanned from port to port? This would include automatically piloting the vessel out of a harbour, avoiding floating and submerged obstacles on route, weather routeing and piloting the vessel into another harbour. Undeniably futuristic? The man made constellation called GPS (Global Positioning System) will be fully operational by the mid 1990's. The anticipated 24 hour global coverage and the accuracy of GPS will bring the concept of automatic navigation closer to reality.

CONVENTIONAL AUTOPILOTS

Few ships of any size are built today without an autopilot. Shipowners recognize this equipment as an investment with a return on capital for many reasons. Reduced manpower, improved fuel economy, accurate course keeping and less wear on machinery are most of the features found in a modern autopilot.

Automatic ship steering was introduced many years ago before control theory was applied to the design of autopilots (synonymous with autohelm or gyropilot). The invention of the gyrocompass initiated the development of the autopilot; its accuracy led to its success. In 1922 both Minorsky [1] and Sperry [2] produced papers on automatic devices. Minorsky treated the problem of automatic steering mathematically, whereas Sperry considered it as a practical problem involving a gyrocompass. These two papers could be regarded as rather dated. Nevertheless, both papers contributed towards the development of the modern autopilot.

Very early autopilots were based on mechanical construction and were able to provide rudimentary control of the rudder; today it is known as PROPORTIONAL control. It is so called because the rudder is moved by an amount proportional to the positional error from the course line or proportional to the heading error. The positional error is the distance off the centre line and the heading error is the difference between the true heading and the desired heading. The control law for proportional control can be represented by the following equation:-

$$\delta_d = a \ \psi_e \qquad\qquad (1)$$

Where δ_d = demanded rudder angle,
             a   = a constant,
and ψ_e = course error or heading error.

In practical terms the measured heading signal from the gyrocompass is compared with the desired heading and the error is input to the controller. The controller output then drives the rudder servo.

Proportional control was adequate for the guidance of small craft such as torpedoes but unsatisfactory for the steering of large ships. This type of control would cause the vessel to continue to oscillate either side of the required course; the steering gear would be constantly hunting to keep the ship on the correct mean course. The vessel would eventually reach its destination but expensive wear in the rudder gears and abnormal high fuel consumption restricted their use as course keeping devices. Both Minorsky and Sperry were aware of this problem. Heavy seas also resulted in excessive working of the steering gear. Attempts were made to reduce rudder activity due to the rough seas. For example, Nomoto [3] suggested a method in 1960. Early autopilots had other problems. The hydraulic telemotor unit, a device used to control the movement of the rudder, was reported to malfunction because of leakage [4]. An electrical system now replaces this inefficient device.

The controller concept since 1922 has hardly changed; developing technology only changed the hardware of autopilots from purely mechanical devices to electronic systems. Proportional autopilots based on mechanical construction were used in ships up to about 1950. Further control terms other than the use of the 'heading error' were beginning to emerge. Luke and West [5] mentioned the commercial use of an autopilot in 1951 based on the control term called 'rate of change of heading'. This derivative signal had the effect of damping the sinusoidal motion of the vessel. The correct combination of both proportional and derivative terms produced reasonable control of the vessel.

However, almost all conventional, marine autopilots by 1980 were usually based on the simple PROPORTIONAL-INTEGRAL-DERIVATIVE (PID) controller systems. Equation (1) is now modified to include the proportions of the integral and derivative terms; the proportions are governed by the values of the constants b and c. Equation (2), below, represents the PID control equation.

$$\delta_d = a\ \psi_e + b\int \psi_e\,dt + c\ \dot{\psi}_e \qquad (2)$$

Where $\dot{\psi}_e$ = rate of change of the course or
heading error.

The addition of the integral term allowed the
course of the vessel to be maintained in the presence
of steady state disturbances such as a cross wind.
Such systems take a signal proportional to the error
between the actual and the desired course as the
controlling input. This heading error signal is also
electronically integrated producing integral control
and differentiated producing derivative control data.
The proportional, integral and derivative control
signals are electronically blended together to
produce a single control signal. To keep the ship on
course, proportions of each of the three signals can
be adjusted manually by a control panel containing
three electrical potentiometers. Attempts to keep the
ship on course was normally performed by trial and
error: in fact this was one of its major criticisms.
Nevertheless, PID autopilots did maintain a straight
line course through the action of the ship's rudder.
In this respect, the composite control signal of PID
autopilots was far superior to the single signal
associated with proportional controlled autopilots.
Unfortunately, PID autopilots had a few irritating
problems. This will be considered next.

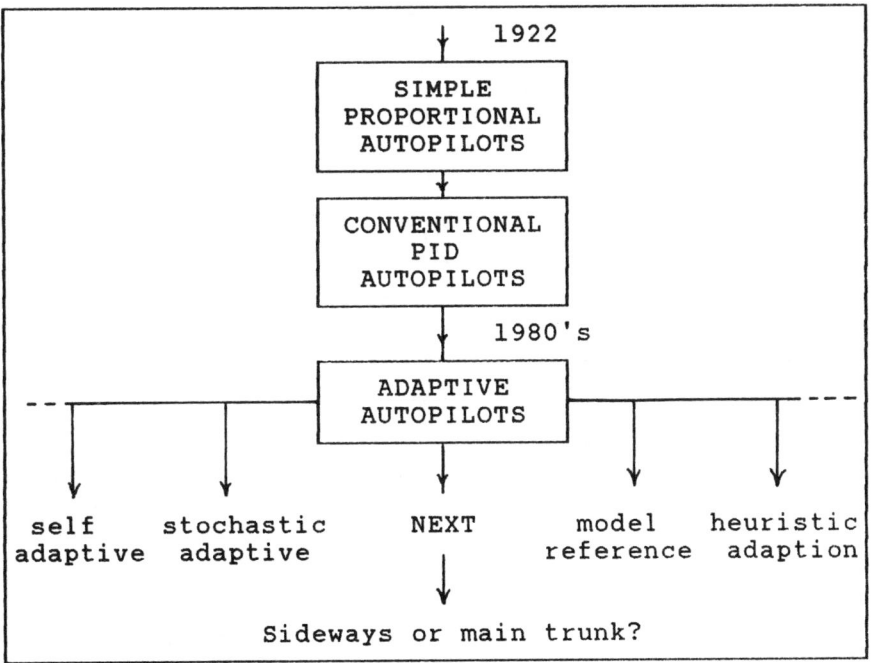

Figure 2. Development of marine autopilots.

Conventional, PID autopilots require manual adjustments to compensate for changes in the ship's environment; settings are seldom optimal for that ship. Adjustments for variations such as sea waves, wind and currents are tedious and time consuming. Furthermore, autopilots perform badly in rough seas as analysed by Blanke in 1981 [6]. On such occasions manual steering is often used in place of the autopilot, yet such circumstances require the use of automatic control. Further problems arise because of changes in the dynamics of the ship; for instance variations in the speed, draught or water depth. Consequently, this explained the growing interest towards autopilots that could automatically adjust or ADAPT themselves to these changes.

ADAPTIVE AUTOPILOTS

Development towards marine adaptive autopilots was rapid in the late 1970's and early 1980's. In 1984 at least three different companies were marketing adaptive autopilots. Interestingly, the origins of adaptive control began much earlier in the early 1950's in the area of aviation. However, early adaptive control was plagued with problems; control theory was in its infancy and microelectronics did not exist. Today, adaptive autopilots make use of complicated control laws and require the functions of the microcomputer. Despite their complexity, adaptive control for ships provides several benefits. Improved fuel economy, increase speed of vessel, reduced steering, reduced manual settings to compensate for wind, waves, currents, speed, trim, draught and water depth. Adaptive control also improves safety and makes the ship operation more convenient in all weather conditions.

There has been much research effort into the design of adaptive autopilots and their control algorithms. In 1975 Oldenburg [7] proposed to add adaption heuristically to ordinary PID autopilots; a similar approach was taken by Sugimoto and Kojima in 1978 [8]. A wide class of adaptive autopilots were also beginning to emerge according to modern adaptive and stochastic control techniques. Stochastic adaptive systems were proposed by Merlo and Tiano in 1975 [9]; Astrom in 1976 [10] and Brink et al in 1978 [11] had also introduced a stochastic approach to the analysis and control of the motion of a ship. Another approach, known as 'self tuning adaptive control', had been investigated by Kallstrom and Astrom in 1977

[12]. Two different autopilots were looked at in their paper. The simplest used only heading measurements, while the more complex system was provided with a Kalman filtering of heading, sway velocity and yaw rate. The Kalman filtering was used to obtain reliable estimates of the three afore mentioned parameters. Self-tuning adaptive control was also considered in a paper by Brink and Tiano in 1981 [13]. This paper describes the simulation of a supertanker and a second generation containership. Results indicate that this type of adaptive autopilot was a feasible and efficient solution for automatic steering of a ship in main operational navigation situations.

However, in 1976 the research of Amerongen [14] was committed to the implementation of an adaptive autopilot based on model reference techniques. The main requirement for such systems is an accurate mathematical model of the ship's steering dynamics and of the external environment. The ideal model is subjected to the same inputs as the actual ship. Inputs of the actual system such as the ship's heading, speed, rate of turn and rudder angle will be fed into a computer containing details of the model. If the response of the stored mathematical model differs from that of the actual system, the error between the two responses is subsequently used to adjust parameters of the real ship. For example, the computer may calculate the rudder angle required to minimise the course keeping errors in the optimum manner. The adaptive aspect in the system will account for changes in weather, water depth and changes in the ship loading condition. A useful side effect of mathematical modelling is the ability to predict the path of the actual ship.

Research has been undertaken into the effects of the ship's natural yaw action in relation to the course to be steered. It has been found that a straight course is not necessarily the most economical and the ship's natural yaw action should not be smoothed out. The added resistance due to steering on a straight course has been analysed by Norrbin [15]. Disturbance levels and load conditions were shown by Astrom [16] to be factors in this process; adaptive control can minimize this loss. Reduced drag will reduce the thrust of the vessel leading to fuel savings and speed increase. Fuel savings of 1-3% and speed increases of 0.5-1.5% were observed by Amerongen in 1984 [17]. The speed increases were mainly due to smoother rudder movements. Consequently, there would also be less wear and tear of the steering equipment, reinforcing the advantages of adaptive control. Earlier in 1979 Kallstrom et al [18] also confirmed a reduction in drag and the corresponding economic benefits of adaptive steering for tankers. Recently, Katebi and Byrne [19] suggested an autopilot using the LQG (linear quadratic Gaussian) approach. It too minimised the added resistance due to steering. Additionally, an improvement in the course keeping performance in all weather conditions is suggested. However, in 1982 both Clarke [20] and Reid [21] have questioned the efficiency of adaptive autopilots.

AUTOPILOTS AND THE SEARCH FOR ECONOMY

Consumption of fuel is one of the major costs in vessel operation. Potentially high oil prices due to the Iraq invasion of Kuwait will initiate yet another interest in fuel saving autopilots. Emphasis will continue on the production of a generation of autopilots with energy saving capabilities.

Fuel consumption is essentially affected by the resistance on the vessel due to movement in its environment. Hull design and engine characteristics are assumed to be fixed. Resistance increases due to ship motions and external influences such as wind and waves can increase the consumption of fuel. Long periods of yawing motion due to an autopilot system can also increase resistance to vessels. This undesirable movement was reported by Motora in 1953 [22]; resistance increase could be as much as 10%. Later in 1966 Nomoto and Motoyama [23] stated if a ship was left to yaw naturally, it would suffer fewer propulsion losses than a vessel undergoing corrective application of the rudder. They forecasted losses of between 2% and 20%. Essentially, rudder movements produce additional drag. However, resistance due to hull inertia was also considered in this paper. Thus, autopilot design was beginning to be based on the concept of "added resistance due to steering". Koyama [24] later argued that these drag forces could not be totally eliminated. He suggested that the correct selection of the proportional and derivative gain controls on the autopilot could minimize rudder resistance. Motora [25], and Motora and Koyama [26] considered the use of a PERFORMANCE INDEX for the measurement of propulsion losses. This expression consisted of two terms. One term was attributed to the excess distance covered by yawing of the vessel; the second term was linked to the rudder resistance. The performance index is sometimes referred to as the LOSS FUNCTION or more importantly as the COST

FUNCTION. This is because propulsion losses give a measure of the cost of the fuel used. The simple performance index, J, is normally expressed as:

$$J = \frac{1}{T} \int_{0}^{T} \psi_e^2 + \lambda \, \delta_a^2 \; dt \qquad (3)$$

or

$$J = \frac{1}{T} \int_{0}^{T} (\psi_a - \psi_d)^2 + \lambda \, \delta_a^2 \; dt \qquad (4)$$

Where

ψ_e = course error (heading error),
ψ_a = actual heading,
ψ_d = desired heading,
λ = weighting factor,
δ_a = actual rudder angle.
T = time over which the optimization is carried out
and $\psi_e = \psi_a \sim \psi_d$.

Thus, the autopilot controls are adjusted to minimize the cost function, J. The optimization of J is carried out during a time interval, T. The interval might be the time to complete an oceanic phase of the voyage, or the total time from departure to arrival ports. According to Koyama the weighting factor, λ, should be set between 8 and 10; its value will depend on the size of the ship. Norbin [15] in 1972 proposed a performance index similar in form to the performance index suggested by Koyama. But, Norbin suggested a value of only 0.1 for the weighting factor. This suggested that resistance due to rudder motions (δ_a term) played a less important part than the resistive forces due to the yawing motions (ψ_e term). The difference produced much controversy. In 1980 Amerongen and Van Nauta Lemke [27] suggested that λ should range from 0.1 to

4. The former value should apply for small ships in a calm sea or for large ships; the latter value should be for small ships in high seas. Observations were based on towing-tank experiments and on full scale trials. Some authors have questioned the validity of equation (3); the optimization of this expression may not lead to maximum economy. Marshall and Broome [28], Clarke [29], Blanke [6], Reid [30], Kallstrom [31], Lim and Forsyth [32] all employed cost functions of different forms. For example, Clarke [29] suggested that a whole family of cost functions exist and are of the form:

$$J = \frac{1}{T} \int_0^T a\, \dot{\psi}_e^2 + b\, \dot{\psi}_a^2 + c\, \delta_a^2 \qquad (5)$$

Where a, b and c are constants, and

$$\frac{d(\psi_a)}{dt} = \dot{\psi}_a = \text{rate of change of the actual heading}$$

Sometimes $\dot{\psi}_a$ is replaced by $\dot{\psi}_e$, rate of change of the course error, since $\psi_e = \psi_a \sim \psi_d$,

thus

$$\dot{\psi}_e = \dot{\psi}_a$$

The term $\dot{\psi}_e^2$ was again concerned with the extra distance covered by the vessel, while the $\dot{\psi}_a^2$ and δ_a^2 terms were concerned with its resistance. The weighting factors a, b and c were dependent on a number of components such as engine conditions and ship type. In a computer simulation Clarke also showed that the optimisation of this cost function could produce improvements in fuel economy of several per cent.

CONCLUSION

Emphasis is on the production of a generation of autopilots with energy saving capabilities. Unfortunately, there is no direct relation between controller settings and fuel consumed. Hence, much of the research today revolves around the minimisation of a COST FUNCTION, but not necessarily of the forms mentioned above. This function normally takes the form of a weighted sum of variables related to excess distance covered in yawing and resistance due to the movement of the steering mechanism. The control algorithm also attempts to maintain the desired course. But is this the correct procedure? What is the alternative?

Another interesting development is where an autopilot is not only used to control the heading of a ship, but is used to reduce the roll motion as well. The rudder roll stabilization (RRS) system has been mentioned in several recent papers [33, 34, 35, 36, 37]. This is an example where autopilot design has produced a unique feature.

Without doubt, the availability of relatively inexpensive, fast digital microcomputers, the development of modern adaptive control and optimal theories have given the impetus to produce more sophisticated controllers. Autopilot design in the future will continue to improve steering characteristics through possibly the use of accurate filtering and modelling; faster processors will hasten this development. It is certain, however, that marginal savings in fuel consumption and accuracy in steering will continue to improve.

REFERENCES

1) MINORSKY, N. Directional Stability of Automatically Steered Bodies. Journal of the American Society of Naval Engineers, Vol.42, No:2, 1922.

2) SPERRY, E. Automatic Steering Presented at the thirtieth general meeting of the Society of Naval Architects and Marine Engineers, New York, Nov. 8-9, 1922.

3) NOMOTO, K. Directional stability of automatically steered ships with particular reference to their bad performance in rough seas. Proc. of the first Symposium on Ship Manoeuvrability. DTMB Report No: 1461, Oct. 1960.

4) TETLEY L., CALCUTT, D. Electronic Aids to Navigation. Edward Arnold ISBN 01713135484.

5)   LUKE R.M. and WEST F. An Integrated Steering System New England Section, SNAME, June, 1960.

6)   BLANKE, M. Ship propulsion losses related to automatic steering and prime mover control, PhD Thesis, Technical University of Denmark, 1981.

7)   OLDENBURG, J. Experiment with a new adaptive autopilot intended for controlled turns as well as for straight course steering. Proceedings of the 4th Ship Control Systems Symposium, The Hague, 1975.

8)   SUGIMOTO, A. and KOJIMA, T. A new autopilot system with condition adaptivity. Proceedings of the 5th Ship Control Systems Symposium, Annapolis, Maryland, U.S.A., 1978.

9)   MERLO, P. and TIANO, A. Experiments about computer controlled ship steering. Semana International sobre la Automatica en la Marina, Barcelona, Spain, 1975.

10)   ASTROM, K.J. and KALLSTROM, C.G. Identification of ship steering dynamics. Automatica Vol.12, No:9, 1976.

11)   BRINK A.W., BAAS G.E., TIANO, A. and VOLTA, E. Adaptive automatic course-keeping control of a supertanker and a containership - a simulation study. Proceedings of the 5th Ship Control Systems Symposium, Annapolis, Maryland, U.S.A., 1978.

12)   KALLSTROM, C.G. et al. Adaptive autopilots for steering of large tankers. Technical Report TFRT 3145, Lund Institute of Technology, 1977.

13)   BRINK, A.W. and TIANO, A. Self tuning adaptive control of large ships in non-stationary conditions. International Shipbuilding Progress, Vol. 28, No:323, pp 162-178, 1981.

14)   AMERONGEN J.van and UDINK TEN CATE A.J. Model reference adaptive autopilots for ships. Automatica Vol.11, pp 441-449, 1975.

15)   NORRBIN, N.H. On the added resistance due to steering on a straight course 13th International Towing Tank Conference 1972, Berlin/Hamburg.

16) ASTROM, K.J. Why use adaptive techniques for steering large tankers? Dept. of Automatic Control, Lund Institute of Technology, Lund, Sweden, 1977.

17) AMERONGEN, J.van. Adaptive steering of Ships - A Model Reference Approach. Automatica Vol.20, No:1, pp 3-14, 1984.

18) KALLSTROM, K.J. et al. Adaptive Autopilots for Tankers. Automatica, Vol.15, No:3, pp 241-254, 1979.

19) KATEBI, M.R. and BYRNE J.C. LQG adaptive ship autopilot. Transactions of the Institute of Measurement and Control, Vol.10, No:4, pp 187-197, 1988.

20) CLARKE, D. Do autopilots save fuel? Transactions IMarE (C), Vol.94, Paper C97, pp 15-19, 1982.

21) REID, R.E. Identification and minimization of propulsion losses related to ship steering. Report No: MA-RD-940-82038, University of Illinois at Urbana-Champaign, U.S.A., 1982.

22) MOTORA, S. On the automatic steering and yawing of ships in rough seas. J. Soc. Nav. Archit. Japan, Vol.94, 1953.

23) Nomoto, K. and Motoyama T. Loss of Propulsive Power caused by Yawing with Particular Reference to Automatic Steering, Jap. Shipbldg. and Marine Eng., Vol.120, 1966.

24) KOYAMA, T. On the optimum automatic steering system of ships at sea. Journal of Soc. Nav. Arch. Japan, Vol.122, 1967.

25) MOTORA, S. On the automatic steering and yawing of ships in rough seas. J. Soc. Nav. Archit. Japan, Vol.122, Dec 1967.

26) MOTORA, S. and KOYAMA, T. Some aspects of automatic steering of ships, Jap. Shipbldg. and Marine Eng., Vol.3, July 1968.

27) AMERONGEN, J.van and NAUTA LEMKE, H.R.van. Criteria for optimum steering of ships. Symposium on Ship Steering Automatic Control, Genoa, 1980.

28) MARSHALL, L. and BROOME, D.R. A cost function indicator for optimal ship course keeping. Automation for safety in shipping and off-shore petroleum operations. I.F.I.P., pp 103-108, 1980.

29) CLARKE, D. Development of a cost function for autopilot optimisation. Proc. Symposium on Ship Steering Automatic Control, Genoa, Italy, June 1980.

30) REID, R.E. A proposal for performance criteria for propulsion losses due to ship steering. Journal of Dynamic Systems, Measurement and Control. Vol.104, June 1982.

31) KALLSTROM, C.G. Identification and adaptive control applied to ship steering. PhD thesis. Technical University of Denmark, 1982.

32) LIM, C.C. and FORSYTH, W. Autopilot for ship design. Proc. I.E.E., Vol.130, No:6, Nov. 1983.

33) AMERONGEN, J.van et al. Roll stabilization of ships by means of the rudder. Proceedings of the Third Yale Workshop on Applications of Adaptive Systems Theory, New Haven, Conn., U.S.A., 1983.

34) AMERONGEN J.van et al. Model tests and full-scale trials with a rudder-role stabilization system. Proceedings of the Seventh Ship Control Systems Symposium, Bath, U.K., 1984.

35) AMERONGEN J.van et al. Recent Developments in Automatic Steering of Ships. The Journal of Navigation, Vol.39, No:3, pp 349-362, 1986.

36) AMERONGEN J.van et al. Adaptive control aspects of a rudder roll stabilization system. Selected Papers from the 10th Triennial World Congress of the International Federation of Automatic Control, Munich, West Germany. Vol.6, pp 221-225, 1987.

37) VAN DER KLUGT P.G.M., VAN NAUTA LEMKE H.R. Rudder roll stabilization and ship design; a control point of view. IEE Colloquium on 'Control in the Marine Industry' (Digest no: 19), pp 2/1-3, 1988.

The Role of a Mathematical Model for Improvement of Marine Navigation

M.J. Dove, K.M. Miller and C.T. Stockel
Ship Control Group, Polytechnic South West, Plymouth, United Kingdom

ABSTRACT

This paper describes part of an ongoing research programme in automatic navigation and guidance. In particular it looks at efforts to improve the accuracy of the the mathematical model used in an integrated navigation system. The paper commences by giving a brief explanation of the reasons for the development of integrated systems. There then follows a description of the Kalman-Bucy filter being developed for use in a marine navigation system, together with a description of the problems of accurately modelling a vessel in changing circumstances. It is suggested that the accuracy of the model and its inputs are important factors in the use of filtering techniques for marine navigation. To overcome the problems of updating the model under operational conditions system identification techniques are introduced, by augmenting the state vector with sway and yaw coefficients derived from the hydrodynamic coefficients. This, in turn, involves extending the matrix which represents the ship. The paper concludes by discussing some of the results obtained.

INTRODUCTION

In the process of bringing a vessel safely to her berth there is great emphasis on the skills of the navigator. These skills are the traditional ones of navigation, seamanship and ship handling; methods which have evolved over centuries and which, in the main, have been carried out without the aid of sophisticated microcomputers and automatic control systems. In this world of traditional navigation the Dead Reckoning or DR position is defined as a calculation of the vessel's position using only the course and speed through the water. This is the least accurate method of recording position. By adding an estimate of the set, rate and drift of the current the Estimated Position, or EP, is obtained. Providing the estimate of tidal flow is reasonably correct, an improved accuracy of position is obtained. Finally , using a compass, chronometer and sextant, or electronic navigation aids, the navigator is able to plot the vessel's position with sufficient accuracy to ensure safe passage between ports. This, the most accurate position available, is known as the Fix.

That none of these skills have been lost is shown in numerous feats of navigation which are frequently reported by the media. So why does the mariner need electronic navigation aids, integrated navigation systems, adaptive autipilots and other systems dependent upon the power of the microprocessor? Is there an argument for increased automation on the bridge? There are several factors which suggest that there is a requirement for moves in this direction, without completely eliminating the mariner from the command loop. This is the case in avionics, where the pilot retains ultimate control of his aircraft, even though automatic navigation and landing systems are being installed in the latest generation of airliners. It would be reasonable to suppose that the travelling public would wish this to continue.

FACTORS IN THE DEVELOPMENT OF AUTOMATIC NAVIGATION

Ship owners and operators have, by the very nature of their business, been conservative. Tradition dies hard and, in 1945, there were none of the incentives which faced the aircraft industries . Ship design was stable, diesel engines were being widely fitted, equipment was largely satisfactory and efficient, and there were no spectacular disasters such as those which dogged the development of the world's first commercial jet airliner, the De Haviland Comet. Things remained that way for twenty years or more.

But by 1970 it had become apparent that the advent of large supertankers and fast container ships, operating in increased traffic density, would require modern navigation systems. Larger, faster vessels have greater momentum and hence require longer stopping distances and greater diameter turning circles. This all requires more sea room when increased draft means the ship may have less space in which to manouevre . These demands, coupled with the dramatic achievements in the world of electronics, paved the way for the systems available today. There was also a widely held view that international shipping was not operated as safely as it might be , with the result that more accidents occurred than were acceptable. To the extent that even well found ships were not being equipped with the aids available to them, it could be argued they were being developed in advance of their demand. Furthermore advanced navigation aids were expensive compared with the more traditional systems available, and there were no definable standards against which to measure improvements in safety [1].

There were, and still are, difficulties in retaining high calibre trained staff at sea. There was, and still is, a decease in job satisfaction. Furthermore the huge oil price increases in the early seventies were a major factor in increased operating costs, leading to a need for optimal operation of ships. Increasing traffic density, particularly in waters such as the Straits of Singapore, was another contributory factor. Finally environmental factors started to emerge as early as the 1960's. For example the cost of clearing up the environment after the Torrey Canon ran aground was in excess of the value of ship and cargo combined [2], and this accident saw a huge public outcry at the damage caused to wildlife and the UK coastline. Finally the automation process itself leads to further decreased job satisafaction for the highly trained personel who may wish to remain at sea.

By 1975 there were a variety of electronic navigation aids available, but none was completely acceptable because no single system covered all aspects of every voyage. In ships fitted with several sensors there was, therefore, the possibility of too much information becoming available, so that one person was increasingly unable to handle, and interpret, the increased information flow whilst undertaking all the other duties of the Officer of the Watch. Even the advent of the Global Positioning System (GPS) has not yet completely solved this problem.

INTEGRATION OF NAVIGATIONAL DATA

The appearance of Automatic Radar Plotting Aids on the bridge paved the way for integration of data from different sources, and several multi-task systems have been developed in Europe, Japan and the USA. The ship is also acted upon by disturbances such as wind, tide and current. These disturbances may be random, as in a gust of wind for example. As the vessel moves sensors measure the position and velocity, but these measurements may be noisy, that is they contain random errors. Use of a Kalman filter will minimise such noise [3] . The filter is a recursive algorithm which estimates the values of the variables of a stochastic system from measurements which contain randomly fluctuating noise. Optimal filtering, using a Kalman-Bucy filter, is a stochastic technique which combines noise corrupted measurements of a dynamic system with other known information about the system, in order to obtain best estimates of the variables, or states, which govern the system.

Central to the development of a good software package to be used in a Kalman-Bucy filter is an accurate mathematical model of the system. Research at Polytechnic South West has shown that the accuracy of filtered data depends very much on the accuracy of this model and its inputs. The remainder of this paper describes some of the work being undertaken to overcome the problems associated with modelling ships in service.

THE MATHEMATICAL MODEL

The model used in this investigation is based upon a Eulerian set of equations of motion. The forces and moments are derived in the usual way, as originally given by Abkowitz [4], with a modular approach as presented by Tapp [5] for use in marine simulation. Forces and moments are decomposed into contributions associated with the system elements, for example hull, propeller, rudder and disturbance terms. In order to model the behaviour of the hull it was first necessary to evaluate the derivatives used in the equations of motion. Some of the difficulties encountered in obtaining these coefficients are described by Chudley [6].

THE BASIC KALMAN FILTER

The theory of the Kalman-Bucy filter is well established and the equations used in the research described in this paper are given by Miller [7]. The filter used is an extended Kalman filter, that is the non-linear system process is linearized about the most recent optimal estimate, while the measurement process is linear and the errors are Gaussian. As a ship constitutes a non-linear system, when parameters such as large alterations of course and/or speed, shallow water effects, and trim are considered there must be some limitation to the technique. The linearization process assumes constant course and speed during each sample period. This is reasonable provided sample times are small when compared with such factors as ship time constants and time between waypoints.

⌒ DECCA   ------ FILTER   ———MODEL

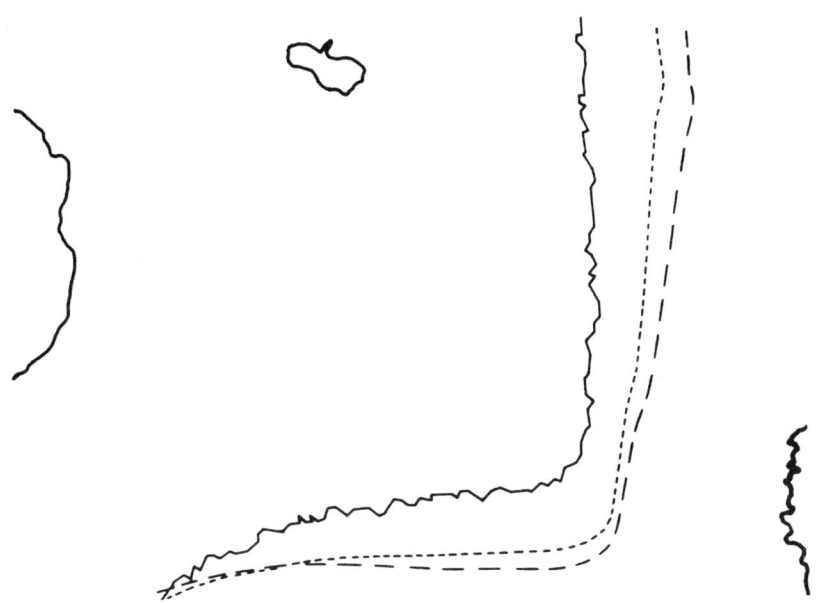

Figure 1.   Estimated Positions With No Disturbance Input

In this basic filter control signals are the only inputs to the model. The model position output can thus be considered as a sophisticated DR . The measured and model values of position and velocity are then combined in accordance with the weighting provided by the filter. A weighting of 1 results in an output of measurement only for the corresponding element in the state vector, 0 gives prediction only, whilst any intermediate value results in a combination, rather like a combination of the DR and Fix.

During trials in Plymouth Sound several runs were made while plotting measured, model and filtered positions. The results for one run are given in figure 1. The model is seen to deviate from the measured track, and this is largely due to disturbances such as wind and tide and also errors in the hydrodynamic coefficients obtained for the vessel. The filtered estimate takes some weighted mean between the measured and model tracks as expected. These results confirmed conclusions from earlier work undertaken in the Solent [8], and highlighted the need to improve the model and its inputs.

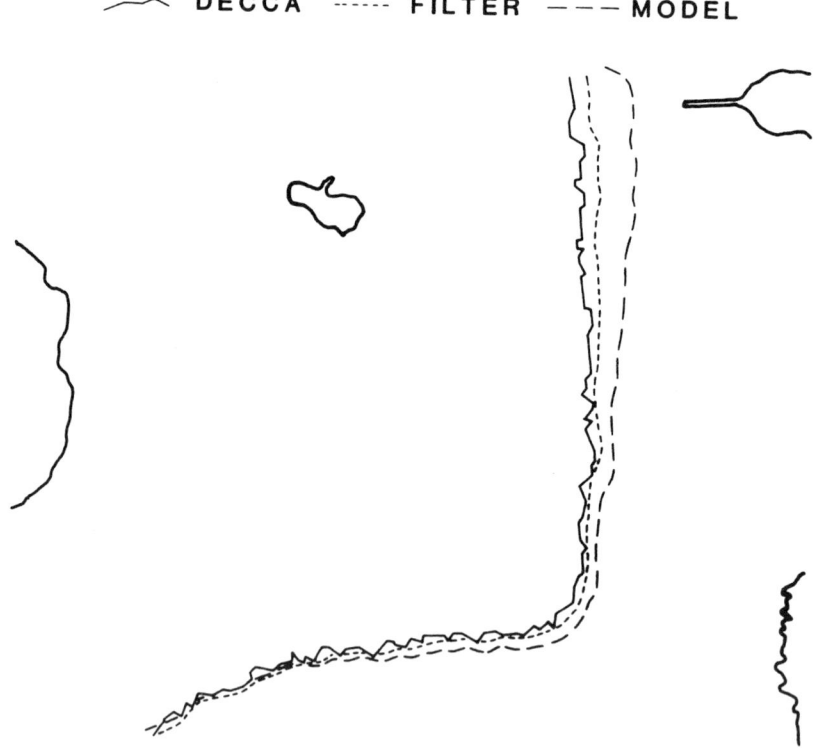

Figure 2. Estimated Positions With Disturbance Input

DISTURBANCE ESTIMATION

The total disturbance effect can be estimated from the difference between position vectors obtained from the smoothed Fix and the DR. Placing a window around a number of previous pairs of vectors and taking the mean value improves the estimate of the disturbance vector.

This can then be used in conjunction with the next DR position to find the EP, or estimated position, of the vessel. Suppose that the filtered output suggests the vessel has moved distances x_1 and y_1 on the grid between fixes, whereas the model gives displacements x_2 and y_2. Any difference between x_1 and x_2 and y_1 and y_2 may be assumed to be due to external forces and errors in the model. The computer program was modified to compute this disturbance vector on each cycle. The prevous twenty values being averaged to give combined values for current and wind, which were then applied to the model equations on the next cycle. Furthermore the variances are calculated and applied to the Kalman filter equations.

With the addition of disturbance inputs to the model the data set used to obtain the plots shown in figure 1 was rerun. The modified plot in figure 2 shows that the model track (DR) follows the measured track closely until the turn is completed, after which it starts to deviate, suggesting that the model is still not adequate for the system. The filtered track is smoothed and close to the measured track during the first leg of the passage, but after the turn it also starts to deviate. The addition of disturbance inputs to the model has thus improved the filter output, but improvement to the model is still required if the Kalman-Bucy filter is to provide the basis of an integrated system.

SYSTEM IDENTIFICATION

Much has been written of the difficulties of obtaining an accurate mathematical model of a ship. Experience has shown however that a good mathematical model is required in a maritime optimal navigation system, and the accuracy of the filter is still further improved if it acquires both control and disturbance inputs, even though the disturbance inputs may only be estimates of the true values. There is the additional problem of changing hydrodynamic coefficients as circumstances change. For example, as the underwater surface of the hull is fouled with growth, when the vesel enters shallow water after an oceanic passage, or when the velocity is changed. This means that the some values in the system matrix will require updating. This may be difficult during routine commercial operations. It certainly would do little to enhance the sale of a filter based integrated navigation system if the potential owner were informed that the vessel would have to be periodiodically taken out of service to update the model. Established techniques for parameter identification, based upon various optimization criteria, would require new algorithms to be included into the Kalman filter recursive loop. These methods are usually time consuming and consequently are unsuitable for real-time applications. However Gelb [9] proposes a method which can be easily implemented into the existing Kalman filter loop. This method has been applied to the navigation system under development, optimizing parameters on the minimum variance estimation algorithms already in use.

AUGMENTING THE STATE VECTOR

Let the unknown parameters be denoted by a vector **a**, having dynamics defined by the differential equation:

$$\dot{\mathbf{a}} = \mathbf{0} \tag{1}$$

with non-linear equations of motion, the system process can now be written:

$$\dot{\mathbf{x}} = \mathbf{f}(\hat{\mathbf{x}}, \mathbf{a}) + \mathbf{Gu} \tag{2}$$

where **G** is the control matrix, **u** the control vector, **x** the state vector, and both **a** and $\hat{\mathbf{x}}$ are to be estimated from the noisy measurement data. Combining **x** and **a** into a composite state vector denoted $\mathbf{x}^*$ such that:

$$\dot{\mathbf{x}}^* = \begin{bmatrix} \dot{\mathbf{x}} \\ \dot{\mathbf{a}} \end{bmatrix} = \begin{bmatrix} \mathbf{f}(\hat{\mathbf{x}}, \mathbf{a}) + \mathbf{Gu} \\ \mathbf{0} \end{bmatrix} \tag{3}$$

and applying this system process to the extended Kalman filter routine yields estimates for both states and unknown parameters.

THE MODELLING PROCESS

Selecting the vector **a** to contain hydrodynamic coefficients for the vessel gives a large dimension augmented state vector and a large transition matrix. This formulation would then lead to cumbersome computations, defeating one of the prime objectives of this research. Furthermore, due to cross coupling, some coefficients cannot be isolated and are therefore unidentifiable. An alternative method was suggested by Robbins [10] who applied this method of parameter identification to aircraft, but used simplified mathematical models.

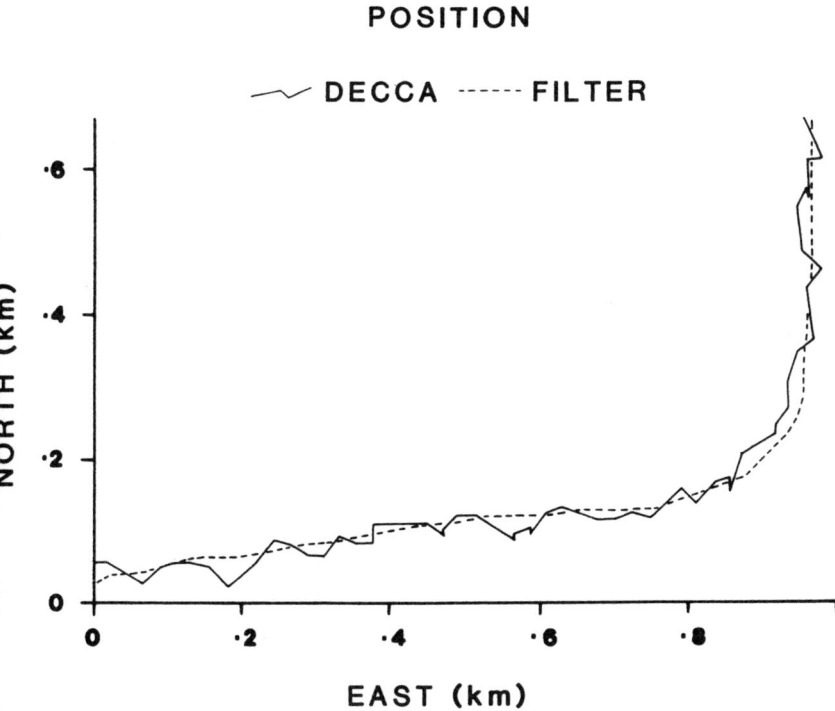

Figure 3.   Comparison of Measured and Filtered Positions with SI.

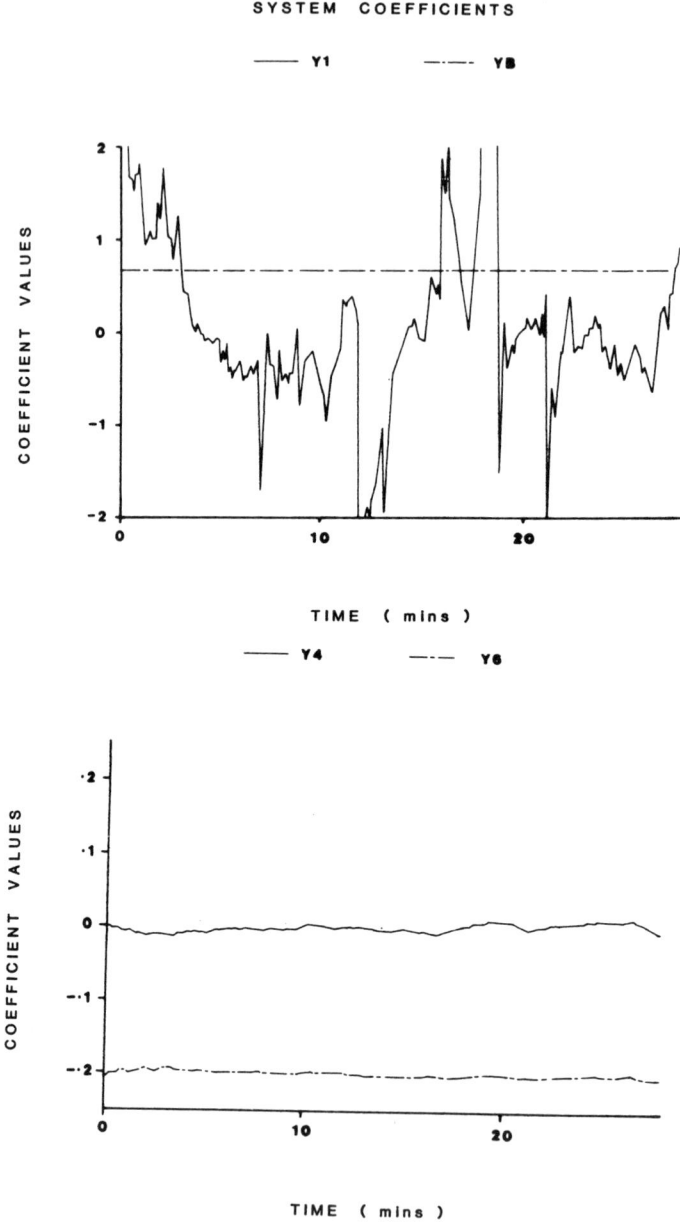

Figure 4.   Sway System Coefficients.

In order to keep the dimensions of the augmented state vector to minimum, maintain identifiability of parameters and to retain sparse population of the transition matrix, it is necessary to identify the components of the latter directly. Initially only sway and yaw terms are considered, as these use the least accurate coefficients and were seen to give poorer results than the surge term. The augmented state vector can be written as:

$$\mathbf{x}^* = (\ \delta,\ n,\ x,\ u,\ y,\ v,\ \psi,\ r,\ Y_1,\ Y_4,\ Y_6,\ Y_8,\ N_1,\ N_4,\ N_6,\ N_8\ )^T \qquad (4)$$

In this paper X and Y are the forces of surge and sway and N is the yaw moment, x, y and ψ are the corresponding displacements of the vessel and u, v and r are their derivatives. The rudder angle is denoted by δ, the propeller revolutions by n and the vessel moves on a grid (x_0, y_0) defined on the earth's surface, where x_0 is the direction of true North, giving a reference from which heading, ψ, is measured.

These constants are dependent on the vessel states, and hence, assuming a slow transition time of the vessel in comparison to cycle time of the Kalman recursive loop, the parameters to be identified may be taken as having dynamics given by equation 1. Space precludes a detailed explanation in this paper, but the developement of the equations is given by Miller et al [11].

The original eight states of the measurement vector can be obtained as before but the augmented states are obtained from the previous optimal estimates. The previous set of values for the sway and yaw coefficients are used in an iterative manner to obtain new values .

SYSTEM IDENTIFICATION RESULTS

Data sets used in earlier work were rerun using the filter algorithm with system identification. A typical track plot is shown in figure 3, from which it can be seen that the filtered track follows closely the true position of the vessel, showing a considerable improvement over the two previous cases shown in figure 1 and 2. Figure 4 shows the identified system sway coefficients. The rudder term Y_1 is seen to be noisy. This term would be expected to reduce to zero when travelling in a straight line and increase during the use of rudders in a turn, which is seen to occur.

Further noise is probably due to noisy rudder measurements. Y_4, the surge term is close to zero. This term influences the turning characteristics with speed and over the 6 to 7 knot speed range used during the trial has little influence.

CONCLUSIONS

A technique to overcome coefficient inaccuracies and variations by incorporating them in to the state vector has been introduced. Trials were undertaken for sway and yaw terms only as these were considered to be the most inaccurate and widely varying, but the method can also be applied to the surge terms. Resulting track plots of filtered position showed the vessel to remain within 20 metres of the demanded track. The filtering algorithm in use is intended to cope with random errors only, and fixed errors must be evaluated and either removed prior to filtering, or an allowance must be made for them. In this way the noise reduction achieved is a better assessment of performance. The ability of the system to maintain a position central to the noise of the position fixing system demonstrated that the filter was producing accurate outputs of displacement in the three degrees of freedom considered. It was shown previously [12] that inaccuraccy in one, or all, of these outputs led to drift from true track. As values for velocity are fed back into the modelling process, any inaccuracy in their estimates leads to cumulative errors in the displacement outputs. The turn rate does however remain noisy and an improvement could be achieved by the use of a gyro input, which was not available in the test vessel.

REFERENCES

1   Maybourn R and Mateer W E, *The Pay-off from Improved Navigational Aids*, J Nav, Vol 27, No 2, pp 133–157, 1974

2   Stratton A and Silver W E, *Operational Research and Cost Benefit Analysis of Navigation with Particular Reference to Marine Accidents*, J Nav, Vol 23, No 4, 1970.

3   Dove M J and Miller K M, *Kalman Filters in Navigation Systems*, J Nav, Vol 42, No 2, pp 255–267, 1989.

4   Abkowitz M A, *Stability and Motion Control of Ocean Vehicles*, Massachusetts Institute of Technology, 1964.

5   Tapp N J, *A Non-Dimensional Mathematical Model For Use In Marine Simulators*, MPhil Thesis, CNAA, Plymouth Polytechnic, UK, Sept 1989.

6   Chudley J, Dove M J and Tapp N J, *A Review of Mathematical Models Used in Ship Manoeuvres*, Proceedings of CADMO 91, Wessex Institute of Technology, Florida, 15-17 Jan 1991.

7   Miller K M, A Navigation and Automatic Collision Avoidnce System *for Marine Vehicles*, PhD Thesis, CNAA, Polytechnic South West, Plymouth, UK, 1990.

8   Burns R S, Dove M J, Evison J L, Stockel C T, Tapp N J, *Improving the Accuracy of Own Ship's Position on the Viewnav Electronic Chart*, Proceedings of a Seminar on Navigating with Electronic Charts, The Nautical Institute and The Royal Institute of Navigation, London, December 1985.

9   Gelb A, *Applied Optimal Estimation*, The M.I.T. Press, 1988

10   Robins A J, T*he Extended Kalman Filter And Its Use In Estimating Aerodynamic Derivatives*, Aerospace Dynamics, Issue 9, pp 16-19, Sept 1982.

11   Dove M J, Chudley J, and Stockel C T, *The Use of a Mathemetical Model in a Track Guidance System*, Proceedings of the Ninth Ship Control Systems Symposium, Department of the Navy, Naval Sea Systems Command, Washington D C, !0-14 September 1990.

12   Dove M J and Miller K M, *On the Role of the Mathematical Model for Improvement of Marine Navigation*, Proceedings of the Third European Simulation Congress, Edinburgh, Scotland, 5-8 September 1989.

Safety Assessment of High Speed Vessel Traffics Using Computer Simulation

M. Numano, K. Okuzumi, J. Fukuto and Y. Murayama
System Engineering Division, Ship Research Institute, Ministry of Transport, 6-38-1, Shinkawa, Mitaka, Tokyo 181, Japan

ABSTRACT

Marine traffic has grown, as has the need for high speed marine transportation. In response to this, various types of high speed vessels have been built. In a congested sea area, the high speed vessel would make a safety maneuver among many high speed vessels and ordinary speed vessels. Safety assessment should therefore be required before the practical use of high speed vessels in the congested area. The safety assessment is considered as rather subjective, because the traffic safety is affected by various factors and contains some ambiguity.

In the present paper, the computer simulation of high speed marine traffic is discussed and the procedure of the safety assessment using the simulation results is proposed. A simulation system for the evaluation of the intelligent ship has been already constructed at the Ship Research Institute. The simulation system is modified to evaluate the high speed vessel traffic. An example of the safety evaluation is shown with simulation results.

INTRODUCTION

Marine traffic for passengers and cargoes is highly congested in a cosmopolite such as Tokyo. The water front is also developed both for business and pleasure traffic. However, recent progress in ship building technology makes it possible to sail high speed vessels.

Standards for high speed vessels are now being investigated in Japan, incorporating standards for ship structure, fire protection and traffic safety.

Above all, the evaluation of high speed vessels as components of the marine traffic is difficult, although the mechanical performance can be easily evaluated. It is necessary to recognise the need for high speed marine traffic to acquire the social acceptance about its safety and availability.

In the present paper, the procedure for the evaluation of high speed vessel traffic using computer simulation is proposed and discussed. The index of the evaluation, which is essential for the evaluation, is also discussed and the safety margin is proposed as the index.

SAFETY ASSESSMENT OF MARINE TRAFFICS

Procedure of assessment
Individuality of each vessel is one of the features of marine traffic to be compared with land traffic. A view point of the vessel operators, such as captains and pilots is, therefore, suitable for the assessment of the marine traffic. From that point of view, a statistical approach which results in a probability of collision and stranding is not sufficient and a case study based upon the properly selected scenarios is useful. A procedure of the safety assessment is proposed as follows:
(1) Selection of the view point for the evaluation,
(2) Listing all the environmental factors and vessel performance ones which affect the marine traffic,
(3) Selection of the scenarios which isolate an effect of each factor,
(4) Selection of indices for the safety margin corresponding to each factor,
(5) Execution of computer simulation according to those scenarios,
(6) Calculation of evaluation indices from the simulation results,
(7) Final assessment by a subjective evaluation of experts with calculated indices.

View point of evaluation
In the evaluation of the traffics, various view points can be considered. To control traffic, its effectiveness and safety are expressed as statistical values. On the other hand, to operate each traffic component, which is one vessel of the marine traffic, effectiveness and safety are expressed as the integration of those in all cases. In marine traffic, an operator such as a captain and a pilot, is responsible for all affairs related to their own ship. The view point of the operator is, therefore, essential for the evaluation of the marine traffic.

Safety margin
Evaluation indices which represent the safety of the evaluated system should be decided. The indices are calculated from some examination results. The safety can be defined as the opposition of the danger. In case of the marine traffic in a highly congested sea area, the danger is represented by the collision danger. The safety of the marine traffic is, therefore, often evaluated by collision probability. It is useful for the statistical approach to evaluate by collision probability.

The safety assessment of the marine traffics, however, requires the evaluation at the present state. Mechanisms of the occurrence of collision danger and keeping navigation safety by avoiding dangers should be cleared. Every system which is properly designed keeps its safety in ordinary conditions. A static performance of the system is evaluated by a document examination. When various environmental factors disturb the system performance, the system compensates the disturbance and keeps its safety dynamically. If each disturbance of the external factors is assumed to be so small that it can be regarded as perturbation, the ability to compensate each disturbance can be evaluated by each respective index, which can be summed up as the total safety margin. Each index has respective value between the lower, necessary, value and the upper, sufficient, value. The total safety margin can therefore be defined as the weighted logical summation of all indices according to the mechanism of the compensation.[2,3]

COMPUTER SIMULATION OF MARINE TRAFFICS

Simulation model
The environmental condition includes a natural condition, a geographic one and a traffic one. The natural condition includes weather, sea state, wind and current. The geographic condition includes water depth, coast lines and remarked objects. The traffic condition includes vessel traffic, regal traffic routes and various vessel traffic services. Each condition has respective accuracy for the evaluation of the collision avoidance which is the chief objective of the marine traffic in a congested sea area.

The vessel condition is also one of the most effective conditions of the evaluation. Each vessel has not only its maneuverability and its outlook, but also its ability for collecting information and decision- making, which is the skill of the vessel operator.

The accuracy of each model is decided according to the simulation scenario. A simple model of maneuverability is sufficient for the safety assessment of marine traffic. In the scenario, some perturbations on some evaluation factors are set and the results of the simulation lead to respective evaluation indices.

Simulation system
To perform the evaluation simulation, a flexible computer simulation system should be constructed. In Ship Research Institute, a computer simulation system "SISANAM (Simulation System for Automatic Ship Navigation and Vessel Traffic Management)" has been constructed for the evaluation of the intelligent ship.[1,4] The simulation system performs a real time simulation of vessel traffics in Tokyo Bay with visual monitors of a sight from a bridge, a RADAR/ARPA (Automatic RADAR Plotting Aids) etc.. The system is explained shortly in the appendix. For the manner in which high speed vessel traffic are treated in the present paper, the simulation system had to be slightly modified. First, the sight from a bridge is improved for the use of the ship operation because the old SISANAM had not an adequate

man-machine-interface for the vessel operation. The added vessel operation device is shown in Fig.1. Secondly,the simulation period,which means a time interval for changing each state variables in the simulation is shortened because more rapid change of the state variables is required to realize a high speed vessel simulation. Although a hard ware restriction such as a cpu speed, a drawing speed etc. prevents the sufficient improvement, some results of the real time simulation for the high speed vessel traffic can be acquired.

Simulation scenario

In the present paper, only a collision danger is considered for simplicity. In fact, the collision danger is the biggest problem in a highly congested sea area. A scenario is considered to be a sequence of the occurrence of the dangerous situation and the recovery of it. The occurence of the dangerous situation is started at an encounter between several vessels. The encounter condition includes an encounter angle, a recognized distance, an approaching speed, types and number of encounter vessels etc.. As the recovery of the collision danger requires adequate information about the encounter, a natural environmental condition which affects the visual information and onboard devices which support the vessel operator collecting the information.

The scenarios are selected to clear the effect of the encounter condition and the information about the encouter.

Simulation results

Some vessel operation simulations are performed to evaluate the effects of the information about the encounter condition on the collision avoidance. An example is shown in Fig.2, where absolute vessel trajectories and time histories of vessel state variables of two cases operated by different persons are shown. In the figure, time taken to recognise a target vessel and that of starting collision avoidance maneuver are indicated with solid lines. Those simulations are a part of the series of an examination which consists of an examination using a radar simulator, that using a night visual simulator and a day visual simulator. In the simulation mentioned in the present paper, our computer simulation system is used as a day visual simulator which means no support of a RADAR/ARPA device.

The simulation results are evaluated both from the view point of the operator of the high speed vessel and the operator of the target vessel which is an ordinary speed vessel.

As the simulations performed coincide with the preparation of the integrated examination planned in 1991, only a qualitative evaluation has been made. The results show that a high speed vessel can clear various encounters easier than an ordinary speed vessel. In spite of the high speed vessel being easy to operate, the evaluation from the view point of the operator of the encountered ordinary speed one is not so good that the ordinary speed one couldn't avoid collision by itself. It is because there is some ambiguity in the behavior of the high speed vessel and a distance of a point of approach is too short for the ordinary speed one.

Indices of the safety margin should represent margin in time and distance in the collision avoidance maneuver from the view point both of a high speed vessel and of an ordinary speed one.

REQUIREMENTS FOR HIGH SPEED VESSELS

A high speed vessel should be designed to clear various navigation environments which include traffic conditions and natural conditions. It has sufficient functions of maneuverability, surveillance and exchange of information as well as an ordinary speed one, which is required because of its own safety as well as sufficient strength of a hull and other onboard facilities.

An operation manual should be also established for a high speed vessel, according to which its navigation causes no menace of other ordinary speed ones and its behavior can be easily understood by other vessel operators, which is required because of the whole traffic safety.

CONCLUDING REMARKS

High speed vessels will be used in marine traffic more popularly than now. Safety assessment of high speed vessel traffic is necessary before their practical use. The safety assessment includes evaluation of collision danger and smoothness of traffics. High speed vessels should not menace or disturb other vessels as well as avoiding collision. A statistical approach is often adopted, where several kinds of probabilities and statistics are estimated. An assessment by the statistical approach is not sufficient because it doesn't include dynamics explicitly in the traffics. It is, therefore, useful to investigate typical cases by a computer simulation.

The investigation by a computer simulation can be modeled as follows:
(1) Environments and functions in vessels are simulated to a sufficient degree,
(2) A behavior of a vessel in collision danger is simulated, where the effects of environmental factors, vessel performance and operators are taken into account,
(3) Scenarios are selected according to which effect of each factor is obviously cleared,
(4) Simulation results are analyzed and respective indices are calculated,
(5) Mechanisms of the occurrence of danger and the recovery of it is cleared and the mechanism for the compensation of each disturbance is also cleared.

Logical summation of all indices according to the mechanisms expresses a total safety margin of the vessel traffic,
(6) Final assessment is made subjectively by experts of vessel operation with the aid of the calculated indices.

As the simulations for the assessment are now being performed, a quantitative evaluation hasn't been obtained yet. As a qualitative evaluation, a safety margin can be regarded as sufficient time from the view point of a high speed vessel operator and sufficient distance from that of an ordinary one.

ACKNOWLEDGMENT

Several research projects about high speed vessel traffics are running in Japan; a research on navigation safety of a high speed vessel under the Ship Building Research Association of Japan and a research on a vessel for Tokyo Water Front under the Ship and Ocean Foundation etc.. We are studying in cooperation with those projects. We are grateful to Prof. Imazu, Tokyo University of Mercantile Marine, and Prof. Hara and Prof. Inoue, Kobe University of Mercantile Marine for their useful suggestion and discussion at the meeting of those committees. We are also grateful to our colleagues for their efforts on constructing and improving the computer simulation system.

REFERENCES

1. Numano, M. Real Time Simulation System for Automatic Ship Navigation, Proceedings of 4th Int. Conf. of Marine Simulation, pp. 350-358, 1987.
2. Numano, M., et al. Automatic Ship Navigation and Safety Evaluation by Computer Simulation, Proceedings of 1st Int. Conf. on Safety in Tokyo, pp. 74-79, 1989.
3. Numano, M., et al. Concepts and Safety evaluation of Intelligent Systems in Future Ship Navigation, Proceedings of Joint Int. Conf. on Marine Simulation and Ship Maneuverability, pp. 209-215, 1990.
4. Kaneko, F., et al. Computer Simulation of Ship Navigation in Realistic Marine Traffic Flow, Proceedings of Joint Int. Conf. on Marine Simulation and Ship Maneuverability, pp. 219-226, 1990.

APPENDIX

Computer simulation system, modified for high speed vessel simulation

<u>Hard ware composition</u> A host computer of the system is a multi- process one with a vessel operation terminal and a RADAR/ARPA monitor. Two graphic work stations are connected through network, which show a sight from a bridge of one of the vessels in the simulation. The sight covers 60 degrees horizontally and 24 degrees vertically. One can operate a vessel by a rudder angle and a vessel velocity order using a steering wheel and a lever, respectively. The composition of the simulation system is illustrated in Fig.3.

<u>Soft ware characteristics</u> A simulation is governed by a set of data called a simulation scenario. The scenario decides various environmental conditions and vessel performances etc. by constructing respective data and using respective processes. A simulated world is expressed as state variables in a global memory section which is renewed by respective processes at respective intervals. Processes included in the simulation perform respective roles such as realizing vessel motions etc. and are initiated and driven at different time intervals according to the class of each process. In this simulation system, data and program are completely separated. An object oriented construction of the simulation can be, therefore, attained using a simulation scenario and a set of processes, which enables a flexible simulation. The processes in the simulation are classified as shown in Fig.4.

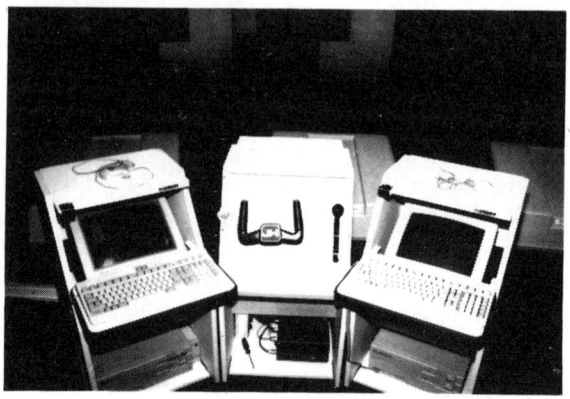

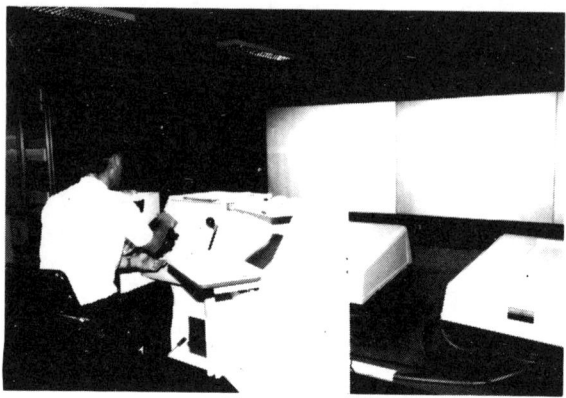

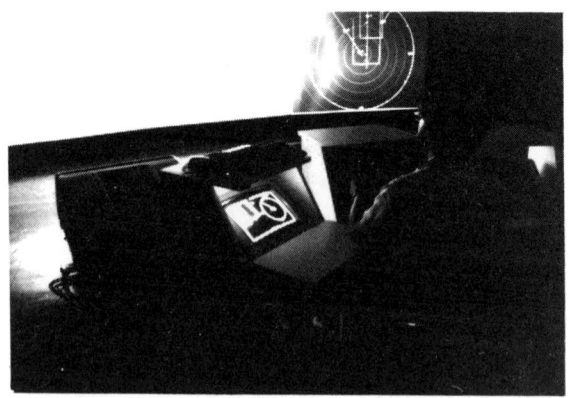

Fig. 1   Photographs of vessel operation
devices

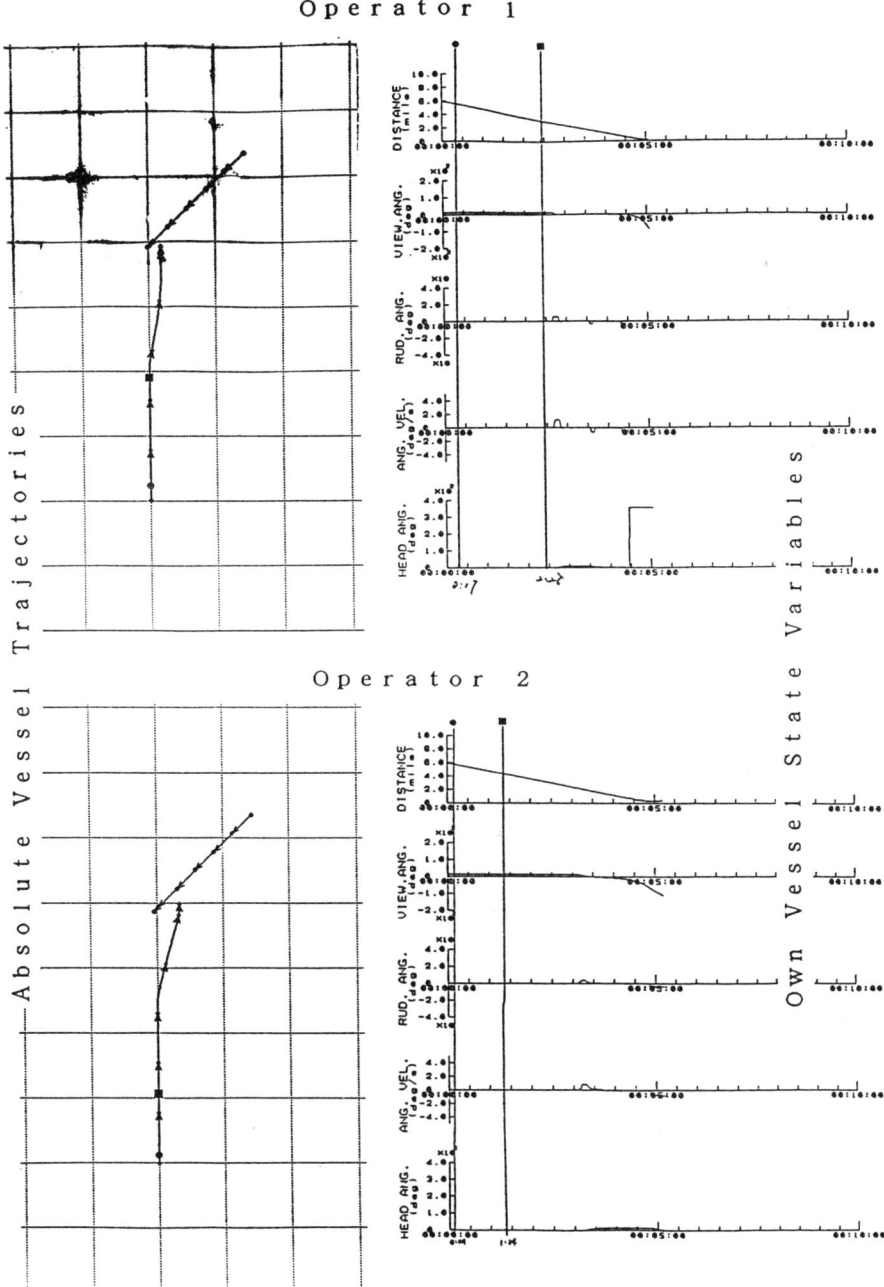

Fig. 2   Example of computer simulation

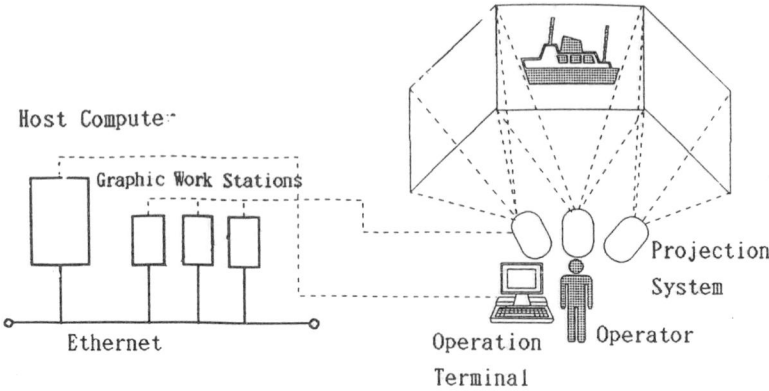

Fig. 3   Illustration of composition
          of SISANAM

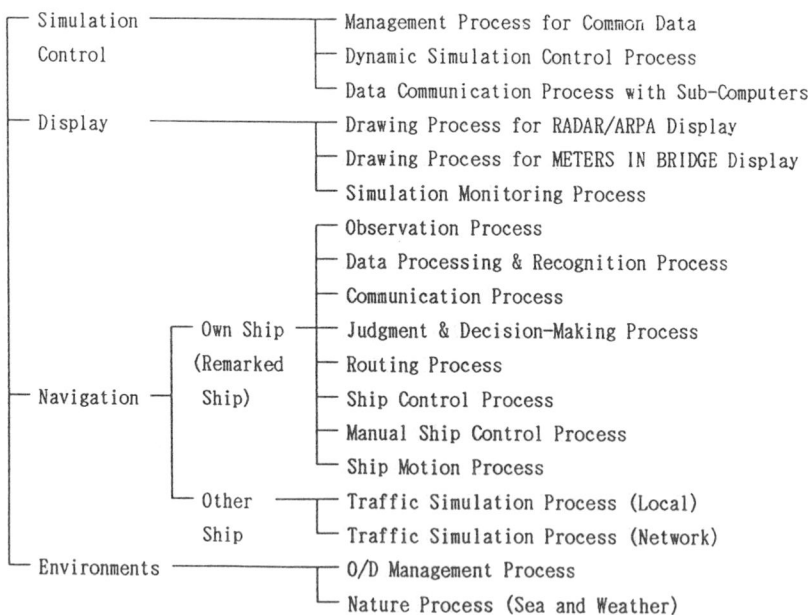

Fig. 4   Classification of processes
          in SISANAM

SECTION 6: SHIP PROPULSION

Numerical Prediction of Effects of Hull Variation on Propeller Performance

D. Hally and D.J. Noble

Defence Research Establishment Atlantic, Dartmouth, N.S., Canada

ABSTRACT

This paper will describe a system for analysing propeller performance given only the description of the geometry of the ship hull and its propellers. The calculation is broken into three distinct stages: representation of the hull, calculation of the nominal flow into the propeller plane, and analysis of the propeller performance.

The hull representation used allows simple modifications of the hull geometry to be made so that the system provides an excellent tool for predicting the effects of hull geometry on propeller performance. As an example of the utility of the system, results of numerical experiments to show the effects of hull geometry on back sheet cavitation will be described.

INTRODUCTION

For the past decade, the Hydronautics Section at the Defence Research Establishment Atlantic (DREA) has been developing computer codes with the purpose of predicting propeller performance; these include codes for describing the ship and propeller geometry, calculating the flow about the hull (the nominal wake), and predicting various measures of propeller performance such as thrust, efficiency, cavitation extent, and cavitation inception speeds. Only recently have these been integrated into a single package so that propeller performance can be predicted given only offsets for ship and propeller. In this paper the system will be described with the focus on prediction of sheet cavitation.

Cavitation forms on the surface of marine propellers at points where the local flow velocity is sufficiently high to cause the local pressure to drop below the saturation vapour pressure of sea water at the ambient temperature. As a propeller blade rotates through a non-uniform vessel wake, the blade surface pressures can vary considerably over portions of a revolution, causing a rapid growth and collapse of the cavitation bubble volumes that are formed on the blades.

Naval propeller designers are concerned with reducing these cavitation volume fluctuations which can generate propeller blade erosion, vibration of nearby struts, rudders and hull panels, as well as increases in the levels of radiated noise. Rapid variations of cavitation volumes are encouraged by increased non-uniformity of the flow in the wake deficit region behind ship hulls, particularly with the increased propeller rotation speeds and reduced propeller tip immersion depths that have been common practice in the design of high speed surface ships.

The accuracy of the DREA cavitation prediction codes will be illustrated using predictions of the extent of sheet cavitation on the propellers of the DREA research vessel CFAV Quest. Quest is an ideal ship for this purpose as measurements of her wake and of the cavitation on her propellers have been made at both model- and full-scale.

The addition of approximately five metres of parallel mid-body (PMB) has been proposed as part of Quest's scheduled mid-life refit. How will this affect the cavitation on her propellers? Because the method of hull representation allows simple modifications of hull geometry to be made, the DREA package of propeller codes provides an excellent tool for answering such questions. A series of numerical experiments have been performed which predict that the inclusion of PMB will cause a significant increase in cavitation.

HULL REPRESENTATION

The hull geometry is defined using the hull representation system HLLSRF[1] developed at DREA. Unlike most hull representation systems which are intended for ship design and fairing, HLLSRF was designed specifically for use in hydrodynamic calculations. Using tensor product B-splines, it provides a regular coordinate system covering the whole hull. The coordinate system provides the means for performing a wide variety of hydrodynamic calculations, from simple evaluation of sectional properties for strip theory, to solution of partial differential equations describing the viscous flow around the hull. HLLSRF has been designed to maintain the regularity of this coordinate system so that numerical errors may be avoided. To facilitate flow calculations, an additional regular coordinate system is provided which covers only the portion of the hull below the waterline.

A useful feature of the hull representation scheme is that variations of existing hulls may be represented by specifying how they differ from their parent. For example, the length of the hull could be changed simply by specifying the new length in a modification file. Similarly, the position of the propellers can be altered, or the waterline can be raised and lowered. Using a method similar to that of Lackenby[2,3] in which the hull sections are displaced fore or aft, the form of the hull can also be modified to change its block coefficients or its longitudinal centre of buoyancy. A special transformation of this sort allows the inclusion of parallel mid-body.

FLOW CALCULATIONS

The nominal wake is calculated using the suite of programs HLLFLO, also developed at DREA. Using a HLLSRF representation of a ship, HLLFLO calculates the potential flow around it using a constant source panel method[4]. The control points are placed on hull sections so that the velocity field at the hull surface can be interpolated using B-splines. The potential flow may then be evaluated anywhere on the hull so that subsequent flow calculations are independent of the choice of hull panelling.

With the potential flow known, the flow in the boundary layer is determined using the program BLAYER[5], which uses an integral method based on the momentum integral equations and Head's entrainment equation. A choice of velocity profiles is allowed: a power law profile with Mager's cross stream assumption, or a three-dimensional version of Coles' profile. The boundary layer equations are solved in the coordinate system provided by the HLLSRF hull representation. An explicit marching procedure with weighted upstream differencing is used; the Courant-Freidrichs-Lewy condition is strictly enforced to preserve stability. Typically 15 to 20 cross-stream points are used and 50 to 70 streamwise points.

While it is well-documented that solutions to the boundary layer equations tend to overpredict the velocity deficit at the stern, boundary layer solutions perform quite well for naval ships such as destroyers, especially at full-scale Reynolds number. When it is propeller performance that is of interest, it is also a great advantage that the propellers on these ships are somewhat upstream of the stern itself. The principal advantage of the boundary layer methods over more sophisticated solvers is their extreme efficiency.

In Figure 1, BLAYER predictions of the axial velocities in the nominal wake at Quest's propeller plane are compared with data taken at both model-scale[6] (Reynolds number based on ship length, $R_e = 1.7 \times 10^7$) and full-scale[7] ($R_e = 5.6 \times 10^8$). The model scale velocities were measured with a pitot tube behind a towed ship model without propellers fitted. The full-scale velocities were measured with an LDV system, slightly upstream of the operating propellers; therefore, the full-scale wake includes the effect of propeller induction. The measurements at both scales include the wake from the propeller shaft bossing; it appears as a narrow region of large velocity defect near one o'clock on the propeller disk. At model-scale the wake is somewhat overpredicted by BLAYER. At full-scale, the contours near twelve o'clock, outside the influence of the propeller shaft bossing, agree well with the predictions. Although not shown in Figure 1, at both scales, the tangential velocity components are also predicted well in regions outside the influence of the wake of the shaft bossing.

It is worth noting that Quest is not nearly as slender as a destroyer, so that the numerical predictions of the propeller wake should be somewhat better for destroyers than for Quest.

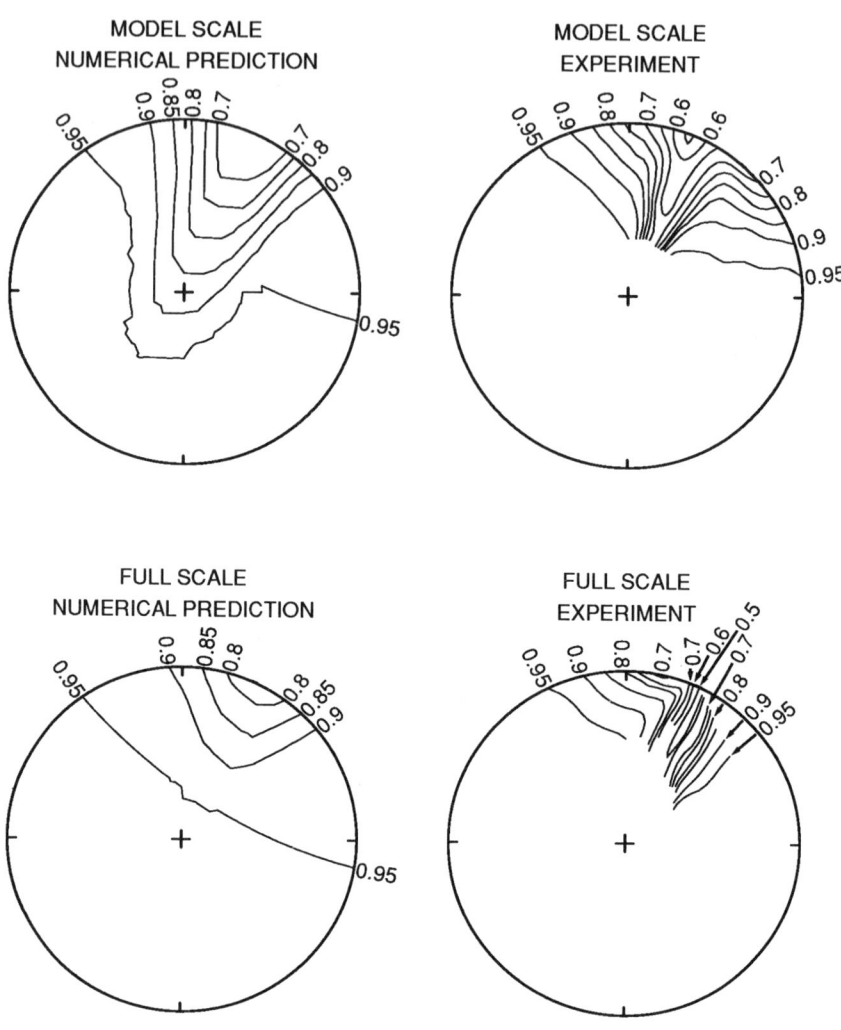

Figure 1: *Comparison of predicted and measured axial velocity contours in the propeller plane of CFAV Quest (port side looking forward). The contours are normalized using the ship speed.*

PROPELLER CALCULATIONS

The development of sheet cavitation on the back or upstream side of the propeller blade is calculated by the DREA program CAVITY,[8] which is based on a lifting surface procedure developed by Lee[9] at the Massachusetts Institute of Technology. Lee's method is a direct extension of a combined vortex lattice and thickness source segment singularity procedure developed at MIT by Kerwin and Lee[10] for prediction of the net unsteady hydrodynamic forces and moments on marine propeller blades.

Sheet cavitation is a particular type of vapour bubble that develops from the blade leading edge, and has the appearance of a smooth continuous layer of vapour. Except for clusters of small bubbles or clouds of vapour that have been observed to form and occasionally detach from the blade at the trailing end of the sheet cavity, the main sheet cavity volume remains attached to the blade as it rotates. Assuming the mean distribution of vessel wake velocity over the propeller plane does not change substantially over several propeller revolutions, this sheet cavity extent and volume will have a mean periodic variation for a constant propeller rotation rate.

To obtain a numerical solution to this problem, the propeller blades and their downstream wakes are represented by a connecting grid of discrete straight line vortex segments located on the blade and wake mean surfaces. The method does not include modelling of the propeller hub or the blade root fillets which fair the blade surfaces into the hub. Vortices are used to induce a circulatory potential flow both around the blades and in their wakes. Discrete line source segments are also used to represent the potential flow contributions of blade and cavitation thickness distributions to the propeller induced flow. The blade thickness source strengths are calculated at the outset from a specified blade section thickness distribution. The strengths of the unknown line vortex and cavity thickness line source segments are determined by satisfying appropriate boundary conditions and conservation laws at discrete time steps as the blades rotate.

The singularity strengths are the solution of a system of linear equations. At a set of discrete control points on the mean blade surface, the equations enforce flow tangency conditions and ensure that the pressure in the fluid adjacent to the cavity is equal to the vapour pressure within the cavity. The net flow is determined as the sum of the induced flow from the singularities and the nominal wake velocities. The fluid pressure along the cavity extent is determined using a form of the Bernoulli equation that is applicable to unsteady flow in blade-fixed coordinates. A cavity closure condition also enforces vanishing thickness at the edges of the cavity volume. The linear equations are formulated in such a way as to simultaneously satisfy mass, vortex and circulation conservation laws.

Starting from a initial fully-wetted or non-cavitating blade solution for vortex and blade thickness source strengths alone, the sheet cavity chordwise extents, cavity thickness source strengths and modified vortex strengths are determined by a trial and error iteration scheme in which trial extents are introduced along individual chordwise strips from the leading edge of the blade. For each fixed blade position, the initial extents are increased or decreased in discrete intervals within each strip until the pressure in the flow adjacent to the cavity equals the specified vapour pressure at a set of control points defined within the cavity extent. During the iteration process, the constant pressure adjacent to the assumed cavity extent is treated as an unknown to be determined along with the vortex and cavity thickness source strengths in the same strip. If the cavity extent needs to be reduced to the leading edge before the pressure adjacent to the cavity drops below the vapour pressure, then that particular chordwise strip of the blade is considered cavitation free. The blade is swept by solving adjacent chordwise strips from hub to tip until a converged cavity is obtained at a given blade position. The cavity thickness distribution is then obtained directly from the cavity thickness source strengths that are calculated within each strip.

Reference 8 discusses some modifications made at DREA to the distribution of singularities on the blades and propeller wake, in order to improve predictions of sheet cavitation development over a wide range of propeller operating conditions. Kennedy, Noble and Casgrain[11] have correlated propulsion and cavitation performance data obtained with DREA lifting surface codes with experimental data for a variety of propeller geometries.

For a given propeller geometry, r.p.m., vessel speed and mean vessel wake distribution over the propeller rotation plane, this numerical method provides sheet cavitation chordwise extents and thickness distributions at discrete blade positions or time intervals over one revolution. Cavity thickness, measured normal to the blade surface, is defined on discrete fixed-radius sections of a blade, and is calculated at discrete points extending from the blade leading edge to the cavity trailing edge along section chordlines. These chordlines lie along fixed pitch helices at each radius, and appear as circular arcs in a projected blade view looking normal to the rotation plane. To obtain a periodic time history of total cavitation volume, cavity thickness distributions are integrated along the chordlines to obtain cavity cross-sectional areas, which are then integrated radially to obtain the overall volume at each blade position.

Figure 2 compares cavitation extents measured on Quest's propellers at model scale, with predictions from CAVITY using the wake at the top left of Figure 1. The cavitation is shown at six different rotation angles of the blade. The excellent agreement is undoubtedly somewhat fortuitous, since the overprediction of the wake deficit is compensated by the lack of the extra velocity defect in the the wake of the propeller shaft bossing. Nevertheless, such agreement is extremely encouraging.

NUMERICAL EXPERIMENTS

In order to determine the effects on cavitation production of the addition of parallel mid-body to Quest, a series of numerical experiments were performed. In the first, different amounts of PMB were added and the extent of cavitation

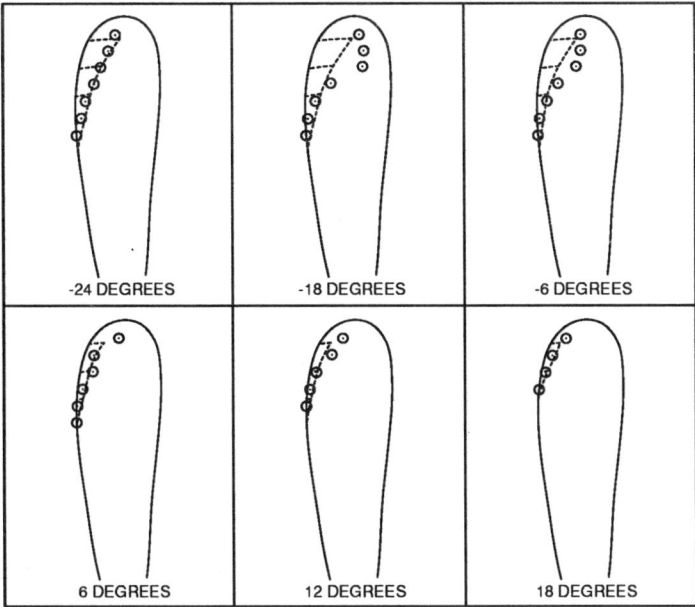

Figure 2: *The extent of sheet cavitation at model scale: the circles show the extent of cavitation measured, the dotted lines the extent predicted. Zero degrees is vertically upwards and the angles increase counterclockwise relative to Figure 1.*

determined. In the second, Quest was simply stretched, by increasing station spacing proportionately, to determine whether the changes caused by the PMB were primarily due to the increased length. An addition of PMB causes an increase in the fore and aft block coefficients. In a third series of tests, the form of Quest's hull was modified by increasing her fore and aft block coefficients to equal those achieved by addition of PMB. The hull modification used the methods described in Reference 3 as implemented in HLLSRF. Figure 3 shows the curve of sectional areas for the variants of Quest increased in length by 10% for each of the three series of tests.

The calculations were run at full-scale. In all cases, the propeller orientation remained invariant. In Series 3, when the block coefficients were altered by moving ship sections, the propellers were moved along with the sections. This retained, as closely as possible, the relative orientation of the propellers and the hull: e.g. the propeller/hull clearance remained the same.

The propellers were assumed to operate at the same advance coefficient in all cases; from the perspective of the program CAVITY, the only input which changed was the definition of the nominal wake. While this is not realistic (e.g. for fixed r.p.m. the advance coefficient would decrease due to the increased resistance of a longer ship), it best illustrates the effects of the nominal wake

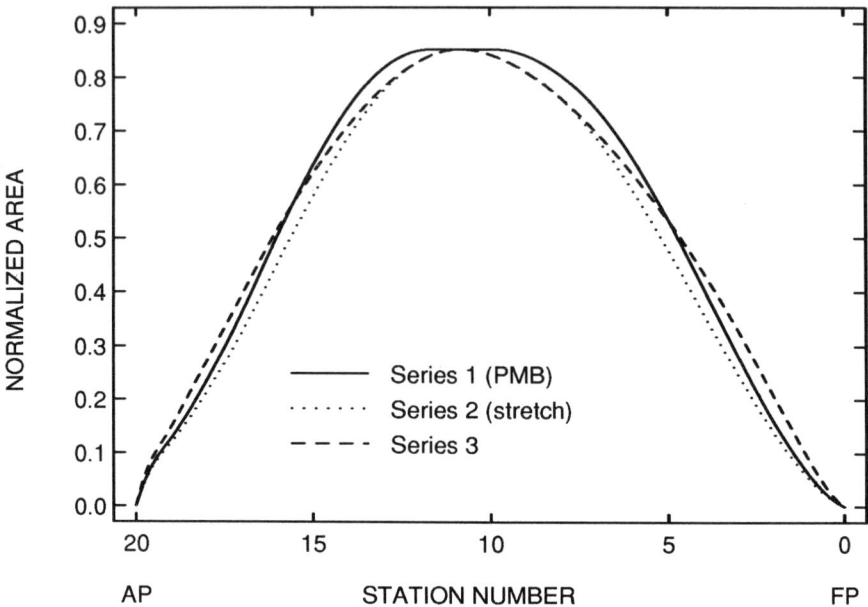

Figure 3: *Sectional area curves for three variants of Quest, each of length 78.8m.*

on the cavitation.

Care was taken to avoid numerical artefacts of the discretization schemes in the different programs. In the potential flow calculations, the same panelling scheme was used on each hull. When parallel mid-body was introduced, the panelling over the forebody and the stern remained the same, while additional panels were introduced to cover the PMB. The same step sizes were used in all boundary layer integrations.

For each run, two quantities were used as measures of the amount of cavitation: the maximum volume of the cavity over a propeller revolution; and the maximum chordwise extent of the cavity at 94% of the propeller tip radius. Figures 4 and 5 summarize the results of the tests; the maximum cavity volume and maximum chordwise extent are each plotted against the hull length. The three curves on each figure represent the three series of hull variants. The common point on the curves, at a ship length of 71.6m, is the unmodified Quest hull.

The predicted cavitation increased slightly as the hull was lengthened (Series 2) due to the small thickening of the boundary layer with ship length; however, this was partially offset by less variation in the potential flow because the longer ship is relatively more slender.

In contrast, the effect of including parallel mid-body (Series 1) was much

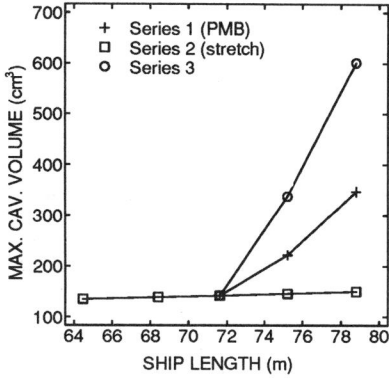
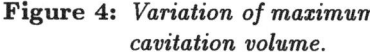

Figure 4: *Variation of maximum cavitation volume.*

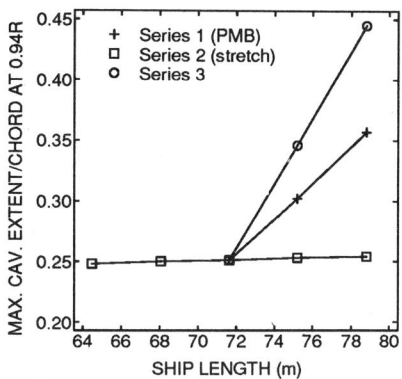

Figure 5: *Variation of maximum cavitation extent at 94% radius.*

larger. Two effects combined to increase the boundary layer wake defect considerably, causing a significant increase in the cavitation levels. Firstly, the increased length causes an increase in boundary layer thickness. Secondly, the parallel mid-body causes a significant augmentation of the maximum speed of the potential flow near the end of the PMB, with a corresponding increase in the adverse pressure gradient over the stern. The result is a rapid increase in the thickness of the boundary layer as the extent of PMB is increased.

The results of Series 3 showed an even greater increase in cavitation with ship length. This can be traced to three separate effects. First, the adverse pressure gradients are even more severe near the stern. Second, the propellers have been shifted further aft to maintain their relative placement with respect to the hull. Third, the boundary layer peak shifted towards the centreplane; this moved the region of velocity defect away from the hub toward the tip where it has a greater effect on sheet cavitation production.

These calculations show clearly that the cavitation is very sensitive to changes in block coefficient, whether by inclusion of PMB or by other means. The increase in cavitation after Quest's refit should be easily measurable, and will provide a good test of the accuracy of the system at full-scale.

FUTURE WORK

Development of the system is continuing on a number of fronts. A significant shortcoming of the current system is that it does not account for the presence of the free surface. This will be remedied shortly.

While the boundary layer methods for predicting the viscous flow work quite well for naval ships, more sophisticated methods are required for greater accuracy and generality. The possible use of Navier-Stokes solvers for the flow near the stern is being investigated. While these would slow the calculations considerably, they allow the inclusion of complicated effects like the wake of the propeller shaft bossing or extension of use to fuller merchant ship forms. For the propellers themselves, use of Navier-Stokes solvers might lead to improved predictions of tip vortex strengths and corresponding pressures, together with the associated cavitation.

The cavitation prediction code is currently being improved by distributing singularities over panels on the blade surfaces and propeller hub, rather than on the mean surface alone. This method will allow prediction of sheet cavitation developing on both sides of the blade simultaneously, as well as predictions of cavitation near the blade root and hub.

More work needs to be done on the integration of the programs to reduce redundant input, and to make the programs truly independent of external input. For example, it should be possible to use hull resistance estimates provided by the flow codes to determine the operating advance coefficient for the propeller. Currently, this is not done.

Currently, when run on a VAX 6600, one run through the full system takes about ten minutes in real time. This performance will degrade significantly when free surface effects are added and even more if sophisticated viscous flow solvers are introduced.

CONCLUDING REMARKS

A system for analysing propeller performance given only the description of the geometry of the ship hull and its propellers has been described, and its performance illustrated using experimental data from CFAV Quest. The system is fast, and works well for naval ships for which the boundary layer method for calculating the viscous flow is sufficient. Due to the ability of HLLSRF to modify hulls, the system is particularly useful for examining the effects of hull variation on propeller performance.

REFERENCES

1. D. Hally, "The HLLSRF Hull Representation System," DREA Report 89/102, 1989.

2. H. Lackenby, "On the Systematic Geometrical Variation of Ship Forms," *Trans. INA* **92**, 289 (1950).

3. D. Hally, "On the Systematic Variation of Hull Representations for Computers," *Trans. RINA* **130**, 77–83 (1988).

4. D. Hally, "POTFLO: A Suite of Programs for Calculating Potential Flow about Ship Hulls," DREA Technical Memorandum 89/210, 1989.

5. D. Hally, "An Integral Method for the Calculation of Boundary Layer Growth on a Ship Hull," DREA Report 85/107, 1985.

6. J. Holtrop and J. Th. Ligtelijn, "Performance test and wake survey for the research vessel Quest; Configuration I," Report 02870-1-VT, NSMB, 1979.

7. C. Norris and N. Leeming, "Full-scale wake measurements CFAV Quest," Report W1106, BSRA Technical Services, 1984.

8. D. J. Noble, "Numerical Prediction of Unsteady Sheet Cavitation on Marine Propellers," DREA Technical Memorandum 87/203, 1987.

9. C. S. Lee, "Prediction of Transient Cavitation on Marine Propellers by Numerical Lifting-Surface Theory," in *Proceedings of the* 13th *Symposium on Naval Hydrodynamics* (Tokyo, Japan, 1980).

10. J. E. Kerwin and C. S. Lee, "Prediction of Steady and Unsteady Marine Propeller Performance by Numerical Lifting-Surface Theory," *Trans. SNAME* **86** (1978).

11. J. L. Kennedy, D. J. Noble, and C. M. Casgrain, "Evaluation of Some Propeller Analysis Methods," in *Proceedings of the 21st American Towing Tank Conference* (Washington, D.C., 1986).

Model Propeller Measurement Using a 3D Laser Scanner

T. Randell

National Research Council, Institute for Marine Dynamics, St. John's, Newfoundland, Canada

ABSTRACT

In this paper a method for measuring the accuracy of small model propellers using a 3D laser scanner is described. The system uses computer processing to extract propeller section shape from the laser scan images. From these sections detailed measures of a number of design components can be obtained. An example is presented and some results obtained for that propeller are given.

INTRODUCTION

An important part of any manufacturing process is the need to determine if the part produced is within the required tolerances. Depending upon the type of part being measured, as well as the type of measurements required, this task can range from relatively simple to very complex. The marine propeller is an example of a complex part which is often difficult to measure. This difficulty is mainly because of the propellers highly contoured surfaces combined with a design method which is somewhat complicated. Traditionally propellers were measured using a device known as a pitchometer. This mechanical device measures the depth from a reference plane to the blade surface at a fixed radius and angle which are set by the operator. Figure 1 illustrates such a device. To use this device the design data for the propeller must be converted to a table of radius, angle and depth settings. The operator adjusts the machine to the settings in the table for radius and angle and then takes a depth reading. The depth reading can then be directly compared to the design depth value from the table. There are several problems with this method of propeller measurement. The method only allows blade offsets to be measured and does not provide a good means for measuring the accuracy of the various geometric components of the design independently. The number of blade offsets measured must usually be kept to a small number for each blade section. Usually only points that correspond exactly to design data offsets are measured. The radius of the measuring tool tip must be compensated for by some means since this is a contact measurement.

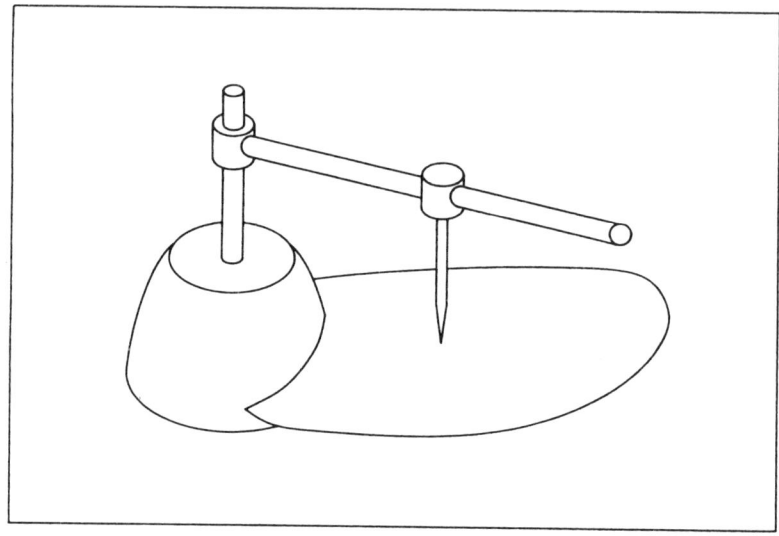

Pitchometer
Figure 1

A more recent method used to measure small propellers involves using a coordinate measuring machine, or CMM. The CMM is programmed to take a series of measurements on the blade surface. The measurements are then processed and converted to numbers which can be compared to the design data. This method can have the advantage of requiring minimum operator intervention during measurement, depending upon the sophistication of the CMM and the program which is driving it. Often the original design data must be used to develop the CMM program so that the required measurements can be based on some knowledge of the shape to be measured. Further processing of the measured data is also required to allow for compensation for the measuring probe tip and data extraction.

The purpose of this project was to investigate the feasibility of another method for measuring the accuracy of the model propellers used at the Institute for Marine Dynamics. This method involved using a laser scanner to scan a model propeller. A number of programs were developed to process the scan data and produce propeller measurements that could be compared directly to the design data. This method yielded an extremely large number of measurements over the blade surface in a very short time without any contact with the actual propeller. The large amount of data provided a very accurate model of the propellers surface which was processed by computer to extract measurements which corresponded directly to the various design elements. The 3D laser scanner which was used is located at NRC's Division of Electrical Engineering in Ottawa. No modifications were made to the scanner in order to perform the measurements. The operator requirements for scanning the blade were kept as simple as possible and all the data processing was performed using computers at the Institute for Marine Dynamics in St. John's.

BLADE DESIGN GEOMETRY

In general terms the geometry of a marine propeller blade is defined by using a series of radial spaced sections. The propeller blade has two sides, the pressure side, which is referred to as the face of the blade and the suction side, which is referred to as the back of the blade. A blade section typically has an airfoil type shape and is defined by the face and back curves and the leading and trailing edges. The face and back curves are usually defined by offsets that are distributed over the length of the curve and are measured with respect to the pitch datum line. The pitch datum line often corresponds to the line defining the face of the blade where the face is a straight line. The offsets are distributed in a nonuniform manner so that more definition is provided near the edges where there is greatest curvature. The offsets are measured from the point of maximum thickness on the section such that the same number of offsets define the section shape from the point of maximum thickness to the leading edge as do to the trailing edge. The shape and radius of the leading and trailing edges for each section are also generally provided. There are typically 8 to 10 sections used to define the shape of the blade which spaced at regular intervals along the radius of the blade. the spacing is expressed as a percentage of the blade radius and the sections are more closely spaced near the tip of the blade where the greatest change in blade shape occurs and greater detail is required. Figure 2 (c) shows a plot of 10 blade sections used to define a propeller blade.

After defining the sections shape the blade itself is described by locating each section in space with respect to the other sections. The sections are located in terms of three parameters, skew, rake and pitch.

Skew

The skew of a propeller blade is define by a line passing through the points of maximum thickness for each blade section and ending at the tip of the propeller. This line, referred to as the skew line, is measured with respect to the generator line. The generator line is a straight line which starts a the centre of the propeller hub and is perpendicular to the axis of rotation of the propeller. The skew of a particular blade section is the distance of the point of maximum thickness for the section from the generator line. In defining a blade the skew is applied by shifting each of the sections by this distance with respect to the generator line. Figure 2 (c) shows a plot of the sections for a propeller with the skew line show and each of the sections offset by the required amount from the generator line.

Rake

The rake of a propeller blade refers to the offset of each section in the direction which is perpendicular to the pitch datum line and the generator line. The rake is often expressed as an angle between the skew line of the propeller and the generator line when viewed in the plane formed by the generator line and the axis of rotation of the propeller. Figure 2 (b) shows the rake as applied to a propeller blade. To apply rake the sections are offset by an amount DY where

$$DY = Radius \times Tan\ (\ Theta\ )$$

Radius = Radius for section
Theta = Rake angle for propeller blade

Pitch

The pitch of a propeller blade refers to the "twist of the blade". The pitch is defined by the helical path that the blade section must follow when the propeller is rotating and advancing. The pitch of a section is the distance which the blade section would advance in one rotation of the propeller, assuming no slippage. For each blade section the pitch can be converted to a pitch angle. that angle, Phi, is defined as

$$Tan(Phi) = Pitch/(2 \times Pi \times Radius)$$

The pitch for a section is applied by rotating the section by the pitch angle about the point defined by the intersection of the pitch datum line and the generator line. Figure 2 (a) shows a blade with the pitch applied to the sections.

After the Skew, rake and pitch have been applied to the sections the final step to completing the blades form is to "wrap" the blade sections. The sections of the propeller are defined in their developed form as shown in figure 2 (c). In reality, however, the sections are not planar but rather line on the surface of the cylinder defined by the radius of the propeller at the section and the axis of rotation of the propeller. Figure 2 (a) illustrates how the sections appear as arcs when viewed looking down the axis of rotation of a propeller.

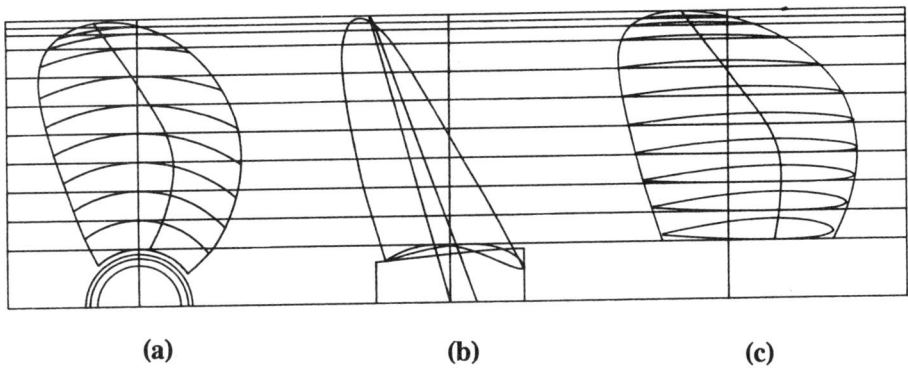

(a)   (b)   (c)

Propeller Design Diagram
Figure 2

Having defined the basic elements of the propellers geometry the objective of scanning the propeller was to be able to obtain measured values for each component of the design geometry. Buy this means not only can a measure of the overall accuracy of the propeller be obtained but the source of the error can be isolated as being an error in section shape, skew, rake, or pitch. By processing the scan image data the blade section shape at any radial position on the propeller can be obtained. By extracting the sections at the same radial locations as used in the design direct comparisons can be made between these sections and the design data.

3D LASER SCANNER

The 3D Laser scanner which was used to measure the propeller model was developed at NRC's Division of Electrical Engineering. The device, shown in figure 3, uses a Synchronized scanning technique which allows the position of the mirror which projects the laser beam onto the object being measured to be constantly aligned or "synchronized" with the mirror which detects the reflected beam during a scan. In order to scan an entire object the scanner is indexed along a axis which is octagonal to the scan direction for each successive scan. The result is a full three dimensional image of the object being scanned.

3D Laser Scanner
Figure 3

There are several inherent difficulties related to the use of this device. The scanner determines the depth of the object being measured at any instant by determining the location of the reflected beam on a CCD. Since the CCD has a fixed number of elements (pixels) along its length there is a direct relationship established between the depth of measurement required and the resolution of the measurements. A smaller depth of field results in a greater resolution since the fixed number of CCD elements maps to a smaller distance on the measured object. An additional problem with the scanner results from the fact that the emitted and reflected beams form an angle where they meet on the object being measured. This angle is reduced to a minimum by the compact design of the laser scanner head but there are still certain conditions where the shape of the surface causes the reflected light to be blocked from the detector mirror. This condition is referred to as a shadowing effect and usually occurs where there is large slope in the measured surface, such as at edges. Because of this certain areas of an object cannot be measured from a single view position. Finally it should be noted that the scanner must be calibrated prior to measurement in order to ensure accuracy of the measurements. This process involves scanning objects of well known physical dimensions and comparing the output of the scanner to the known values. From this comparison a "calibration table" can be developed which applies the required correction factors to the lasers as measured values. Because of this the laser scanner is limited in its absolute measurement abilities by the resolution of the calibration procedure.

The data file produced as output by the scanner system contains the scan lines in a specific format and is accompanied by a configuration data file which describes the setup for the scan. The data has been processed by the actual scanner system to the extent that any geometric conversions and corrections such as calibration have been applied to the data. It is still required that the data be further processed so that meaningful measurements can be extracted from it.

DATA PROCESSING

The propeller which was scanned for measurement is show in figure 4. The overall diameter of this three bladed propeller is 343 mm. The scanning of the propeller was performed with a scan interval of 0.75 mm. The resolution of the scanner is 0.001 mm with an after calibration accuracy of 0.025 mm. Once both sides of the propeller had been scanned there were a number of data processing steps required before the propeller sections could be extracted and compared to the design sections.

Example Propeller
Figure 4

Initially some filtering of the data is required in order to remove any noise components which appear as "glitches" in the data. This filtering was performed using a median filtering technique. The median filter has the an important property in that it preserves edge detail which would otherwise be lost using a typical averaging filter. The median filter sorts the values found within the given window around the point being processed and determines the median value within the window. That value is then assigned to that point. By this method any point which has a value which is not similar to surrounding points will be "removed" since its value would be at or near the start or end of any sorted set of neighbouring points. There are some conditions in which this technique can produce errors. In the general case of a propeller, where the component surfaces being measured are smooth and have a high degree of continuity , the method works well. Figure 5 is a plot of the scan data for the example propeller used in this paper. The data has been filtered by a median filtering program which performed five passes through the data with a window size of five.

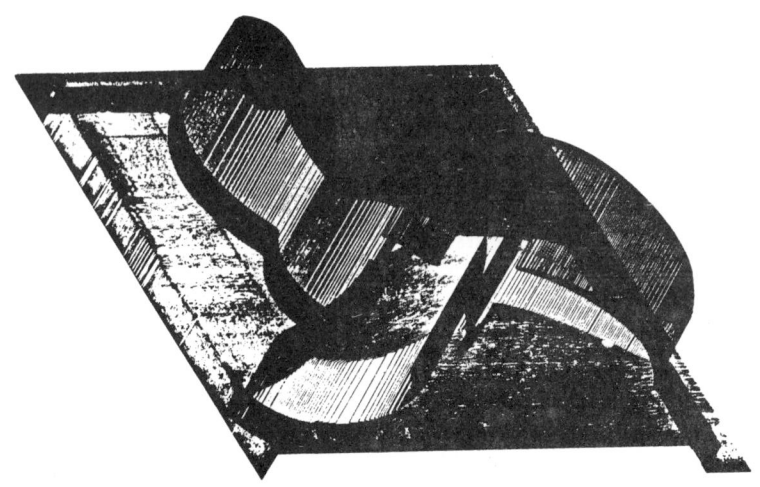

Scan Image of Propeller (Back Side)
Figure 5

Once the data has been filtered the next procedure is to examine a scan of the empty fixture which held the propeller. The fixture device, as illustrated in Figure 6, has a large flat base which is aligned as flat as possible before the propeller is placed in the fixture and scanned. In order to be able to compensate for any small tilt in the fixture the base is scanned without the propeller mounted. By processing the scan data for the base any small angular tilt can be measured and a compensation for the tilt can be applied to the propellers scan data. A computer program examines the four corners of the fixture base and determines the orientation of the normal vector to the base. two angles, one along the scan path and the other octagonal to it are computed. The scan data for the propellers is then "levelled" to compensate for these angles. While this technique adjusts the values of individual scan points it does not adjust their actual locations by distorting the scan grid. Since the degree of adjustment is very small and the overall height of the object from the base is not too large ignoring this distorting effect does not introduce any significant error. It can be seen, however, that it is very important to assure that the fixture base is as flat as possible prior to scanning the actual propellers. The tilt angles measured for the example used in this paper were 0.021 and 0.025 degrees for the angles along the scan path and octagonal to the path respectively. These angles result in a maximum height adjustment of 0.075 mm at the extreme blade tip locations and would require a maximum lateral shift of 0.038 mm at the extreme height of the propeller for data correction

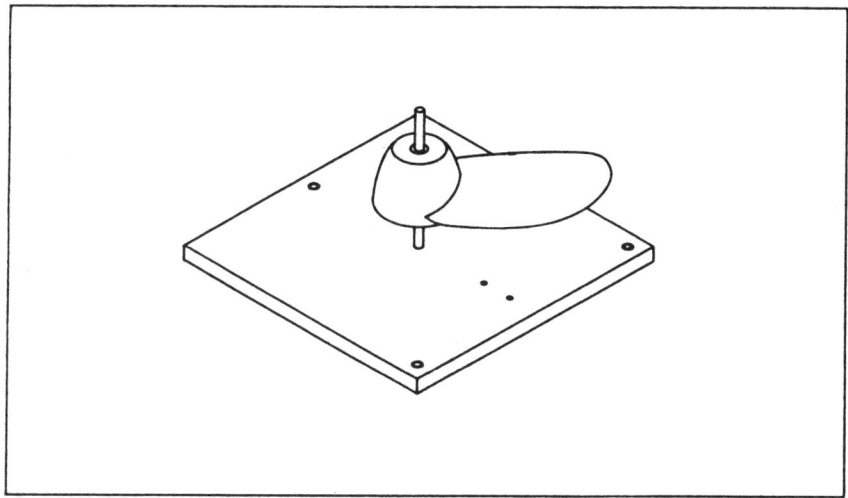

Propeller Scanning Fixture
Figure 6

In order to extract the individual blade sections for the propeller from the scan data it is necessary to convert the data to a cylindrical coordinate system with its origin on the centerline of the propellers rotation axis. The location of the origin must be determined as accurately as possible in order to ensure accuracy of the results. This task involved locating the shaft of the fixture arbour in the scan data and calculating the centre of the arbour. This was done by finding the scan line with the greatest cross section width through the arbour shaft and then by interpolating between this scan section and the surrounding scan sections to find the centre. There is an inherent problem with this technique in that the scan interval spacing and the diameter of the arbour determines the accuracy to which the centre of the arbour can be located. With a scan interval of 0.75 mm the centre of a cylinder of diameter 34.1 mm can be found to an accuracy of approximately 0.030 mm. A more accurate method for locating the centre of the propeller hub would be desirable and a new technique which has been proposed for doing this will be described in the conclusion of this paper.

After the centre of the propeller has been determined the next processing step is to interpolate the cylindrical sections from the scan data. A program interpolates these cylindrical sections based on several input from the user. The user specifies the centre of the cylindrical coordinate system, the radial interval at which the sections are to be interpolated and the desired point spacing along the sections. The program interpolates the sections and writes the data to a file. Figure 7 shows a plot of the cylindrical data interpolated at a radius interval of 8.575 mm and a point spacing of 0.75 mm. The data points in Figure 7 have been interpolated at the same locations as the original propeller design sections. It is these sections which are used for comparison and accuracy measurement. The sections are interpolated in opposite directions for the face and back scan data since the coordinate system with respect to the propeller for each scan are inverted.

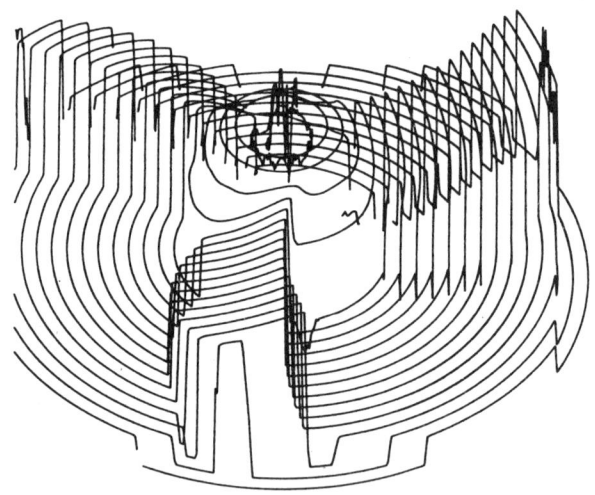

Interpolated Cylindrical Data (Face Side)
Figure 7

The interpolated cylindrical sections are "unwrapped" and displayed using an interactive graphics program which allows the user to view a specific section and chose the range of points to be written to a data file for a particular radius and blade combination. Using the program the section points for all the blade sections for both the face and back scan files are written to data files. The user names each data file so as to identify the particular combination of blade number and radius. The files containing section curves for the face and back of the blade can be merged together to form the final blade section after the data in one of the files is offset to allow for correct alignment. The amount of offset required is the same for all sections and is determined by comparing the heights of the propeller hub section between the Face and Back scan data files with the known hub length for the propeller. The cylindrical coordinate systems used to interpolate the face and back of the propeller must also be aligned to ensure proper section shape. The centre axis of the systems are identical but one must be rotated through a specific angle in order to achieve proper alignment. The required angle is determined based on the alignment pin of the measuring fixture which is located at a known distance form the centre of the propeller hub.

After the final propeller sections have been assembled they can now be compared to the original design sections. An interactive graphics program allows the scan section data to be read in and displayed along with the data representing the original design sections. The original design section shape is created by fitting elliptic splines through the design offsets for each section. The complete set of 10 design sections for the example propeller are shown in Figure 8. The measured sections are compared to the design sections by first displaying the design section curves then the corresponding measured sections for that radius. The design data must be shifted into the location for the section being compared. The amount of shifting required depends upon the number of blades on the propeller and is physically equivalent to rotating the design section around the propeller in order to position it for each individual blade. Once this has been done the two sections can be compared. The design and measured sections for a 40 percent radius are shown in Figure 9. These overlaid plots show very graphically how close the actual section is to the design section.

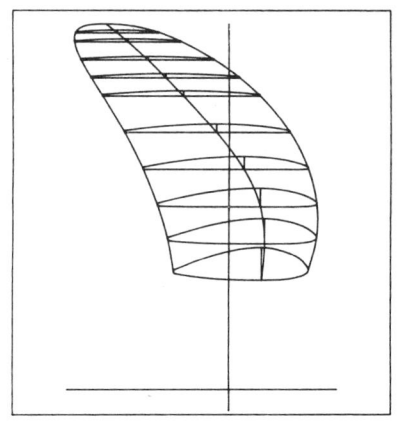

Design Sections
Figure 8

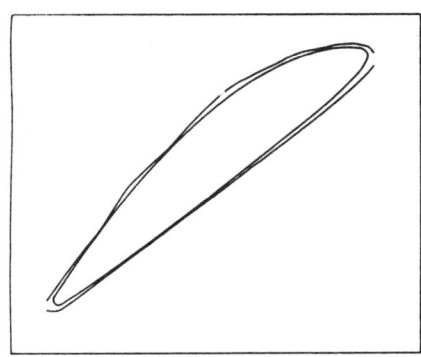

Design and Measured
Section at 40% Radius
Figure 9

In order to determine some quantitative values for the accuracy of the propeller it is required to "move" the measured section until the best fit is found to the design data. This "moving" requires both translation and rotation of the measured data. Once a "best fit" has been found three values are recorded which tell how much translation and rotation was required. these values are Xoffset, Yoffset, Angle. They each provide a measure of the accuracy in terms off skew, rake, and pitch respectively. The amount of Xoffset applied to the data represents the amount of rotation of the measured section around the propellers centerline in order to fit the design data. This measure corresponds to an error in the Skew position for that section. The Rake error is indicated by the amount of Yoffset applied to the section data since this movement is in the direction of the propeller axis. Any rotation which was applied to the data represents an error in the pitch of the section. The actual accuracy of the section shape can be evaluated by comparing the data after the "best fit" has been obtained. Any differences between the contours at this stage represents an error in section shape. This error generally has two forms, the section is undersized when the measured contour is inside the design contour, or it is oversized when the measured contour is outside the design contour.

MEASUREMENT RESULTS

For each of the three blades of the propeller sections were interpolated at 8 locations along the radius. These sections correspond to the design sections of the propeller and are at 40 percent through 80 percent in increments of 10 percent of the overall blade radius and from 85 percent to 95 percent in increments of 5 percent. Each of these sections was interpolated from the scan data and the best fit was made to the design data. The adjustments required to fit the data for blade 1 is tabulated and given in Table 1. The ISO 484/2 Standard for propeller manufacturing tolerances is used as a basis for accessing the results. While this standard is for larger propellers it is much more concise than any type of standard currently established for propeller models.

Table 1   **Section Fitting Offsets for Blade 1**

	40%	50%	60%	70%
DX	−66.692	−83.327	−99.567	−116.151
DY	6.883	7.233	6.982	7.792
	80%	85%	90%	95%
DX	−133.640	−142.073	−150.658	−161.879
DY	8.464	8.865	9.170	9.337

Skew Error

Using the data in the Table 1 values for skew error were calculated by the following relationships.

$$\text{Skew error} = DX - (DX7 \times R/R7)$$

Where DX is the adjustment required to align the measured section with the design section and DX7 is the adjustment required to align the 70 percent section of the blade being checked. R is the Radius at the section and R7 is the radius at the 70 percent section. The 70 percent section was chosen as the reference section in accordance with ISO standard 484. This is however somewhat arbitrary since there was no absolute reference position on the propeller.

The results of the skew error calculations for blade 1 are given in Table 2. The skew error measurements are given for each section within the blade with respect to the 70 percent section of that blade. This provides a measure of the individual sections accuracy. The skew for the 70 percent section of all blades are also measured with respect to the 70 percent section of blade 1. This provides a measure of the axial position of each of the blades with respect to each other. This measure is given in degrees and is calculated using the formula.

$$\text{Angular Diff.} = (DX7\text{-}DX7A) \times 180/(0.70 \times R \times pi)$$

Where DX7 is the skew value for the 70 percent section on the blade being measured and DX7A is the reference skew adjustment value for the 70 percent section on Blade 1. R is the overall design radius of the propeller.

The ISO 484/2 standard for manufacturing tolerances specify a tolerance of +/- 1 degree for class S (very high accuracy) and class I (high accuracy) propellers.

Table 2   **Skew error for Blade 1**

	40%	50%	60%	70%
SKEW ERROR	-0.267 %	-0.242 %	-0.005 %	0.0 %
RAKE ERROR	0.73 %	-0.18 %	-1.12 %	-1.30 %
	80%	85%	90%	95%
SKEW ERROR	-0.374 %	-0.406 %	-0.490 %	-1.49 %
RAKE ERROR	-1.47 %	-1.52 %	-1.60 %	-1.72 %

Rake Error

The rake error is calculated by the DY adjustment required to best fit the measured section to the design section. This measurement can be converted to an angular deviation by using the formula...

$$\text{Angular Dev.} = \text{ATAN(DY/R)-THETA}$$

Where THETA is the design rake angle, 5 degrees for the example propeller, and R is the radius of the section.

The DY values are measured with respect to the generator line of the propeller. The Rake angle errors for each section of Blade 1 are given in Table 2 along with the overall blade rake error. In accordance with ISO 484/2 the rake error is measured by taking 3 points along the blade face and determining the deviation in distance from these points to some arbitrary reference plane as a percent of the propeller diameter. The allowable deviation for class S propellers is +/- 0.5 percent and for class II it is +/- 1 percent.

Pitch error

The pitch error is measured for each individual section and from this the mean pitch for the each particular blade is determined and finally the mean pitch for the propeller. Five values were obtained for each section at approximately 15%, 30%, 50%, 70%, and 85% of the section length. For blade sections below 50 percent only the middle three values are used since the face of the blade is more rounded in this area. The error in local pitch is determined as the deviation between the values for a given section. The mean pitch for a given section is taken as the average of the individual values for that section while the blade mean pitch is the average for all the mean pitches for the sections of that blade. The overall propeller mean pitch is taken as the average for all the blades of the propeller.

The ISO tolerances for Local pitch are +/- 1.5% and +/- 2.0% for class S and II respectively. The tolerances for Mean pitch for each section are +/- 1.0 % and +/- 1.5%. These percentages are with respect to the design pitch for the particular section. The tolerances for mean blade pitch are +/- 0.75% and +/- 1.0%. The overall propeller pitch tolerances are +/- 0.5% and +/- 0.75%. For both the mean blade and propeller tolerances the percentages are with respect to the overall mean design pitch of the propeller. The results of the pitch error measurement are given in Table 3.

Table 3   Pitch Error

SECTION MEAN PITCH ERROR (Blade 1, 70%) = -0.8 %

BLADE MEAN PITCH ERROR (Blade 1)   = -1.23 %

PROPELLER MEAN PITCH ERROR   = -1.31 %

Section Shape Error

The accuracy of the section shape can only be assessed after the section has been corrected for skew, rake and pitch errors. This correction results in the "best fit" condition and therefore any further inaccuracies between the scanned section and the design section can be termed as section shape error. The ISO 484 standard specifies tolerances for section thickness and section length. These tolerances are +2 % and -1% for class S propellers and +2.5 % and -1.5 % for class II. The measurement of the blade thickness can be extracted from the scan data by "unpitching" the sections and intersecting the upper and lower curves of the section at the point of maximum thickness. The results for thickness measurement for the blade 1 are given in Table 4. The measurement of the section length cannot be measured as accurately as would be desired with this current method. The sampling interval of the laser scanner limits the accuracy of measurements which are perpendicular to the scanners beam in the same manner as was discussed for locating the centre for the bore of the propeller. The shadowing effect also results in a loss of definition at the edges of the blade section. The net result of this is that the section length can only be measured to an accuracy of +/- 1 mm. This accuracy could be improved by increasing the sampling interval of the laser scanner. Table 4 also contains the measurements of section lengths as could be best obtained from the scan data.

Table 4   **Section Shape Error for Blade 1**

	40%	50%	60%	70%
WIDTH ERROR	16.3 %	13.0 %	18.2 %	23.2 %
LENGTH ERROR	5.0 %	0.5 %	0.8 %	0.1 %

	80%	85%	90%	95%
WIDTH ERROR	23.7 %	16.0 %	9.0 %	3.0 %
LENGTH ERROR	-0.7 %	-1.2 %	-0.6 %	1.0 %

CONCLUSION

This first attempt at measuring a model propeller using the laser scanner proved successful in achieving most of the desired results. Perhaps more than anything it provided a learning experience from which more accurate and simplified techniques can be developed for future usage. Some modifications have already been made to the measurement fixture which holds the propeller in place under the laser scanner. A set of cone shaped attachments have been manufactured which will allow for much more accurate location of the centerline of the propeller. These cones ,shown in Figure 10, should make locating the tip of the cones a depth measurement rather than measurement in the plane orthogonal to the scanner beam. By this means the accuracy of locating the centre is based on the scanners depth of field rather then the scan interval, which is less accurate. The use of two of these cones also allows for a more accurate measure of any small tilt of the base since a vector joining the tips of both cones will be the same as that of the axis of rotation of the propeller.

In order to provide a more accurate means of aligning the face and back sections from the two separate scans of the propeller a small locating device, shown in Figure 11 has been made. This device is attached to the tip of the propeller and remains in place for both scans. By ensuring that the ball on the device has the correct spherical shape when the scans are aligned a high degree of accuracy can be achieved.

It is planned to use these new devices to scan a set of model propellers which have recently been made using a sand casting process. In addition to measuring the final finished propellers and comparing them to the design data the rough castings before finishing will also be scanned along with the wood pattern used to produce the castings. Buy making a overall surface differential comparison between these it should be possible to see the casting shrinkages and distortions that occurred in fine detail since the laser scanner can provide very accurate results in this type of relative measurement operation where calibration error is not significant. As a further improvement a program called Veri-foil, which is used for checking turbine impeller sections, will be examined to see if it can be adapted to allow for a more automated "best fit" of the section shapes to the design sections.

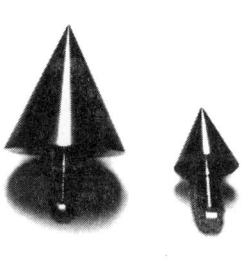

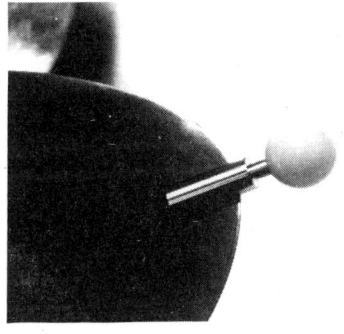

Locating Cones
Figure 10

Scan Alignment Device
Figure 11

REFERENCES

1. International Organization for Standardization
    Shipbuilding - Ship Screw Propeller - Manufacturing Tolerances
    Part 2: Propellers of Diameter Between 0.80 and 2.5 M inclusive.

2. Standard Procedure for Resistance and Propulsion Experiments
    with Ship Models, 9th ITTC , Formal Discussion.

3. Rioux, M. ,Laser Range Finder Based on Synchronized Scanners,
    Applied Optics, 1 November 1984 / Vol. 23, No. 211

4. Rioux, M. , Becthold G. , Taylor D. , Duggan M. ,
    Design of a Large Depth of View Three-Dimensional Camera for Robot
    Vision, Optical Engineering, 26(12), 1245-1250 (December 1987)

5. Livingstone, F.R. and Rioux, M.
    Development of a Large Field of View 3-D Vision System
    Proceedings of SPIE, International Society of Optical Engineering
    Volume 665, June 1986

6. Brownrigg D.R.K. , The Weighted Median Filter,
    Communications of the ACM, August 1984, Volume 27, Number 8

7. Yang G. and Huang T.S. , Median Filters and Their Applications to
    Image Processing, School of Electrical Engineering,
    Purdue University, TR-EE 80-1, May 1980

8. Principals of Naval Architecture, The Society of Naval
    Architects and Engineers

9. O'Brien T. P. , The Design of Marine Screw Propellers,
    Hutchinson Scientific and Technical, (revised) June 1968

SECTION 7: OFFSHORE STRUCTURES

Numerical and Experimental Study On Fatigue Analysis of Offshore Tubular Joints

K.N. Cho, Y.S. Jang, W.I. Ha(*) and C.D. Jang, S.J. Kang, D.H. Nam(**)

(*) Hyundai Maritime Research Institute, HHI, Ulsan, Korea

(**) Seoul National University, Seoul, Korea

ABSTRACT

In this paper, typical tubular joints' fatigue strength is investigated focusing on the Stress Concentration Factors calculation using Finite Element Methods. For the calculation of the SCF of the members, the joints are modeled using thin shell elements and comprehensive analysis are carried out. Related techniques for the numerical analysis are studied. Experimental studies are performed for the verification and comparison with the numerical analysis results. Model tests of K joints are carried out not only for finding SCF values but also for the calculation of fatigue lives of the joints using specially designed test facilities.

INTRODUCTION

In 1965 an early example of fatigue damage occurred in a triangular semi-submersible drilling rig in the Gulf of Mexico. From that time on, various kinds of fatigue failures of ocean structures have been reported. As a result, the importance of fatigue life estimation at the design stage was recognized and various kinds of analysis approaches have been discussed [1]. Fatigue crack growth results from predominantly dynamic loading, relatively high local stresses, application of high strength steel and fabrication defects. For offshore structures subject to variable loads such as wave loads, a probabilistic approach using spectral analysis methods is preferable. Spectral analysis is a technique capable of relating, in a statistcal manner, cause and effects due to randomly occurring phenomena [2].

In this paper characteristics of the spectral method are studied and the elements of the approach are discussed. The concepts of maximum stress concentrations phenomenon at certain points which are commonly referred to as hot spots and the stresses at these locations "hot spots stresses" are discussed. Various kinds of stress concentration factors formula and the application of FEM for searching for SCFs are studied. Generally, FEM analysis is necessary for complex joints to which known SCF formulas cannot be applied. On the other hand, for simple T, K, joints and similar ones, formulas such as Gibstein's, Kuang's, Wordsworth's, Smedley's, Kinra's and Efthymiou's are applicable depending on the characteristics of the joints considered [3,4].

In the following, components of the spectral fatigue analysis are reviewed and the roles and the characteristics of SCF in the fatigue problems are studied in the view point of numerical analysis and experimental analysis. Comparison of SCF values of typical K joint obtained by different methods are carried out. Also fatigue tests are performed for the joints.

SPECTRAL FATIGUE ANALYSIS

Metal fatigue in welded structures is a complex phenomenon affected by a number of synergistic factors, the most important being the cyclic stress range [5]. For most offshore structures, a spectral fatigue analysis approach, may be performed without any difficulties. In this case, wherein the entire long-term distribution of fatigue stresses is determined in each specific case, considering the characteristics, such as significant wave height and representative wave period, of each sea state and the time spent in it [6,7].

The spectral method applies the theory of stochastic processes for the calculation of the response to environmental loading, especially wave loading. For a particular sea state, spectrum of a response variable is found by combining wave spectrum with transfer function relating wave amplitude to amplitude of response. By integrating the response spectrum, variance of response and spectral moments can be calculated. Once the stress spectrum for a particular point in the structure during a certain sea state is known, predictions of

the stresses experienced at that location can be made.   All
statistcal stress predictions are related to the moments of the
relevant stress spectrum about the origin.   Spectral fatigue
analysis procedures are well introduced [2,7].

Among the major elements of the fatigue analysis, for
example, environmental condition, stress concentration factor,
S-N curve, the most crucial element is found to be the stress
concentration factor.   Typical sensitivity analysis shows that
a doubled SCF increases damage ratio by 21 times [6].   SCF may
be said to be the most sensitive factor in fatigue behavior of
offshore structures.

STRESS CONCENTRATION FACTORS

Stress concentration can be defined as a condition in which a
stress distribution has high localized stresses : usually
induced by an abrupt change in the shape of a member ; in the
vicinity of notches, holes, changes in diameter of a shaft, and
so on, maximum stress is several times greater than where there
is no geometrical discontinuity.   The stress concentration
factor is the ratio of the greatest stress in the region of
stress concentration to the corresponding nominal stress.   The
location of the stress concentrations are called hot spots.

The SCF may be calculated from theory of elasticity by
various methods.   Analytical methods tend to become
mathematically complicated, and are applicable to simple
geometries only.   Finite Element Method is more versatile.   For
3-dimensional case, the Finite Element Method is still
practical, but is very expensive in many cases.   The SCF may be
estimated based on parametric equations or experimental data.
Finite Element Method is used commonly to determine the stress
distribution and hot spot stress especially for complex tubular
joints.   Shell elements traditionally have been used for the
SCF calculation.   However it is difficult to include the
geometry in the weld region in the FEM modeling.   Stress
analysis by FEM is by the way, the most efficient, reliable and
economical tool for detailed stress analysis of tubular joint
with the rapid development of computation skills and modeling
techniques and high speed computers.

There are many formulas for SCF calculation which were
obtained from numerical analysis or experiment.   Several

researchers have suggested empirical approximations for SCFs in
tubular joints [3]. Generally, nondimensional ratios of
geometric parameters are used to allow for the generalization
of the results to many joint sizes.
SCF is assumed as a function of parameter γ, β, τ, ζ, θ
as:

$$SCF = f_1(\gamma) \cdot f_2(\beta) \cdot f_3(\tau) \cdot f_4(\zeta) \cdot f_5(\theta) \tag{1}$$

where $\gamma = D/2T$ chord diameter to thickness ratio
       $\beta = d/D$ brace diameter to chord diameter ratio
       $\tau = t/T$ brace thickness to chord thickness ratio
       $\zeta = g/D$ gap to chord diameter ratio
       θ brace angle with the chord

With the above assumptions for SCF, the actual curve fit
of the SCFs was performed in a graphical manner generally. The
empirical equation in the form shown below is obtained.

$$SCF = a \cdot \gamma^{m_1} \beta^{m_2} \tau^{m_3} \zeta^{m_4} \sin^{m_5}\theta \tag{2}$$

Obviously any set of empirical guidelines must be
restrained within certain limitations to minimize data
dispersion and maintain design applicability. Recent
researches on the SCF formulas are performed mostly by making
use of numerical analysis [8,9].

NUMERICAL APPROACH FOR FATIGUE ANALYSIS

SCF calculation of K joint by FEM
The stress analysis using the FEM is performed on the typical K
joint models. The geometrical particulars of the models are
shown in Table 1.

Table 1   Dimension of K joint models

unit : mm

	D	T	L	d	t	g
Model 1	168.0	7.1	1400.0	76.0	5.2	45.76
Model 2	168.0	10.0	1400.0	76.0	5.2	45.76

In the analysis, the axial load of 5 ton was imposed on

the brace. To provide a satisfactory model for a tubular joint
FEM analysis, the mesh generation was prepared as follows.
1. 3-D thin shell element was used.
2. In the immediate vicinity of the branch-to-chord junctions,
   the dimension of the element was chosen not to exceed 0.75
   $\sqrt{Rt}$ as UKOSRP (United Kingdom Offshore Steels Research
   Project) recommended. Where R is the radius of the
   corresponding member and t is the thickness of the member.
3. Element aspect ratio was chosen not to deviate from
   unity.
4. Mesh generation is made to provide that the location of the
   strain gauge attachment point and the nodal point of FEM
   analysis are to be coincided. Total number of element is
   1256 and total number of nodal point is 1266. Figure 1 and
   Figure 2 show the FEM modeling of the K joint.

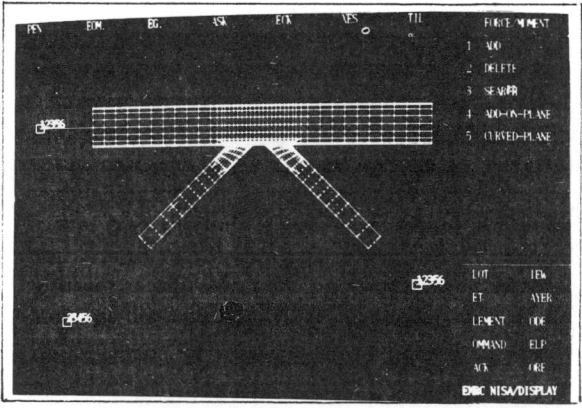

Fig.1 FEM model of K joint, front view

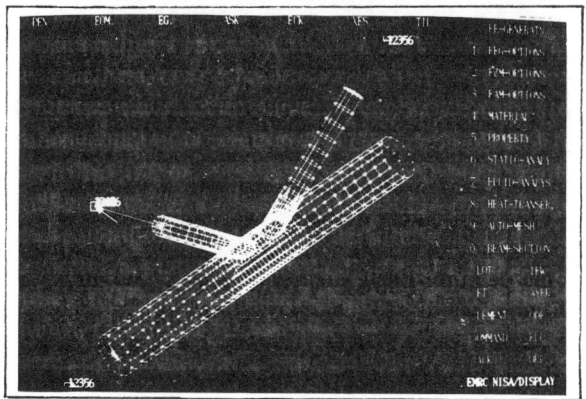

Fig.2 FEM model of K joint, prospective view

Figure 3 and Figure 4 show the mesh configuration and mesh size of joint intersection area of the model.

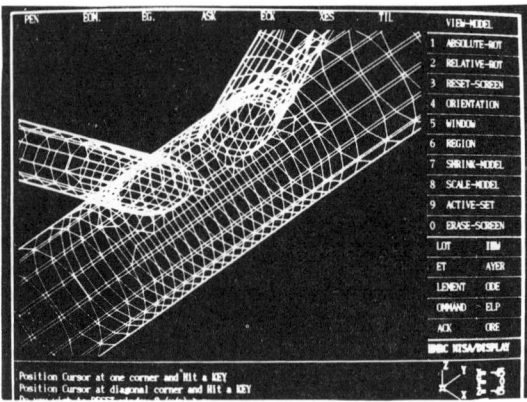

Fig.3   Mesh configuration at joint intersection

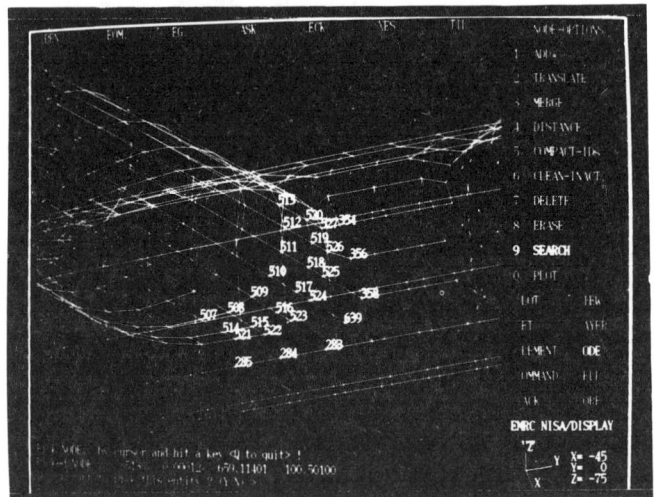

Fig.4   Mesh size at brace-chord intersection

The boundary conditions of the model for FEM analysis is same as those of the model for actual test. Axial load of 5 ton is applied at one brace member with other brace member is free of loading. The chord is in simply supported-free condition. The stress analyses have been carried out on two types of tubular K joints by using the general purpose code for structural analysis NISA II [10]. Figure 5 and Figure 6 show the principal stress and the von-Mises stress distribution with hot spot locations, respectively, for model 1.

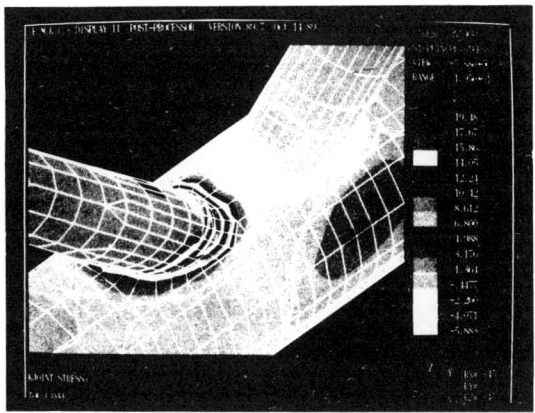

Fig.5   Principal stress distribution at the intersection for model 1

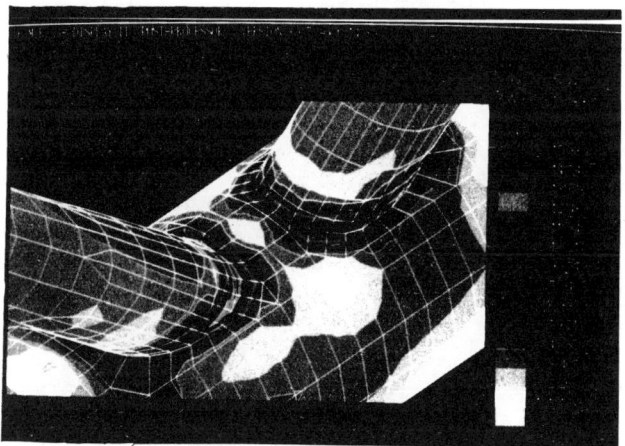

Fig.6   Von-Mises stress distribution at the intersection for model 1

Figure 7 and Figure 8 show the principal stress and the von-Mises stress distribution with hot spot locations respectively, for model 2. Since high local stresses are observed along the intersection lines especially at the saddle point it may be said that the saddle point area is the most possible failure occurrence area eventhough there are no welding defects.

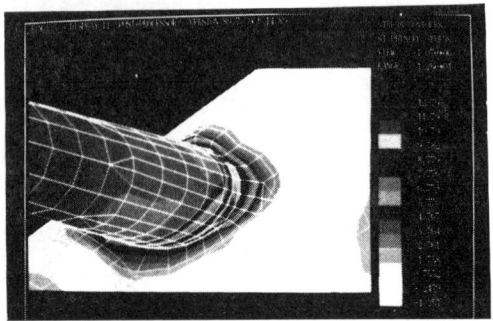

Fig.7   Principal stress distribution at the intersection for
        model 2

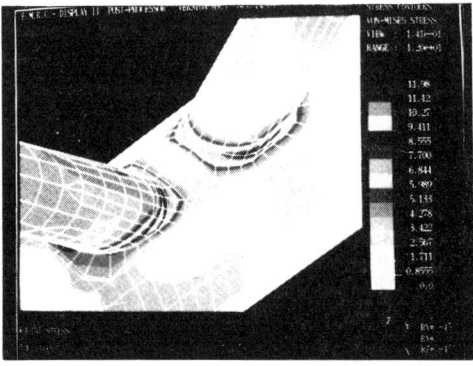

Fig.8   Von-Mises stress distribution at the intersection for
        model 2

Table 2 shows the results of the stress analysis for model
1 and model 2. The location of maximum SCF for model 1 is the
center between saddle point and crown point. The location of
maximum SCF for model 2 is around the saddle point at brace
side.

Table 2   SCF values obtained by FEM

		Chord	Brace	Kuang's formule
Model 1	SCF	4.339	4.213	4.782 (Chord)
	Prin. Stress	18,744	18,200	
Model 2	SCF	2.211	3.229	3.264 (Brace)
	Prin. Stress	9,522	13,949	

Fatigue life estimation based on S-N curve

Miner's Rule is most conveniently applied for the fatigue life estimation. The validity of Miner's Rule for random loading conditions is open to question. However, it is known that a linear accumulation of damage is approximately true in cases where fatigue is predominantly a result of the propagation of initial cracks that are already present [2].

For the simple calculation of fatigue life of the K joint using the SCF value obtained previously, American Welding Society curve for category X is used [1]. There are many S-N curves in existence and they differ greatly, depending on the type of structural detail referred to and the origin of the data. The modified AWS-X curve is used here, since this curve is thought to be most suitable for the case. Using the obtained SCF value of 4.213 for model 1 and 3.229 for model 2 and given axial load of ± 7.5 ton, fatigue life is obtained. The fatigue failure cycle obtained is 11604 for the case of model 1, and 22073 for the case of model 2. These values are thought to be very conservative.

EXPERIMENTAL APPROACH FOR FATIGUE ANALYSIS

SCF calculation of K joint by test

The SCFs of the specimens which have same dimensions as the models for FEM analysis were estimated using strain gauges readings. The extrapolation techniques have been used to interpret the measured results. Test arrangement for K joint is shown in Figure 9. A typical strain gauge layout is shown in Figure 10. To measure the values of SCF, strain gauges were attached at the crown and saddle points of both chord and brace side. To get the stress at the weld toe, the linear extrapolation method of maximum principal stresses is employed. Figure 11 shows the way of extrapolation of hot spot stress. Table 3 shows the results obtained by the tests and FEM.

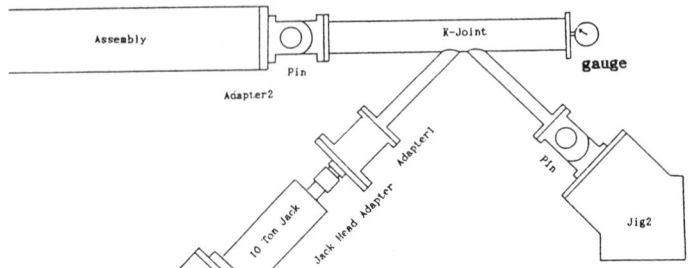

Fig.9 Fatigue test arrangement

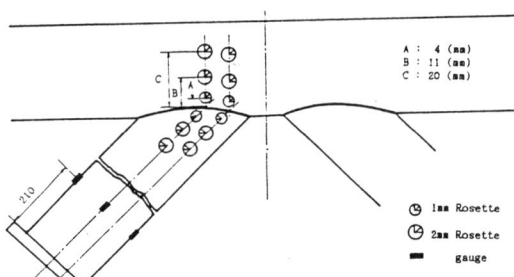

Fig.10 Typical strain gauge layout

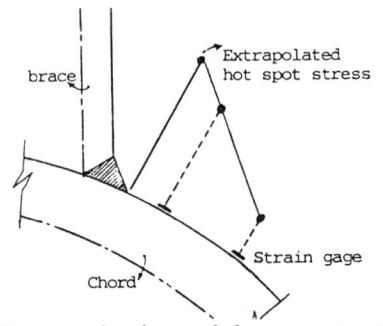

Fig.11 Extrapolation of hot spot stress

Table 3   SCF values obtained by tests and FEM

		Chord	Brace
Model 1	Test	3.780	3.140
	FEM	4.339	4.213
Model 2	Test	2.500	2.840
	FEM	2.211	3.229

Fatigue test of K-joint

Fatigue test were conducted on large scale size specimens. The geometrical particulars of the specimen are same as the model used in FEM analysis. Testing layout is shown in the Figure 9.

Fatigue test were carried out under load controlled conditions. The chord ends were simply supported and free and the axial load of ± 7.5 ton was applied at the end of the brace. The life of thorough thickness cracking was detected with the aid of pressure gauge which is installed at the end of the chord as shown in the Figure 9. The results of the tests under the cyclic loading for Model 1 and Model 2 are summarized in Table 4 with the comparison of the results by S-N curve approach . The testing set up is shown in Photo 1. The failure modes of the specimen for model 1 and model 2 are shown in Photo 2 and Photo 3, respectively.

Table 4   Fatigue life by various methods

unit : cycle

	AWS-X Curve	Fatigue test
Model 1	11,604	306,000
Model 2	22,073	515,300

Photo 1   Testing set up

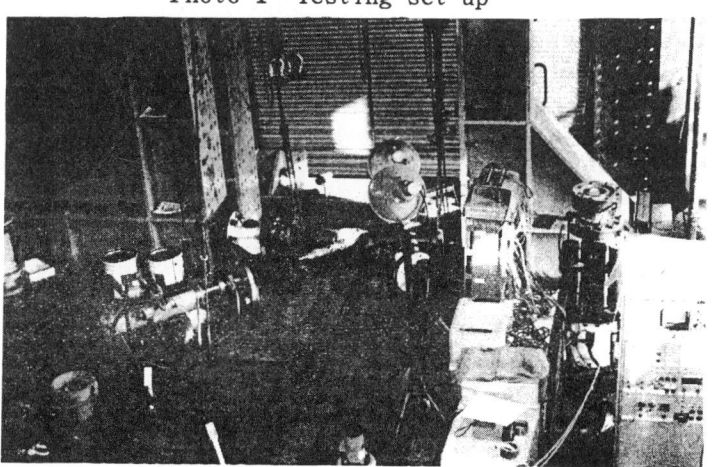

Photo 2   Failure mode for model 1

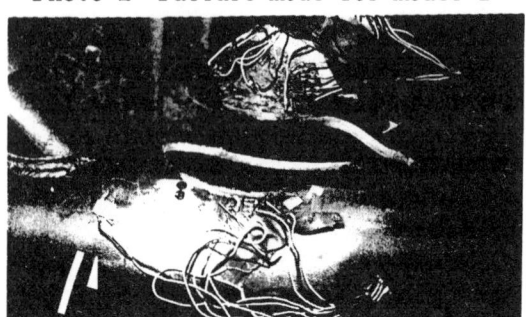

Photo 3   Failure mode for model 2

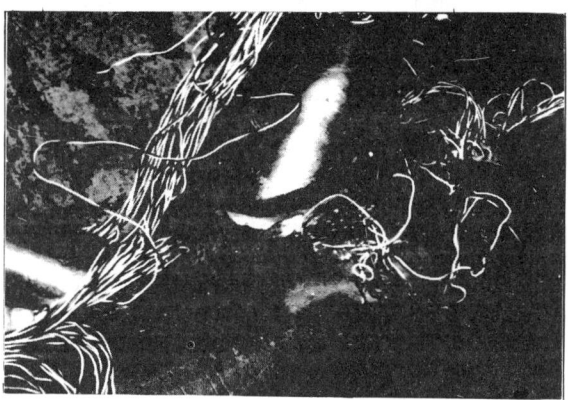

DISCUSSIONS AND CONCLUSIONS

A fully satisfactory solution of fatigue problem has not been achieved yet. Several kinds of approaches are sought. Among these methods, spectral method described here represents physical characteristics more realistically. In the method, stress concentration factor is the most crucial factor in fatigue life calculation. In order to obtain accurate fatigue life calculations, a reasonable estimation of SCF is essential. FEM analysis is necessary for finding SCFs for complex offshore structure joint. Even though the SCF formulas are available, the detailed FEM analysis is strongly recommended. In this paper the SCF values of K joints are obtained by FEM and compared with those obtained by the tests. The values are in good agreement each other in this case.

The experimental approach may be utilized for SCF calculations. The inherent difficulties of experimental approach are exposed. Any set of empirical guidelines must be restrained within certain limitations to minimize data dispersion and maintain design applicability. Recent researches on the SCF formulae are performed mostly by making use of numerical analysis, mainly because of huge amount of time and expenses required for experimental investigations.

Fatigue life estimation based on the spectral analysis may be very conservative. In this paper, fatigue life of the K joint obtained by the spectral method is come out to be more than 20 times conservative comparing with the actual test results. However, the spectral method, with the realistic SCF values obtained by FEM is thought to be main tool for solving the problems.

The main conclusions drawn from the study are as follows:
1. The integrity of offshore structure is critically dependent on the behavior of tubular welded joints which are subjected to stress concentration and thus fatigue. Despite of all the progress made for the solving of stress concentration problems, FEM is the reliable main tool for solving the problems.
2. To provide a satisfactory model for a tubular joint stress analysis, thus for precise SCF value calculations, mesh generations need to be carried out carefully according to the geometry of the member, boundary conditions and loading circumstances.
3. Spectral fatigue analysis gives very conservative results. However the method is by far the most convenient and economical tool for the estimation of the fatigue life of offshore structures.

REFERENCES

[1]. Almar-Naess, A., Fatigue Handbook-Offshore Steel Structures, Tapir, Trondheim, 1985
[2]. Vughts, J.H. and Kinra, R.K., Probabilistic Fatigue Analysis of Fixed Offshore Structures, Proceedings of Offshore Technology Conference, OTC 2608, pp889-906, 1976
[3]. Jang, Y.S, SCF Calculations for Various Tubular Joints, HMRI Report 2S87176, 1989
[4]. Kuang, J.G., Potvin, A.B, and Leick, R.D., Stress Concentration in Tubular Joints, Proceedings of Offshore

Technology Conference, OTC 2205. pp593-612, 1975

[5].   ABS, R & D. Department Technical Report RD-89020F, 1989

[6].   Jang, Y.S., Yi, W.S., Cho, K.N., Transportation Technique Development for ONGC Jacket Structure, Part I. Fatigue Strength, HMRI Report 2S87042, 1987

[7].   Kim, D.Y, Cho, K.N., Basic Design Development of H520/H521 Part 3, Fatigue Analysis.   HMRI Report 2S 87225-3, 1988

[8].   Soh, A.K., Too, H.K., Wong, C.F., SCF Equations for T and K Square Tubular Welded Joints, Proceedings of Seventh International Conference on Offshore Mechanics and Arctic Engineering, 1988

[9].   Hellier, A.K., Connolly, M.P., Dover, W.D., Corderoy, D.J.H., Parametric Equations to Predict the Full Stress Distribution in Tubular Welded Y and T-Joints, Proceedings of the First Pacific/Asia Offshore Mechanics Symposium. 1990

[10]. Engineering Mechanics Research Corporation, NISA II User's manuals. 1989

Evaluation of Fatigue Damage on Fixed Jacket Platforms

T.A.P. Lopes and L.C.M. Meniconi

COPPE/UFRJ and CENPES/PETROBRÁS, Cx. Postal 68508, CEP 21945, Rio de Janeiro, RJ - Brasil

ABSTRACT

This paper presents a procedure to evaluate accumulated fatigue damage on tubular joints of fixed jacket steel platforms based on a numerical–experimental methodology. Experimental results as wave height, displacement and stress spectra, were obtained for the structure during actual measurements. After adjusting the numerical finite element model, the fatigue damage were calculated for some selected joints. The input for this calculation are the power spectral stress densities obtained from the actual dynamic behaviour of the platform.

INTRODUCTION

Reliability of jackets of fixed platforms is related to the frequency and quality of periodic inspections. Nowadays, inspection depends on deep sea divers, which makes it expensive or unpractical due to the increase in jacket number and water depth. To overcome this problem, the researches have given special care to two areas of study: *submarine robotics* and *structural monitoring*. Structural monitoring employs vibration analysis, acoustic emission and strain transducers to make the inspection more selective and to reduce to a minimum the role of sea divers.

This paper presents a procedure which qualifies as structural monitoring by vibration analysis. Systematic measurements of the acceleration upon the platform deck, with the help of an adjusted numerical model of the structure, gives the accumulated fatigue damage on each important tubular joint, permitting to establish a priority as concerns that sort of fault.

Our analysis was centered on the central platform of Garoupa oil field (PGP–1), because of its strategic role as a converging point for the production from the other platforms.

DATA MEASUREMENT AND SIGNAL PROCESSING

The gathering of the mainly dynamic information associated to this paper is made through a system of data collecting shown in Figure 1.   Each signal is monitored every six hours, thus a time series of approximately half an hour duration is obtained in each measuring schedule.

The first element of the system of data collecting are the wave height sensor, strain gages and structural acceleration sensors.   Each of those sensors generates a voltage signal in continuous or alternative current, which multiplied by a conversion factor indicates the value of the measured variable.

The obtained signal is amplified to increase the final resolution. Afterwards, it passes through a filter, which lessens the oscillations above an admitted frequency so that it can eliminate noises that might introduce errors in the process.

It is necessary to take the measurements for long periods of time. From november 1984 to september 1986, 19 measuring campaigns were held on Garoupa platform, almost one monthly campaign during two years.   Each campaign lasted for one week and four daily measurements were taken.   The results were presented by Lopes [1].

The most frequent measurement aimed for records of wave height and acceleration of deck points, but occasionally measurements of the strain were also taken to check the method.   Processing of the time series, by *Fast Fourier Transform* (F.F.T.), gave a group of spectral density graphs as shown in Figure 2.

It is noted that the platform response to the sea waves, expressed by the deck movement, divides into two well defined components.   The first one, called *direct response* or *quasi–static*, is identified by the spectral peak in the same frequency range of the distant sea where the largest part of the waves energy is concentrated.   This component is detected by all sensors.   In the second one, peaks having higher frequency values are shown, where the sea energy decays continuously because it is a local sea region.   In spite of that, the peaks are sharply outlined because they correspond to natural frequencies of the jacket, thus forming the resonant response.

In order to settle the total response for the concerned jacket, we only have to keep the first group of resonant frequencies of the structure from the first to the third one, having in mind that the latter can only be detected by the transversal sensors as its excitation is very weak.

DETERMINATION OF THE NUMERICAL PARAMETERS

The curves of spectral density displacement obtained for the deck of Garoupa Platform are characterized by the dynamic parameters of the structure.   The dynamic parameters of the structure that concerns us are the natural frequencies and modes and damping ratio, which are fundamental to establish the numerical model of the jacket.

It is necessary to establish simplifying hypothesis to make possible the calculation of the parameters.   The structure has a large number of natural frequencies although only the first group is relevant to formulate the modal response to the wave action.

It is usually accepted that each natural mode forms an independent system of one degree of freedom (Brebbia and Orzag [2]) and it is assumed that:

- the modal damping coefficients are uncoupled;
- the structure behaves linearly;
- the damping is small;
- the natural frequencies of interest are wide apart.

The system mass contains the structural mass, the mass on the deck and the additional hydrodynamic mass. The damping coefficients include the structure and soil internal friction, the hydrodynamic viscous friction and friction associated to wave generation from the jacket vibration.

From then on rigidity, mass and damping refer to a certain mode of the structure and are then considered invariable, thus being a linear system.

In our present study, in which we deal with damping ratio lower than 10% and with natural frequencies with a maximum of 1.0 Hz, practically, the natural frequency of a vibration mode is the frequency relative to the corresponding maximum peak in the displacement spectral density function.

The jacket's natural frequencies were established with one hundredth Hertz accuracy and correspond to the following mean values:

- First mode   : 0.30 Hz
- Second mode   : 0.40 Hz
- Third mode   : 0.52 Hz

After determining the natural frequencies, we deal with damping ratio by way of the method of half power spectral bandwidth, which is an approximate one , and uses the following premises:

- as the damping is low, the resonant frequency mixes up with natural frequency; in this case the peak is almost symmetrical in relation to the natural frequency, in the level which corresponds to half of the maximum spectral density, the level being Δf wide; the modal damping ratio is given by:

$$\zeta = \frac{\Delta f}{2f_n} \tag{1}$$

The mean values obtained for the damping are:

- First mode   : 4.5 %
- Second mode   : 4.1 %
- Third mode   : 3.0 %

ASSESSMENT OF DAMAGE BY TRANSFER FUNCTIONS

Consider a linear system with constant parameters as illustrated in Figure 3. According to Bendat and Piersol [3], the system behaviour is represented by the complex frequency functions $H_u(f)$ and $H_v(f)$. The system input are two time series $u(t)$ and $v(t)$ related to random stationary process of null mean.

As a consequence, there is a stationary output $q(t)$ whose measurement is affected by the noise $r(t)$. The following relation among the spectral density functions is valid for the system but, to make it simpler, the frequency variable ω was left out:

$$G_{qq} = (H_u)^2 \, G_{uu} + H_u^* \, H_v \, G_{uv} + H_u \, H_v^* \, G_{vu}$$
$$+ \, (H_v)^2 \, G_{vv} + G_{rr} + H_u^* \, G_{ur} + H_u \, G_{ru}$$
$$+ \, H_v^* \, G_{vr} + H_v \, G_{rv} \qquad (2)$$

The system presented here is the mathematical model of the jacket, considered as linear, so that the random dynamic analysis becomes possible by the ADEP program, from PETROBRÁS. The soil was modeled by linear springs. Airy's wave formulation was employed throughout the analysis, which is linear and can be applied to the natural conditions of the Brazil coast.

The wave forces induced on the structural parts are obtained from Morison's equation, where the drag force is not linear to the particles velocity, becoming linear by the ADEP program.

The inertia and drag coefficients were taken as being 2.0 and 1.2, respectively. It was taken a linear damping equivalent to 4.3 %, which is the mean value taken from modes 1 and 2.

The magnitude $u(t)$ and $v(t)$ shown in Figure 3 are the deck displacements towards x and y obtained from the model. The series $q(t)$ refers to a stress on the section of one part of the model.

The noise $r(t)$ incorporates the effects of non—linearity, non—stationarity and measurement noise. Thus, as defined, the parts concerning noise are null, as we deal with a numerical model.

The real coherence function is defined as:

$$\gamma_{uv}^2 = \frac{|G_{uv}|}{G_{uu} \, G_{vv}} \qquad (3)$$

The image of the coherence function is the interval (0, 1). An univalent coherence function means that two time series contain redundant information, thus one of them can be dismissed. A null coherence function implies on uncoupled time records, with no cause and effect relation between them.

In practical use we find a gradation between the two mentioned extremes.   In this paper, where we deal with displacements in two orthogonal directions, or two distinct degrees of freedom, a low coherence function must be expected. if the sea "enters" from the x–direction of the jacket, the y–direction will have low excitation, unless unexpected factors occur.

This reasoning is based on practice, considering the coherence values evaluated during the two years that the measurements were taken in the Garoupa Platform. Up to the frequency 0.2 Hz, which corresponds to the direct response to the sea, coherence is 0.5 or less. From 0.2 to 0.4 Hz, in the range of the first and second modes this value tends to zero.

In our study, in which we considered only direct action from the sea and the two first modes, the low coherence values allow us to assume the uncoupled hypothesis. Thus, the crossed terms of the former equation are canceled, and it becomes, in the absence of noise:

$$G_{qq} = (H_u)^2 \, G_{uu} + (H_v)^2 \, G_{vv} \qquad (4)$$

and the gain factors $(H_u)^2$ and $(H_v)^2$ are obtained from the model of the structure, taking each direction as a separate linear system.

The random dynamic analysis module of program ADEP uses as input a wave height spectrum divided into a series of bands where width is chosen by the user. To each band is associated a wave height equal to the double of the square root of its area, and a period equal to the inverse of its mean frequency. An analysis on the frequency domain is made for each wave (Lima et all [4]).

We considered a height spectrum of waves that spread to the x–direction of the model, covering the frequency band of interest, from 0.05 to 0.50 Hz with intervals of 0.01 Hz. We obtained from the program the displacement spectrum G_{uu}, as well as the stress spectrum in the section of a selected member, G_{qq}. The gain factor $(H_u)^2$ is obtained by:

$$(H_u)^2 = \frac{G_{qq}}{G_{uu}} \qquad (5)$$

The same procedure is repeated for the y–direction to get $(H_v)^2$.

Stresses that are relevant, for a frame structure as a jacket, are the axial stress and the two bending stresses, referred to the two local axes of the section of each member. The two gain factors from the model are applied to the functions of displacement spectral density obtained from each measurement, so as to arrive to the stresses acting in that period.

Gain factors are determined only once to avoid the costly computational analysis. If there is any important change on the structure, such as on the deck mass, the model should be reviewed.

The described method considers only the influence of the first two modes on the jacket, as we can not simulate on the model a "rotational" spectrum which would correspond to the 3rd mode. Anyway, for the jacket dimension, the influence of the third mode is irrelevant.

After getting the stress spectrum of an extreme section of a tube, we must find the stress spectrum of *hot spot* to calculate the fatigue damage. The current analysis method calculates the *hot spot* stress in eight spots along the tube intersection, set apart 45°, as shown in Figure 4.

$$\sigma^\phi(t) = \frac{FX \; F(t)}{S}$$
$$+ \cos \phi \; \frac{FY \; M_y(t)}{Z} + \sin \phi \; \frac{FZ \; M_z(t)}{Z} \tag{6}$$

Equation (6) is a valid formulation for the time domain, where $F(t)$ represents the axial force and $M_y(t)$ and $M_z(t)$ the bending moments as concern the local axes of the member. Terms FX, FY and FZ are stress concentration factors referring to each force. S and Z are respectively the area and the strenght modulus of the analyzed *brace*.

Relating to the spectral density function, the former equation becomes:

$$G^\phi_{\sigma\sigma} = G_{\sigma_x\sigma_x} + \cos^2 \phi \, G_{\sigma_y\sigma_y} + \sin^2 \phi \, G_{\sigma_z\sigma_z}$$
$$+ \cos \phi \, (G_{\sigma_x\sigma_y} + G^*_{\sigma_x\sigma_y}) + \sin \phi \, (G_{\sigma_z\sigma_z} + G_{\sigma_x\sigma_z})$$
$$+ \sin \phi \cos \phi \, (G_{\sigma_x\sigma_z} + G_{\sigma_y\sigma_z}) \tag{7}$$

As soon as the stress spectra (including the crossed ones) on the outermost section of the *brace* and the factors of stress concentration are known, we have the *hot spot* stresses at any point along the intersection line of the tubular joint. The remaining problem is to have the crossed spectra from the ADEP program which is not prepared for it.

In order to illustrate the adopted procedure, let us analyze the hot spot tension in position 1:

$$G^1_{\sigma\sigma} = G_{\sigma_x\sigma_x} + G_{\sigma_y\sigma_y} + (G_{\sigma_x\sigma_y} + G^*_{\sigma_x\sigma_y}) \tag{8}$$

having in mind the usual notation of crossed spectral density:

$$G_{ab} = C_{ab} - j\,Q_{ab} = |G_{ab}|\,e^{-j\theta}$$
$$= |G_{ab}|\,(\cos\theta - j\,\sin\theta) \tag{9}$$

results

$$G^1_{\sigma\sigma} = G_{\sigma_x\sigma_x} + G_{\sigma_y\sigma_y}$$
$$+ 2\,(G_{\sigma_x\sigma_x}\,G_{\sigma_y\sigma_y})^{\frac{1}{2}}\,(\gamma_{xy}\cos\theta) \tag{10}$$

and a conservative analysis becomes available, supposing that in the above expression the term $\gamma_{xy}\cos\theta$ is maximum and equal to 1.0 in each frequency:

$$G^1_{\sigma\sigma} = G_{\sigma_x\sigma_x} + G_{\sigma_y\sigma_y} + 2\,(G_{\sigma_x\sigma_x}\,G_{\sigma_y\sigma_y})^{\frac{1}{2}} \tag{11}$$

By analogy, in position 3 we have:

$$G^3_{\sigma\sigma} = G_{\sigma_x\sigma_x} + G_{\sigma_z\sigma_z} + 2\,(G_{\sigma_x\sigma_x}\,G_{\sigma_z\sigma_z})^{\frac{1}{2}} \tag{12}$$

The use of the above expressions is strict, because it supposes that the component stresses of the *hot spot* value will always be in phase and in high coherence, giving the highest possible resultant. On the other hand, this results in a conservative analysis which is helpful to compare the damage on two distinct joints.

After calculating the stress spectrum of *hot spot*, the fatigue damage can be obtained from the expression proposed by Chaudhury [5]. The damage is evaluated by means of the characteristics of the curve of power spectral density of stress, the parameters of the adopted SN curve and the period of time related to each measurement. By this way, the fatigue analysis is carried out just in the frequency domain, which is faster than in time domain.

RESULTS AND CONCLUSIONS

In order to get actual results from the method presented on this report, 21 jacket joints were selected, according to three criteria [6]:

— in order to evaluate the damage and its correlation to the position of the joint along the height of the jacket, for the same type of joint, the intersections of the members 5401, 5501, 5601 and 5701 with the jacket leg were selected;

— joints more liable to fatigue according to design were selected, located at the members 2602, 3605, 3606, 3609 and 3610;

—   joints that represent the variety of types of joints on jackets were selected as well.

Figure 5 presents the selected joints, together with their locations.

Among the two years campaigns, the ones from number 6 to 17 were held monthly and correspond to one year of measurement. Table 1 shows the accumulated values of fatigue damage during that year in positions 1 and 3 (see Figure 4) of the selected joints. At the same table it is also presented the inverse of the annual damage, that corresponds to fatigue life of each joint.

Evaluation of the results relative to members 5401, 5501, 5601 and 5701 indicates, roughly, that damage increases with the decrease of depth of the joint along the jacket. According to design, joints which would have a life span lower than 200 years — at members 2602, 3605, 3606, 3609 and 3610 — only the one at member 2602 remains critical. As a matter of fact, this joint presents the shortest life span: 44.2 years. The other ones, according to our calculation, present a life of thousand of years.

The main idea of the present report is to introduce a methodology for reanalysis of the dynamic behaviour of jackets as concerns to fatigue of welded joints and, as a consequence, a guidance for inspection to detect cracks. It seems to us that this goal has been reached as it was verified that the fatigue will not lead to *trespassing cracks* on the tubular joints of PGP–1 platform, installed in Campos Basin, Offshore Brazil, during its operational life. Concerning the design verification, the above statement allows us to infer that the structure is well designed for fatigue damage. However, it should be noted that the joints which accumulated more damage in this study do not match those of the design, except one of them. As a consequence, inspection plans must be reviewed, for they are based mostly on data from design.

REFERENCES

1.   Lopes, T. A. P. — *Instrumentação de Plataformas Fixas de Produção na Bacia de Campos* — PETROBRÁS/CENPES, Rio de Janeiro, 1987

2.   Brebbia, C. A. and Orzag, S. A. — *Vibrations of Engineering Structures* — Springer–Verlag, Berlin, 1985

3.   Bendat, J. S. and Piersol, A. G. — *Random Data Analysis and Measurement Procedures* — John Wiley & Sons, New York, 1986

4.   Lima, E. C., Elwanger, G. and Torres, A. L. — *Subsistema ADEP–DINAL: Manual de Teoria* — PETRÓBRÁS/CENPES, Rio de Janeiro, 1988

5.   Chaudhury, G. — *Spectral Fatigue of Broad–Band Stress Spectrum with One or More Peaks* — Offshore Technology Conference, Paper OTC 5333, 1986

6.   Meniconi, L. C. M., Lopes, T. A. P. and Ebecken, N. F.
F. – *Numerical Experimental Study of Damage Caused by Fatigue on
Jackets of Offshore Structures* – 7th International Symposium on
Offshore Engineering, Rio de Janeiro, 1989

SYMBOLS

ζ	–	Modal damping ratio
G_{qq}	–	Force power spectral density
G_{uu}	–	Displacement auto power spectral density
G_{uv}	–	Displacement cross power spectral density
G_{rr}	–	Noise auto power spectral density
G_{ur}	–	Noise cross power spectral density
$(H_u)^2$	–	Square of magnitude of complex response function
$\sigma^\phi(t)$	–	Stress at position ϕ in time domain
$G^\phi_{\sigma\sigma}$	–	Stress spectral density at position ϕ

TABLE 1 – Annual damage and fatigue life

JOINT Member/Node	ANNUAL DAMAGE Pos. 1 – Pos. 2	FATIGUE LIFE Pos. 1 – Pos. 3
2502/175	8.70 – 3.89	115,000 – 257,000
2502/209	14.8 – 5.2	67,700 – 191,000
2506/181	21.7 – 4.48	46,000 – 223,000
2506/216	21.9 – 8.40	45,600 – 119,000
2602/208	22,600 – 12,100	44.2 – 82.4
3501/179	6.71 – 76.9	149,000 – 13,000
3501/207	20.1 – 442	49,800 – 2,260
3506/205	0.662 – 3.91	1,510,000 – 256,000
3506/212	2.53 – 3.85	395,000 – 260,000
3605/238	5.75 – 5.81	174,000 – 172,000
3606/238	10.8 – 78.1	93,000 – 12,800
3609/239	5.85 – 5.85	171,000 – 171,000
3610/239	11.2 – 86.2	89,000 – 11,600
4609/228	204 – 128	4,900 – 7,810
5401/146	82.0 – 188	12,200 – 5,330
5501/174	56.5 – 361	17,700 – 2,770
5601/207	382 – 9,700	2,620 – 103
5701/240	2,470 – 3,720	405 – 269
5708/270	1,750 – 1,700	572 – 589
6603/208	59.2 – 1.63	16,900 – 612,000
7613/214	800 – 41.7	1,250 – 24,000

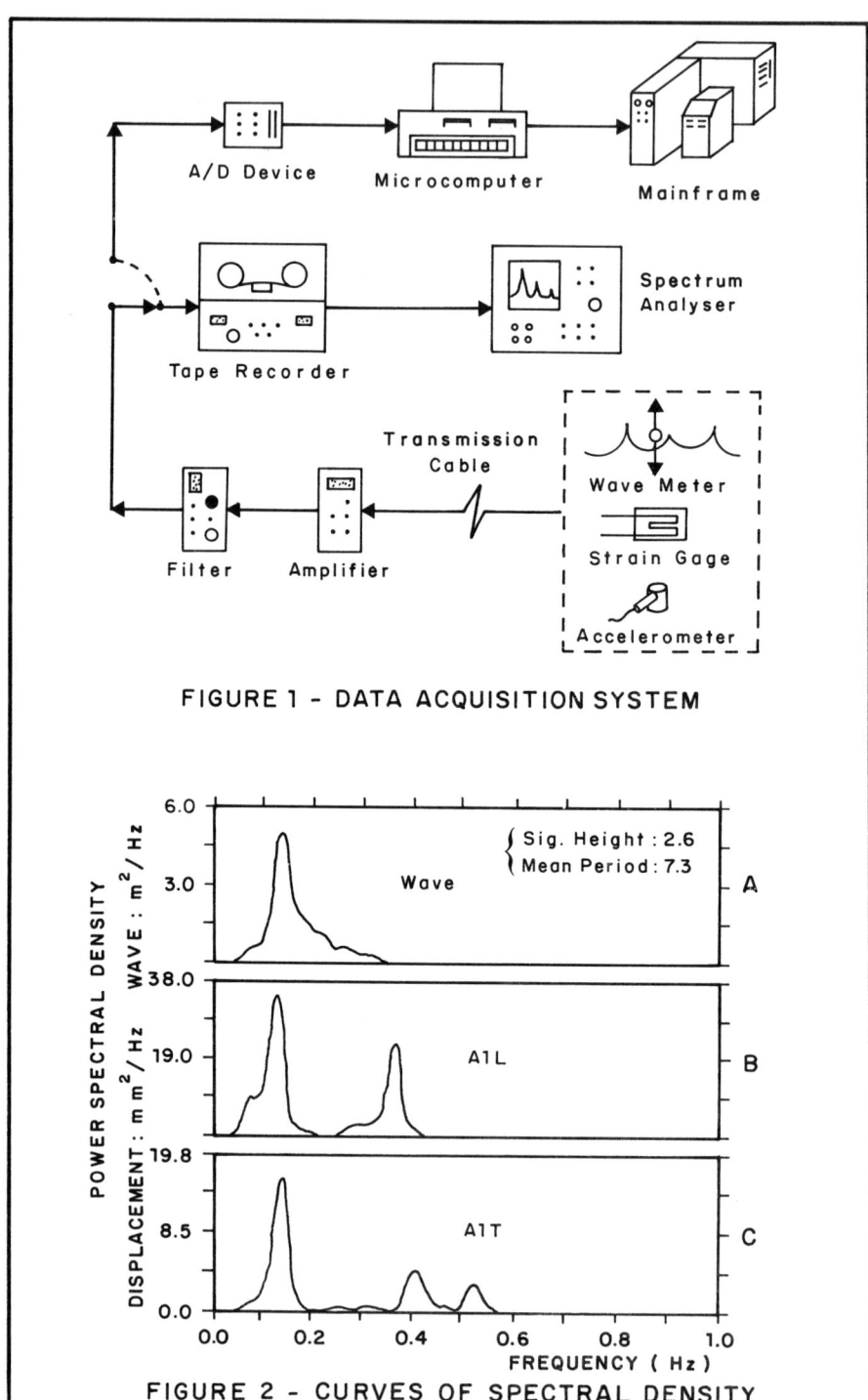

FIGURE 1 - DATA ACQUISITION SYSTEM

FIGURE 2 - CURVES OF SPECTRAL DENSITY

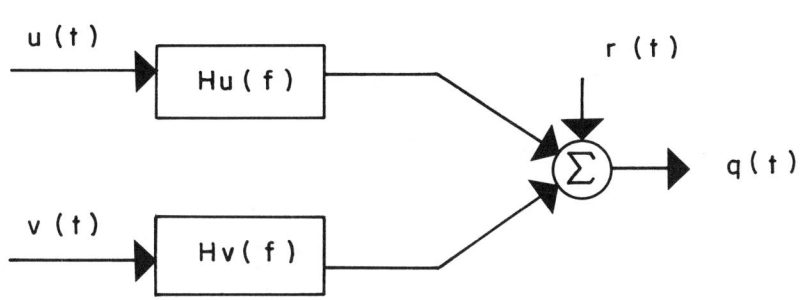

u(t), v(t) - Displacements in x and y directions

q(t)   - Stresses on a selected section

r(t)   - Noise effects

FIGURE 3 -   TWO INPUT AND ONE OUTPUT
             LINEAR SYSTEM

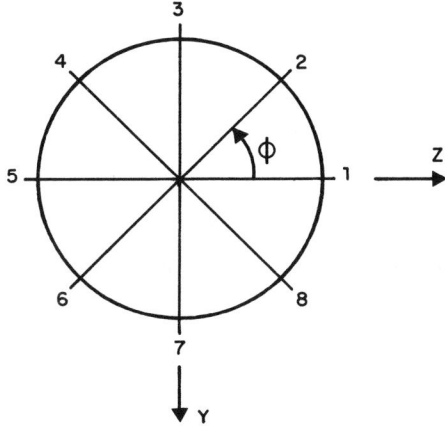

FIGURE 4 - SPOTS ALONG A JOINT FOR
           DAMAGE CALCULATION

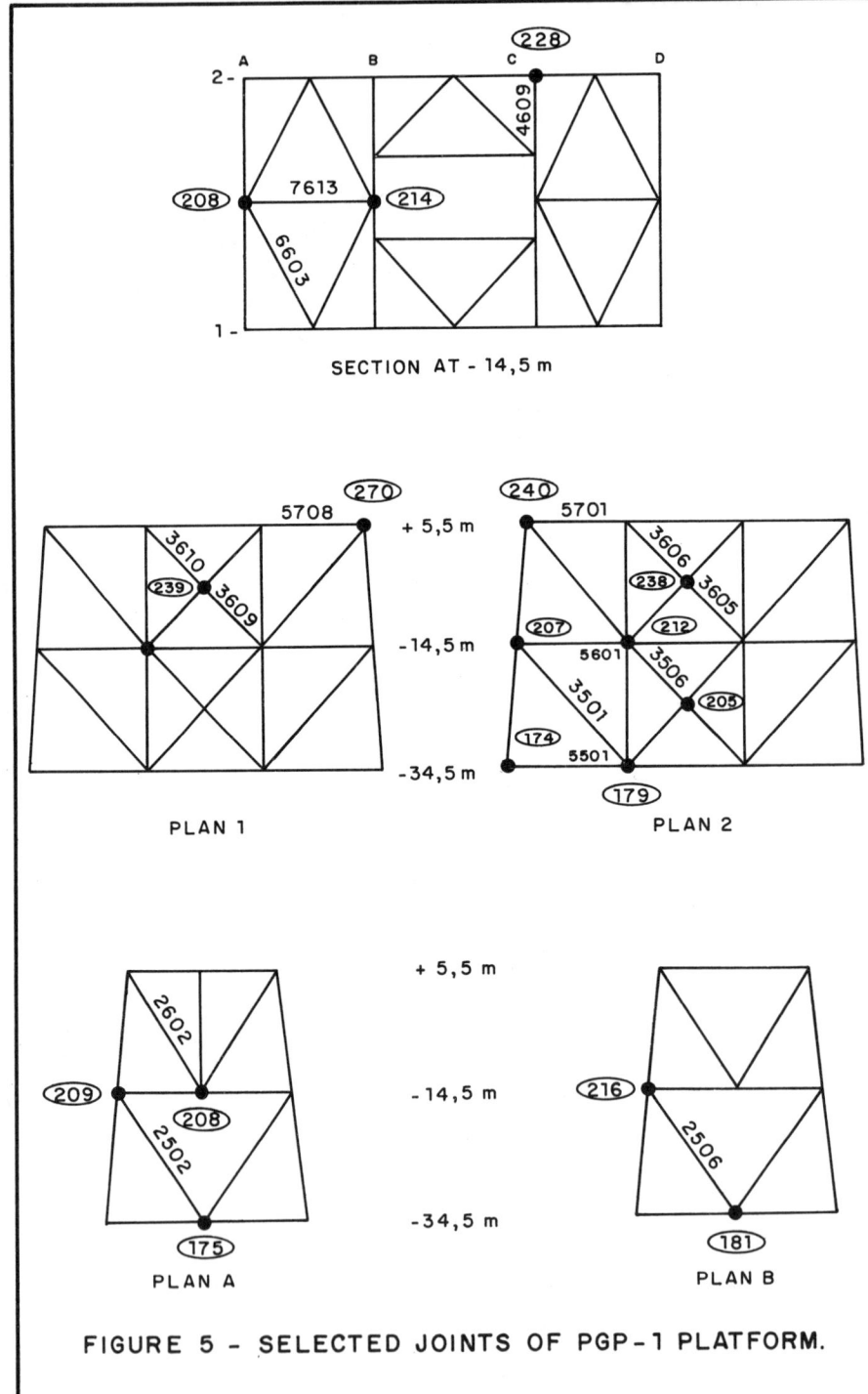

FIGURE 5 - SELECTED JOINTS OF PGP-1 PLATFORM.

Shape Optimization on the Basis of Biological Growth with Special Regard to Slanted Joints

M. Beller(*) and C. Mattheck, J. Schäfer(**)

(*) PREUSSAG Anlagenbau GmbH,

Pipeline-Service, Breslauerstr. 56B,

7500 Karlsruhe, Germany

(**) Kernforschungszentrum Karlsruhe GmbH,

Institute for Material and Solid State Research IV,

7500 Karlsruhe, Germany

ABSTRACT

Biological load-carriers (e.g. wood, bone etc.) selfoptimize by adaptive growth with respect to the most important natural loading condition encountered. It is shown that trees, bone, claws and other natural designs follow the rule of a constant Mises-stress at their surface. A new method named CAO (Computer Aided Optimization) was developed at the Nuclear Research Center Karlsruhe (KfK) [1], which allows the simulation of adaptive growth as well as the shape optimization of engineering components.

The method which was well received by industry will here be applied to slanted joints. In order to illustrate the method it will first be applied to natural designs such as branch-stem-joints in trees. CAO will then be applied to slanted joints of engineering components, as found for instance in offshore framework-structures.

It is shown as a significant result that circular transition contours especially, connecting the jointed members of such structures, cause large notch stresses which can be reduced drastically by shape optimization using CAO. This in turn leads to geometric designs similar to the biological ones.

INTRODUCTION

Biological load carriers are subject to hard competition for energy and living space. As a consequence only the best mechanical construction has a chance to survive which results in an optimum design characterized by minimized weight in conjunction with sufficient mechanical strength. A large number of preliminary studies by Mattheck [1,2] have shown that biological structures adapting their growth with respect to external loads will always grow into a state of constant mechanical stress at their surface. This results in a surface contour where no point of the biological structure is either over- or underloaded. As a consequence no notch stresses will exist, thus resulting in a surface without localized stress peaks. This is a very good "insurance" to have regarding a long fatigue life for a component. Understanding and copying natural design can also be used to improve engineering design. In the following a new structural optimization method will be introduced which makes use of the biological design rule of "adaptive growth". This technique called CAO (Computer Aided Optimization) can be used in the fields of biomechanics and engineering.

COMPUTER AIDED OPTIMIZATION

Nature as a teacher
Nature has had ample time in which to optimize biological structures. Optimized in this context shall imply structural optimization, i.e. reducing or even eliminating stress concentrations for a given geometry and loading situation. Nature is a very good teacher! By observing trees for example it can be seen how nature achieves optimization by the mechanism of *adaptive growth*. In the case of trees this means that material will be added at overloaded regions (in bone for instance material will even be removed at underloaded regions) in order to obtain a more advantageous stress distribution. All this therefore suggests learning from and simulating nature in order to further improve engineering design proposals.

Different optimization techniques
Optimized transition contour lines for beams with a narrowing cross-section have been evaluated by Baud [3] as early as 1934. Baud used an experimental approach in order to obtain transition lines without notch stresses. More recently many attempts of structural optimization using numerical methods are reported in the literature. For reasons of space only two of those shall be mentioned here. Umetanu and Hirai [4] also decided to copy biological growth for structural optimization purposes. Huiskens et.al. [5] reported a similar method. A more detailed comparison of these individual methods is given in [1].

Simulating adaptive growth
The CAO (Computer Aided Optimization) method simulates the natural mechanism of adaptive growth by special application of a commercial FE-code on a computer. The procedure consists of the following steps as shown in Fig.1. Initially a reasonable design proposal has to be made. This is followed by a standard elastic FE-analysis. The FE-run will calculate the Mises stress distribution present in the structure due to the external loading applied. In a subsequent step only the stress values obtained for the surface are used. The main principle of the technique is to set the Mises equivalent stresses formally equal to a fictitious temperature field. This implies that the problem of stress homogenization is solved by transformation into a problem of thermal strains. The resulting temperature field is applied in a further FE-run as the only applied loading, using the thermal expansion or thermal swelling routine provided by most commercial FE-codes. The structure will then "grow" by stress thermal expansion which simulates the mechanism of adaptive growth and will lead to a better structural design regarding the initially applied load. A subsequent standard FE-run will reveal the Mises stress distribution present in the "swollen" structure. The procedure can be repeated until a satisfactory stress distribution is obtained. Alternatively CAO provides the option to remove material from the region of a structure which is underloaded. In that case a minimized weight design can be obtained for a given stress distribution. The main advantages of CAO are:

- Any standard FE-code including a thermal expansion option can be used.

- There is no need for costly postprocessing routines adapted to individual problems.

- A very much improved surface stress distribution with drastically reduced stress concentrations can be obtained within only a few computing cycles, which is a major bonus considering costs.

EXAMPLES

CAO has been applied successfully in the fields of biomechanics and engineering. In the latter especially, problems due to fatigue have been investigated and very much improved geometric designs could be achieved using the method. Numerous examples can be found in [1]. In this paper CAO will be applied to slanted joints as found in engineering structures or tree branches. Structural optimization will result in a more homogeneous stress distribution at the surface of a component and will reduce or even eliminate local stress peaks and result in an increased fatigue resistance. A T-joint design which was optimized using CAO and then manufactured using Computer Aided Manufacturing techniques (CAM) was fatigue tested together with the non-optimized design. The same load parameters were applied to both designs and the optimized design survived 36 times more loading cycles without failure than the non-optimized version [6].

Slanted branch joint
Fig.2 shows a slanted branch joint. This example clearly illustrates how CAO can be used in biomechanics in order to understand and learn from nature. Fig.2a shows a photograph of the section of the tree investigated. The boundary and loading conditions are shown in Fig.2b. The initial non-optimized FE-mesh is shown in Fig.2c. The geometry of this initial mesh was taken from the photograph. The transition between stem and branch however was modelled as a circular notch. The FE-code ABAQUS [7] was used throughout the analysis. The 2D FE-mesh was generated with standard linear, isoparametric elements. The model consisted of plane strain elements incorporating 3626 degrees of freedom. Youngs modulus for wood is taken as $E = 2 \cdot 10^4 \, Nmm^{-2}$ and Poissons ratio as $v = 0.3$. Fig.2d shows the contour plots for the initial and final structure. The optimized contour leads to an elimination of the initial local stress peaks and to a more homogenous stress distribution at the surface, see Fig.2e. The final contour fits the actual tree shape very well.

T- and Y-joint
Two different engineering joints were also investigated, a T-joint geometry where brace and chord are at right angles to each other and a Y-joint (slanted joint) where the angle was 45^0. In order to limit the mesh generation effort for this initial investigation of slanted joints, the brace was modelled as a square tube and fitted onto a narrow plate representing the chord, see Fig.3a. A circular transition contour, as usually found in engineering design, was chosen between pipe and plate. Fig.3a-d shows the T-joint, and Fig.4a-d the Y-joint investigated. For both 3D meshes, 8-node brick elements were used, Youngs modulus was taken as $E = 2 \cdot 10^5 \, Nmm^{-2}$ and Poissons ratio as $v = 0.3$. The T-joint and Y-joint meshes incorporated 3320 and 3544 nodes respectively. Fig.3a shows the FE-mesh, Fig.3b the loading and boundary conditions implied and Fig.3c the intial and optimized contour for the T-joint. The local stress peaks present in the initial geometry were drastically reduced. The Mises stress distribution for the initial and the optimized structure are shown in Fig.3d. Fig.4a-d shows the Y-joint analogous to Fig.3. As can be seen the stress peaks initially present were reduced, resulting in a more homogenous stress distribution, see Fig.4d.

CONCLUSIONS AND OUTLOOK

It was the aim of this paper to show that shape optimization based on biological growth can successfully be applied to improve the stress distribution in T- and Y-joints. It should be noted that even a small geometric change can drastically reduce local stress peaks present in engineering structures usually due to the redirectioning of the force flow in the vicinity of notches. Furthermore a more homogeneous stress distribution can be achieved and the fatigue resistance can be improved. From nature one learns that notches present in biological load carriers are shape optimized and therefore do not cause any notch stresses. The mechanism of adaptive growth models this biological design rule. It does so by simply attaching material to overloaded regions and optionally even removing material from underloaded regions (as happens in bone). The CAO-method simulates adaptive growth by stress controlled thermal expansion. CAO is a straight forward and easy to use method for 2D- and 3D-shape optimization in an engineering environment. This study shows that CAO can be used to optimize T- and Y-joints. In further work T- and Y-joints will be modelled more exactly and optimized for various loading conditions. The authors are aware that structural shape optimization potential for welded joints (simple brace-chord connection welded) is limited. In this case merely the weld contour could be optimized. However CAO is well suited for the optimization of casted T- and Y-joints or nodes. Fabrication of an optimized geometry using casting would not cause any problems.

REFERENCES

1.   Mattheck, C. Engineering Components Grow like Trees, Mat.-wiss. u. Werkstofftech., 21, pp.143-168, 1990.

2.   Mattheck, C. Design and Growth Rules for Biological Structures and Their Application, Fatigue Fract.Engng Mater.Struct., Vol.13, No.5, pp.535-550, 1990.

3.   Baud, R. Beiträge zur Kenntnis der Spannungsverteilung in prismatischen und keilförmigen Konstruktionselementen mit Querschnittsübergängen, Report 29, Schw.Verb. für Materialprüfung i.d. Technik, Zürich, 1934.

4.   Umetani, Y., Hirai, S. An adaptive shape optimization method for structural material using growing-reforming procedure, Proc. of 1975 Joint ISME-ASME, Applied Mechanics Western Conference, pp.359-365, 1975.

5.   Huiskens, R., Weinans, H., Grootenboer, H., Dalstra, M., Fudala, B., Sloof, T. Adaptive bone remodelling theory applied to prosthetic design analysis, J. Biomechanics, 20, pp.1135-1150, 1987.

6.   Mattheck, C., Beller, M., Bethge, K., Erb., D. Computer Aided Optimization of T-Joint Structures By Simulation of Biological Growth, Proc.First Offshore Mechanics Symposium, ISOPE, Trondheim, 1990.

7.   Hibbit, H.D., Karlsson, B.J., Sorensen, E.P. ABAQUS User's Manual, Vers.4.8, Providence, R.I., 1989.

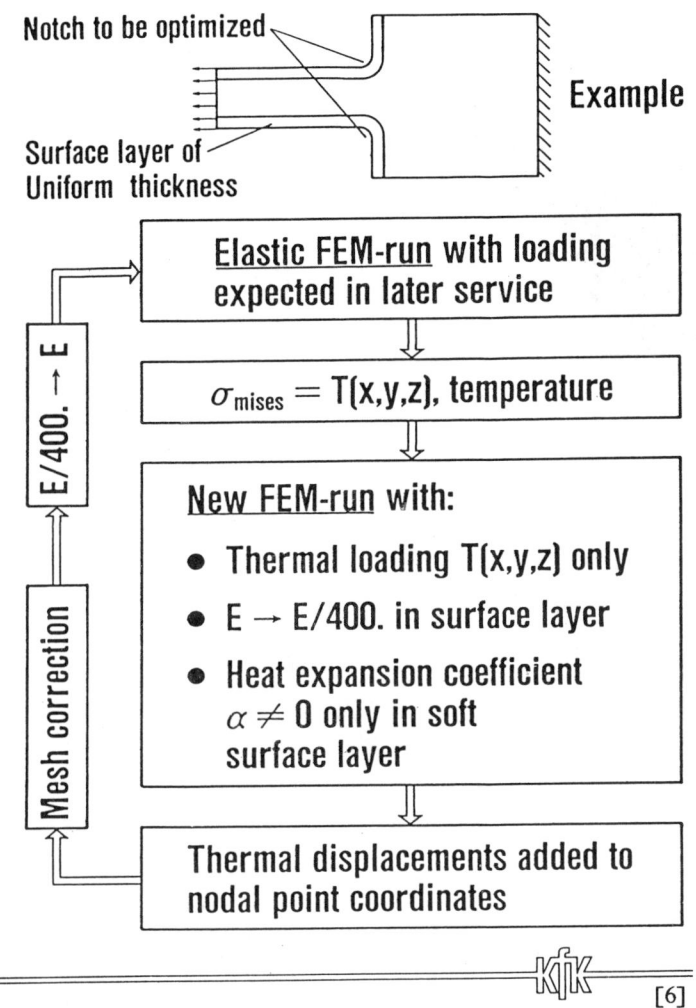

Notch to be optimized

Example

Surface layer of
Uniform thickness

Elastic FEM-run with loading
expected in later service

$\sigma_{mises} = T(x,y,z)$, temperature

New FEM-run with:

- Thermal loading T(x,y,z) only
- E → E/400. in surface layer
- Heat expansion coefficient
  $\alpha \neq 0$ only in soft
  surface layer

Thermal displacements added to
nodal point coordinates

E/400. → E

Mesh correction

[6]

Fig.1 Procedure of CAO

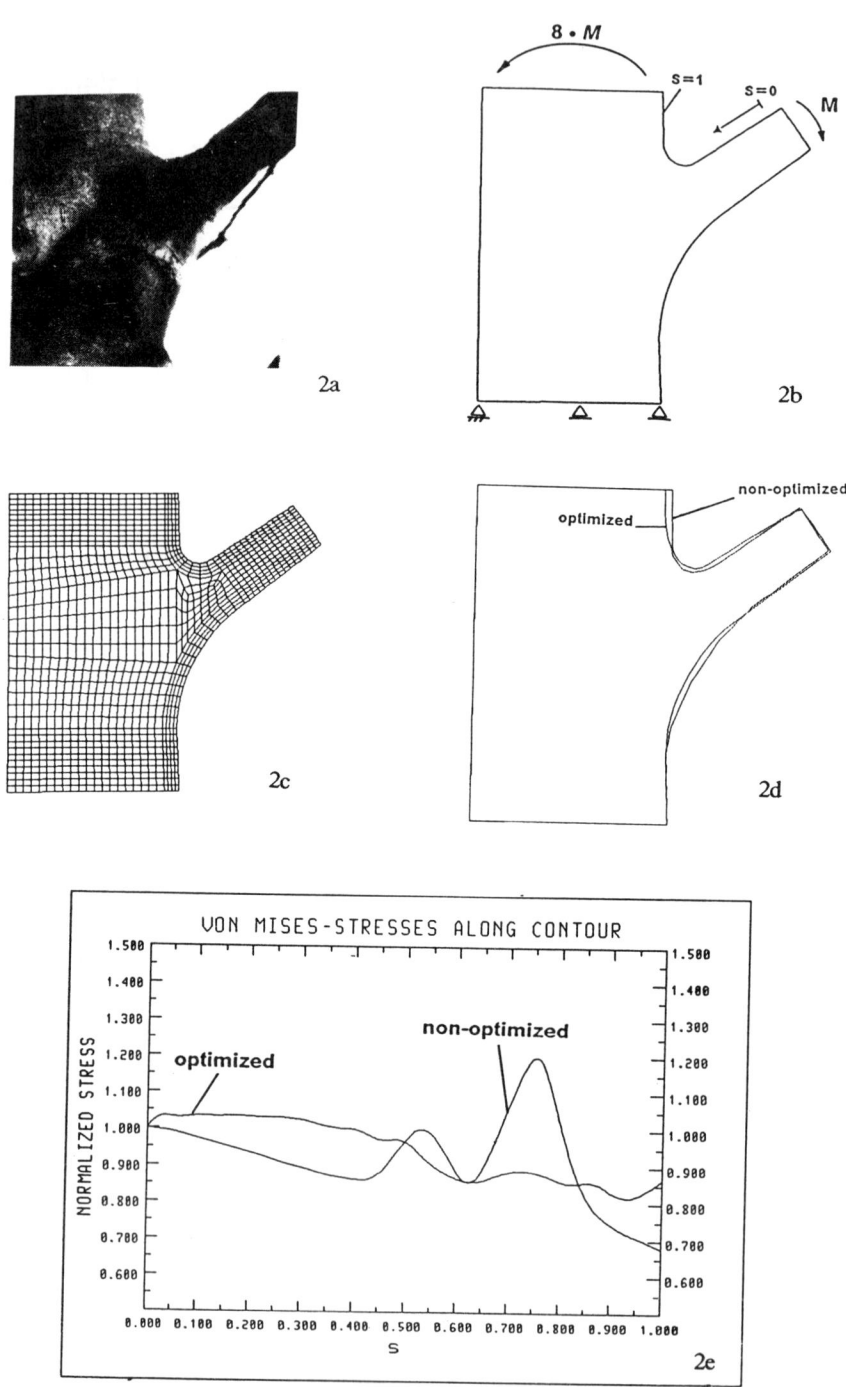

Fig.2a-e Slanted branch joint

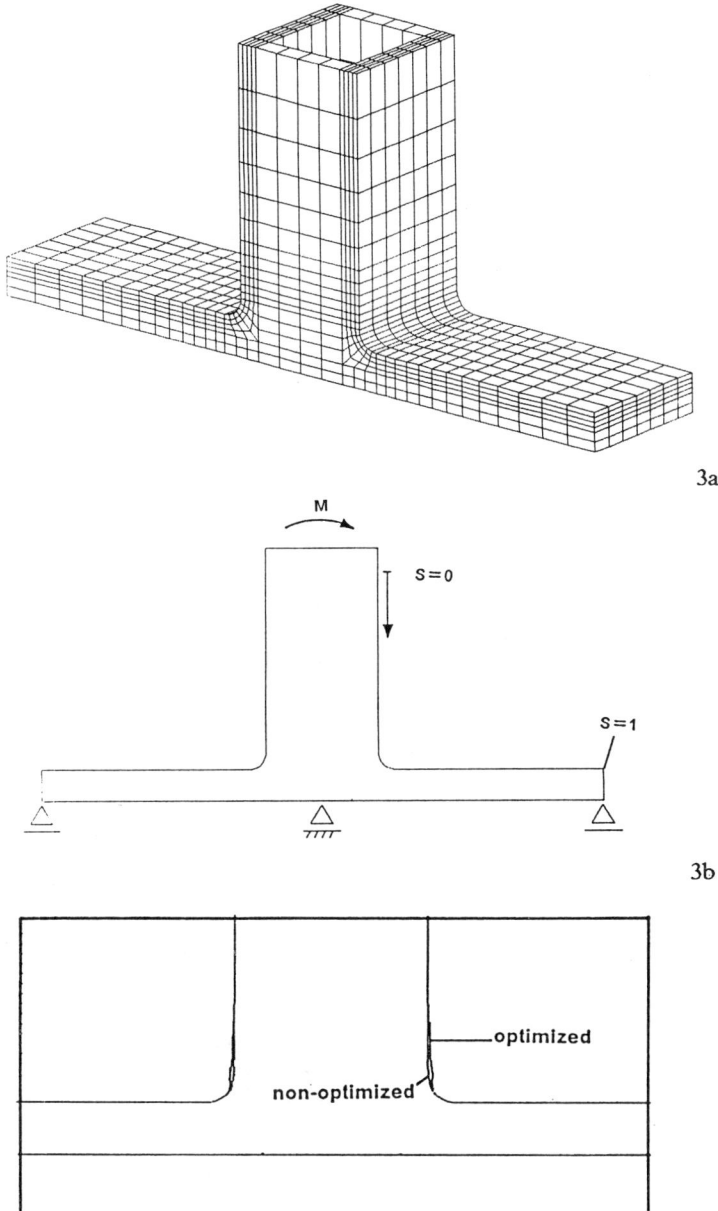

3a

3b

3c

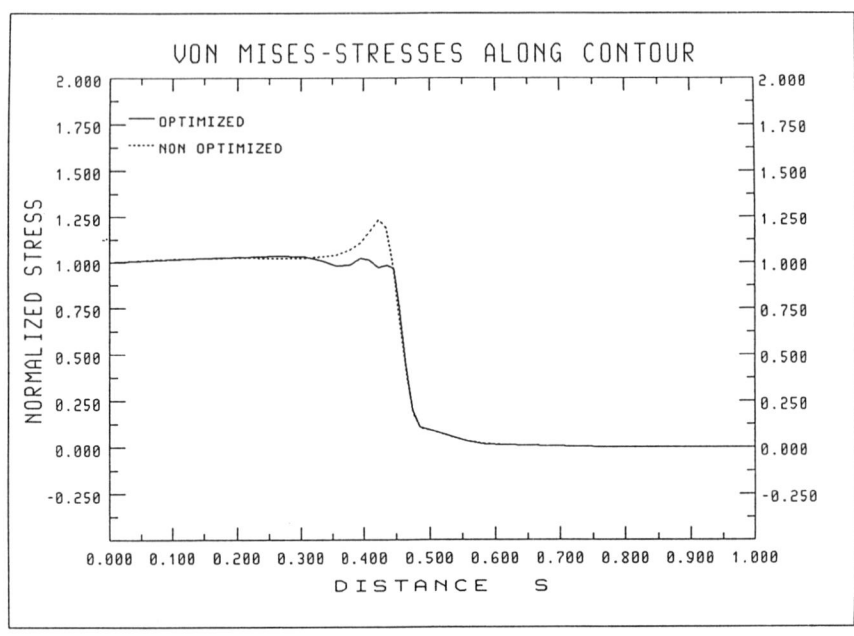

3d

Fig.3a-d T-joint

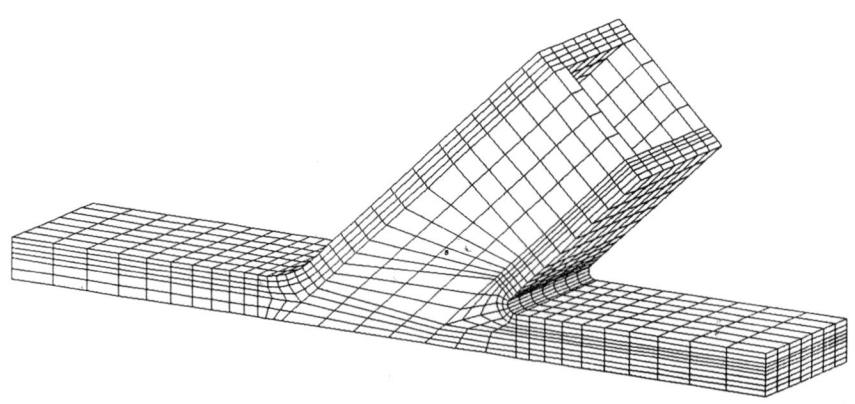

4a

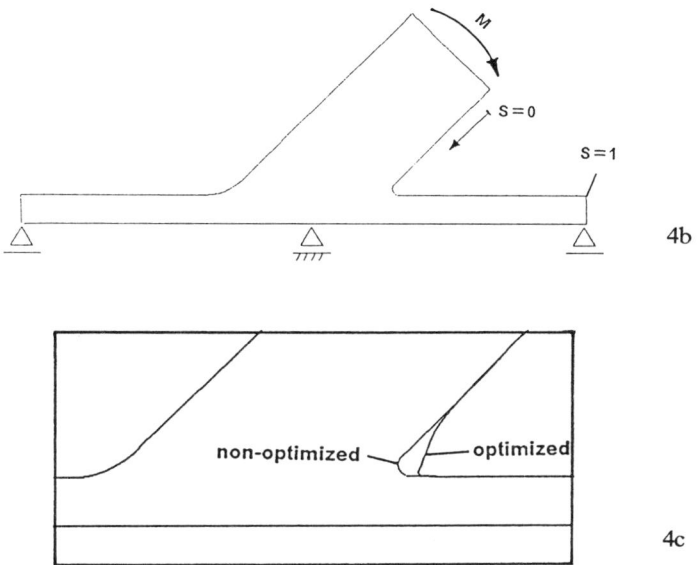

4b

4c

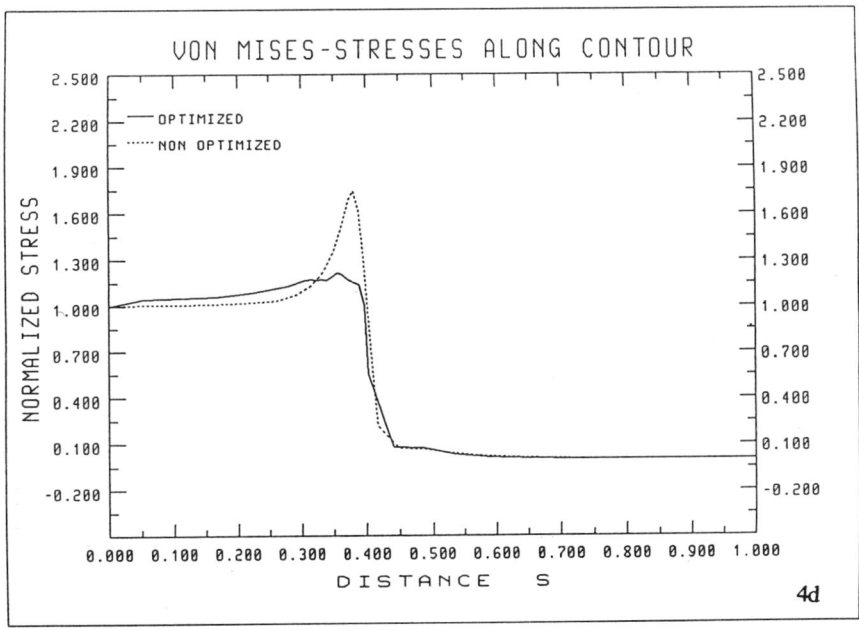

4d

Fig.4a-d Y-joint

Spectral Integrations for Dynamic Responses of Offshore Platforms to Random Waves

Y.H. Chen and D.H. Tsaur

Department of Civil Engineering, National Taiwan University, Taipei, Taiwan

ABSTRACT

The spectral analysis and the corresponding integration formulas of the response spectra of an offshore platform in random waves are presented. Any shape of the in-put wave-height spectrum is described by the piecewise straight lines for approximation. Both the effects of the nonlinear drag force and the modal-damping coupling are taken into account in analysis. The nonlinear drag force is linearized by the method of least square in statistical sense, and the modal-damping coupling is uncoupled by the use of the generalized complex damping. As a consequence the integration formulas of the response spectra for all vibrational modes were derived.

INTRODUCTION

For the spectral analysis to predict the stochastic dynamic responses of an offshore platform, the linearization of the nonlinear drag force [1,2] exerting on an offshore structure is required. By least square method the coefficient of the linearized drag force will as a consequence be a function of the standard deviation of the relative velocity between structure and fluid, if the process of this relative velocity is Gaussian. Therefore an iterative calculation procedure should be carried out until the solution is converged. The generalized complex damping [3,4] is applied to uncouple the modal-damping coupling in order to simplify the calculations. Any shape of the input wave-height spectrum is considered and approximately described by piecewise straight lines, and the integration formulas for the response spectra are derived. Additionally the cross-spectra of modal responses are also included in formulation. A jacket-type offshore platform is presented as an example for demonstration. The stochastic responses of the platform with fixed and pile foundations are calculated by the method described in this paper. The influences of the foundation conditions and the cross-spectra on the responses are also discussed.

GENERALIZED COMPLEX DAMPING AND TRANSFER FUNCTIONS

The equations of motion of an offshore structure modeled by a finite-beam-element and excited by the random sea waves can be expressed in the matrix form as

$$[m]\{\ddot{v}\} + [c]\{\dot{v}\} + [k]\{v\} = \{f(t)\} \tag{1}$$

where $[m], [c]$, and $[k]$ represent the mass, damping, and stiffness matrices, respectively; $\{\ddot{v}\}, \{\dot{v}\}$, and $\{v\}$ represent the structural acceleration, velocity, and displacement vectors, respectively; $\{f(t)\}$ represents the force vector. Both the fluid-structure interaction and the pile foundation effect would be considered. Morison equation is used to calculate the forces exerting on any member of the offshore platform. The matrices $[m], [c], [k]$, and the vector $\{f(t)\}$ can be given by

$$
\begin{aligned}
[m] &= [m'] + C_a[m''] \\
[c] &= [c'] + C_d[c''] + [c'''] \\
[k] &= [k'] + [k'''] \\
\{f(t)\} &= (C_a + 1)[m'']\{\ddot{u}(t)\} + C_d[c'']\{\dot{u}(t)\}
\end{aligned} \tag{2}
$$

where $[m'], [c']$, and $[k']$ represent the structural mass, damping, and stiffness matrices, respectively; $[m'']$ and $[c'']$ are the matrices related to the added mass and the hydrodynamic damping; C_a and C_d represent the added mass and the drag force coefficients; $[c''']$ and $[k''']$ represent the damping and the stiffness matrices due to the pile-foundation effect; $\{\ddot{u}(t)\}$ and $\{\dot{u}(t)\}$ represent the water-particle acceleration and velocity vectors, respectively. The added-mass matrix $[m'']$ for a single structural element can be expressed as

$$[m''] = \rho[V][\lambda] \tag{3}$$

where ρ represents the water density; $[V]$ and $[\lambda]$ represent the lumped displaced volume and the member orientation matrices [4].

If the nonlinear drag force terms in Morison equation have been linearized by the least square method, the hydrodynamic damping matrix $[c'']$ for a single structural element can then be expressed as

$$[c''] = \rho[A][\alpha][\lambda] \tag{4}$$

where $[A]$ and $[\alpha]$ represent the matrices of the lumped projected area of the structural member and the linearized parameter of the nonlinear drag force. More discussion on the linearized parameter will be given in the following paragraph. The matrices $[V], [A]$, and $[\alpha]$ are all diagonal matrices, but the matrix $[\lambda]$ is, in general, not a diagonal matrix.

The pile-foundation effect included in analysis is based on the theoretical studies of the impedance functions of a pile-foundation [7]. The behaviors of soil-pile interactions are modeled mathematically by a series of springs and dashpots along the pile as shown in figures 1.-3. The impedance function can then be determined by using the theory of a continuous beam on elastic springs and dashpots. By performing a steady-state dynamic analysis under a harmonic excitation at the top of the pile, a complex form impedance function can be obtained. The real part and the imaginary part of the impedance function represent the stiffness and the damping due to the soil-pile interactions, respectively. The impedance function of the pile group can be obtained from the impedance function of single pile by using a reduction factor [7], then the matrices $[c''']$ and $[k''']$ can be achieved.

By modes superposition technique, the corresponding normal equations of equation (1) could be written as

$$[M]\{\ddot{y}\} + [C^c]\{\dot{y}\} + [K]\{y\} = \{F(t)\} \tag{5}$$

where $[M]$ and $[K]$ represent the matrices of the generalized mass and stiffness; $[C^c]$ represents the matrix of the coupled damping due to the influence from the hydrodynamic damping and the foundation damping terms; $\{y\}$ and $\{F(t)\}$ represent the generalized coordinates and the generalized forces vectors, respectively.

According to linear gravity wave theory, the motion of the water particle is harmonic, due to an incoming sinusoidal-wave excitation; therefore the generalized forces and coordinates could be written in complex forms as follows

$$\{F(t)\} = \{d + ie\}e^{-i\omega t}$$
$$\{y(t)\} = \{-a + ib\}e^{-i\omega t} \tag{6}$$

where a, b, d, and e are constant, and $i = \sqrt{-1}$.

The coupled normal equation (5) could become uncoupled as follows

$$[M]\{\ddot{y}\} + [C]\{\dot{y}\} + [K]\{y\} = \{F(t)\} \tag{7}$$

where

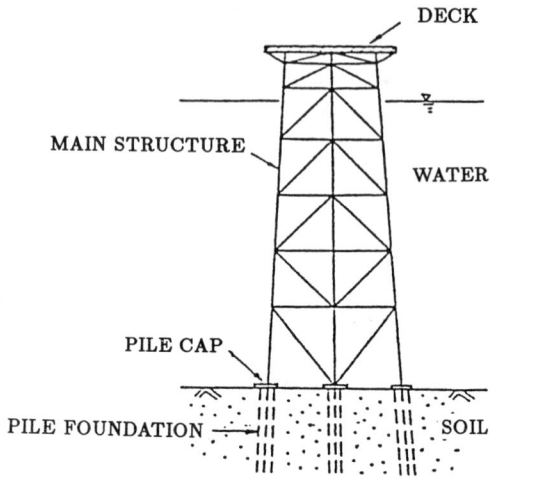

Figure 1. Offshore platform with pile foundation.

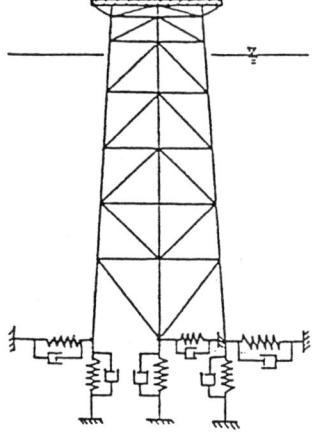

Figure 2. Offshore platform with spring and
dashpot supports.

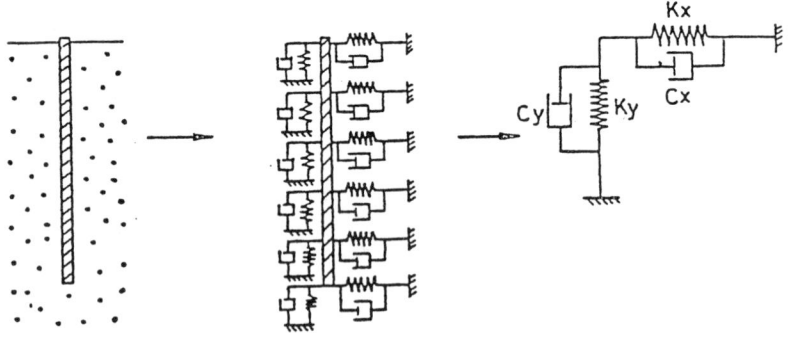

(a) Pile foundation   (b) Continuous beam on spring and   (c) Impedance function
dashpot supports.

Figure 3. Model for pile foundation.

$$[C] = [C_1 + iC_2] \tag{8}$$

$$= \text{generalized complex damping matrix}$$

The generalized complex damping can be determined at resonance for any vibrational mode and given by

$$C_{1p} = \frac{a_p e_p + b_p d_p}{\omega_p (a_p^2 + b_p^2)}$$

$$C_{2p} = \frac{-a_p d_p + b_p e_p}{\omega_p (a_p^2 + b_p^2)} \quad , \qquad p = 1 \sim N \tag{9}$$

where the subscript p denotes the pth mode, N represents the total number of modes of concern, and ω_p represents the natural frequency.

The corresponding generalized complex damping ratio is then defined by

$$\xi_p \equiv \xi_{1p} + i\xi_{2p}$$

$$= \frac{C_p}{2M_p \omega_p} \tag{10}$$

The transfer functions of $y_p(t)$ and $\dot{y}_p(t)$ are given as follows

$$H_{y_p}(\omega) = \frac{1}{M_p \omega_p^2 [(1 - \beta_p^2 + 2\xi_{2p}\beta_p) - 2i\xi_{1p}\beta_p]}$$

$$H_{\dot{y}_p}(\omega) = -i\omega H_{y_p}(\omega) \tag{11}$$

$$= \frac{-i\omega}{M_p \omega_p^2 [(1 - \beta_p^2 + 2\xi_{2p}\beta_p) - 2i\xi_{1p}\beta_p]}$$

where

$$\beta_p = \frac{\omega}{\omega_p} \tag{12}$$

INTEGRATIONS OF RESPONSE SPECTRA

The coefficient of the linearized drag force term in Morison equation is a function of the standard deviation of the relative velocity of the structural member and the water particle, which is unknown in advance and should be calculated iteratively and given by

$$\alpha = \sqrt{\frac{8}{\pi}} \sigma_{\dot{r}_n} \tag{13}$$

where α and $\sigma_{\dot{r}_n}$ represent the coefficient of the linearized drag force and the standard deviation of the relative velocity $\dot{r}_n$, respectively. The subscript n denotes the normal direction of the structural member.

The relative velocity $\dot{r}_n$ is given by

$$\dot{r}_n = \dot{u}_n - \dot{v}_n \tag{14}$$

Therefore, the variance of the relative velocity $\dot{r}_n$ can be expressed as follows

$$\sigma_{\dot{r}_n}^2 = \sigma_{\dot{u}_n}^2 - 2U_{\dot{v}_n\dot{u}_n} + \sigma_{\dot{v}_n}^2 \tag{15}$$

where $\sigma_{\dot{r}_n}^2$, $\sigma_{\dot{u}_n}^2$, and $\sigma_{\dot{v}_n}^2$ represent the variances of $\dot{r}_n$, $\dot{u}_n$, and $\dot{v}_n$, and $U_{\dot{v}_n\dot{u}_n}$ represents the covariance of $\dot{v}_n$ and $\dot{u}_n$, respectively.

For any response $R(t)$ of a linear system, the spectral density function is equal to the sum of the auto-spectral density functions and the cross-spectral density functions of all modes [10,11] such that

$$S_{RR}(\omega) = \sum_p \sum_q S_{R_p R_q}(\omega) \tag{16}$$

where

$$S_{R_p R_q}(\omega) = H_{R_p}^{\star}(\omega) H_{R_q}(\omega) S_{F_p F_q}(\omega) \tag{17}$$

where $H_{R_p}(\omega)$ and $H_{R_q}(\omega)$ are the transfer functions of the response $R(t)$ contributed by the pth mode and the qth mode respectively, those can be obtained from the transfer function $H_{y_p}(\omega)$ by the standard structural analysis. The superscript $\star$ denotes the complex conjugate. $S_{F_p F_q}(\omega)$ is the cross-spectrum of the modal forces F_p and F_q .

The variance of response $R(t)$ can be obtained using the relation

$$\sigma_R^2 = \int_{-\infty}^{\infty} S_{RR}(\omega) d\omega \tag{18}$$

substituting equations (16) and (17) into equation (18), and the integrals could be separated into a finite number of parts, such as

$$\sigma_R^2 \approx \sum_p \sum_q \{ \sum_j \int_{\omega_j}^{\omega_{j+1}} Re[H_{R_p}^\star(\omega) H_{R_q}(\omega) S_{F_p F_q}(\omega)] d\omega \} \tag{19}$$

where $Re[\cdot]$ denotes the real part of the complex value in bracket.

The vectors $\{v\}$, $\{\dot{v}_n\}$, and $\{\dot{u}_n\}$ could be expressed by the following matrix forms

$$\{v\} = [\phi]\{y\}$$
$$\{\dot{v}_n\} = [\lambda]\{\dot{v}\}$$
$$= [\lambda][\phi]\{\dot{y}\} \tag{20}$$
$$\{\dot{u}_n\} = [\lambda]\{\dot{u}\}$$

where $[\phi]$ represents the matrix of the mode shapes.

Hence, the matrices $[\sigma_v^2]$, $[\sigma_{\dot{v}_n}^2]$, and $[U_{\dot{v}_n \dot{u}_n}]$ can be obtained as follows

$$[\sigma_v^2] = [\phi][\sigma_{y_p y_q}^2][\phi]^T$$
$$[\sigma_{\dot{v}_n}^2] = [\lambda][\phi][\sigma_{\dot{y}_p \dot{y}_q}^2][\phi]^T[\lambda]^T \tag{21}$$
$$[U_{\dot{v}_n \dot{u}_n}] = [\lambda][\phi][U_{\dot{y}_p \dot{u}}][\lambda]^T$$

where the superscript T denotes the transformation of the matrix. And where the element $\sigma_{y_p y_q}^2$, $\sigma_{\dot{y}_p \dot{y}_q}^2$, and $U_{\dot{y}_p \dot{u}}$ can be calculated by following relation

$$\sigma_{y_p y_q}^2 \approx \sum_j \int_{\omega_j}^{\omega_{j+1}} Re[H_{y_p}^\star(\omega) H_{y_q}(\omega) S_{F_p F_q}(\omega)] d\omega$$
$$\sigma_{\dot{y}_p \dot{y}_q}^2 \approx \sum_j \int_{\omega_j}^{\omega_{j+1}} Re[H_{\dot{y}_p}^\star(\omega) H_{\dot{y}_q}(\omega) S_{F_p F_q}(\omega)] d\omega \tag{22}$$
$$U_{\dot{y}_p \dot{u}} \approx \sum_j \int_{\omega_j}^{\omega_{j+1}} Re[H_{y_p}^\star(\omega) S_{F_p \dot{u}}(\omega)] d\omega$$

If the input spectra could be approximated by a finite number of piece-wise straight lines, that is

$$S_{F_pF_p}(\omega) \approx r_j + s_j\omega$$

$$S_{F_pF_q}(\omega) \approx \left(r_{R_j} + s_{R_j}\omega\right) + i\left(r_{I_j} + s_{I_j}\omega\right) \quad , \qquad \omega_j \leq \omega \leq \omega_{j+1}$$

$$S_{F_p\dot{u}}(\omega) \approx \left(r'_{R_j} + s'_{R_j}\omega\right) + i\left(r'_{I_j} + s'_{I_j}\omega\right) \tag{23}$$

The integrals of the right-hand side of equation (22) can be derived easily by use of the transfer functions given in equation (11); therefore equation (22) can be rewritten as follows

if $p = q$, then

$$\sigma^2_{y_p y_p} \approx \frac{1}{M_p^2 \omega_p^3}[\sum_j (\hat{C}_1 \hat{I}_0 + \hat{C}_2 \hat{I}_1)_j]$$

$$\sigma^2_{\dot{y}_p \dot{y}_p} \approx \frac{1}{M_p^2 \omega_p}[\sum_j (\hat{C}_1 \hat{I}_2 + \hat{C}_2 \hat{I}_3)_j] \tag{24}$$

if $p \neq q$, then

$$\sigma^2_{y_p y_q} \approx \frac{1}{M_p M_q \omega_p \omega_q^2}[\sum_j (C_1 I_0 + C_2 I_1 + C_3 I_2 + C_4 I_3 + C_5 I_4 + C_6 I_5)_j]$$

$$\sigma^2_{\dot{y}_p \dot{y}_q} \approx \frac{\beta}{M_p M_q \omega_q}[\sum_j (C_1 I_2 + C_2 I_3 + C_3 I_4 + C_4 I_5 + C_5 I_6 + C_6 I_7)_j] \tag{25}$$

and

$$U_{\dot{y}_p \dot{u}} \approx \frac{1}{M_p \omega_p}[\sum_j (C'_1 \hat{I}_0 + C'_2 \hat{I}_1 + C'_3 \hat{I}_2 + C'_4 \hat{I}_3)_j] \tag{26}$$

where M_p and M_q represent the generalized mass, and ω_p and ω_q represent the natural frequencies for the pth and qth modes, respectively, $\beta = \omega_p/\omega_q$. The constants $\hat{C}_1 \sim \hat{C}_2$, $C_1 \sim C_6$, and $C'_1 \sim C'_4$ are given in Appendix A, integrals $\hat{I}_0 \sim \hat{I}_3$ and $I_0 \sim I_7$ are given in Appendix B.

APPLICATION AND DISCUSSION

A jacket-type offshore platform as shown in figure 4. is used as an example for application. The size of all steel tubular members are given in Table 1. The finite-beam-element

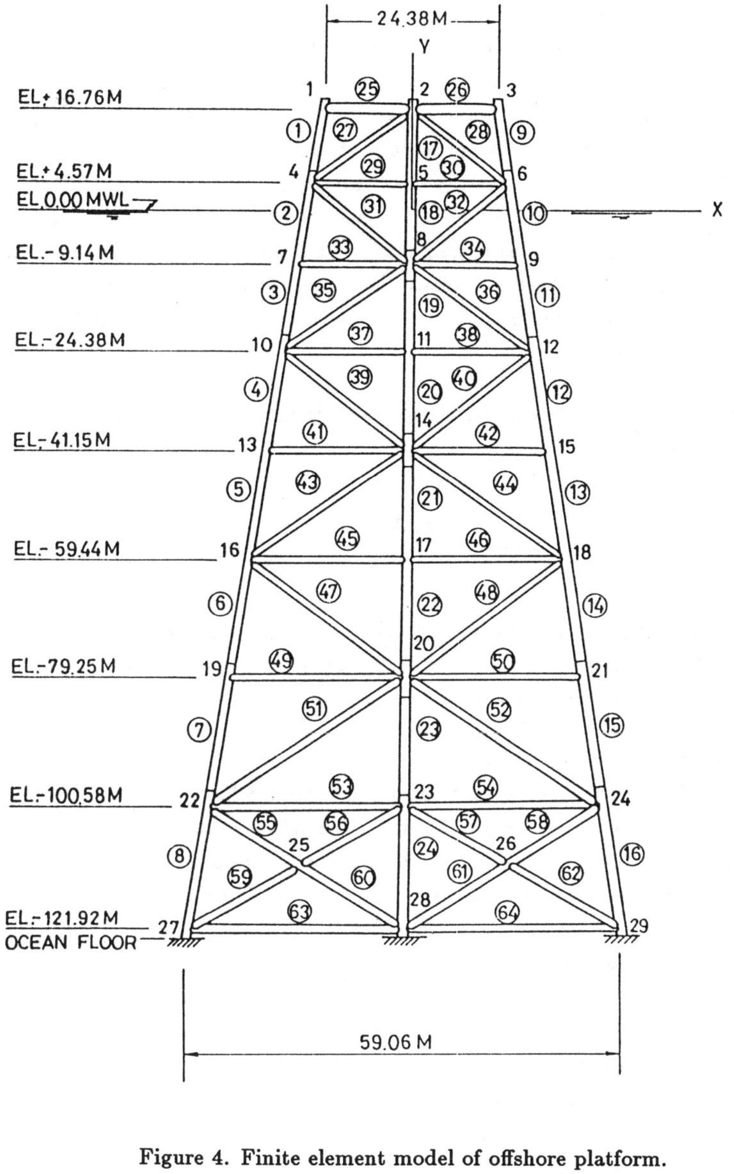

Figure 4. Finite element model of offshore platform.

model of the platform is also shown in figure 4. , resulting in 29 nodes and 64 elements. The lumped deck masses are given in Table 2. The height of the platform is 138.68m, and the water depth is 121.92m. The size of all piles of the foundation of this offshore platform are 101.6cm outside diameter and 2.54cm in thickness, and the depth of penetration into soil is 60.96m. The properties of steel are as follows: the mass density is $7.8 \times 10^3 \text{kg/m}^3$, modulus of elasticity $2.1 \times 10^{11} \text{N/m}^2$, poisson ratio 0.25, respectively. The mass density of water is $1.029 \times 10^3 \text{kg/m}^3$. The soil domain is assumed to be homogeneous, isotropic, and elastic half space. The properties of soil are as follows: the mass density is $1.79 \times 10^3 \text{kg/m}^3$, poisson ratio 0.5, shear modulus $9.576 \times 10^7 \text{N/m}^2$, respectively. The added-mass coefficient C_a and the drag-force coefficient C_d are assumed 1.0 and 0.7, respectively.

Pierson-Moskowitz wave-height spectrum is used to described a random sea, which has the following form

$$S_{\eta\eta}(\omega) = \frac{0.0081g^2}{\omega^5} exp(-\frac{3.11}{\omega^4 H_{1/3}^2}) \qquad (27)$$

where g represents the gravitational acceleration; $H_{1/3}$ represents the significant wave height, and $H_{1/3}$=18.3m (60ft) is used for the calculation. The total duration is assumed to be 4 hours to determine the extreme responses.

The lowest five modes are included for the modes superposition to determine responses. The lowest five natural frequencies corresponding to the lowest five modes of the offshore structure are shown in Table 3. The 1st, 3rd, and 4th modes are the major modes (X-X modes), the 2nd and 5th modes are the minor modes (Y-Y modes). The structural damping ratios are assumed to be 0.02 for all modes. Table 4. shows the generalized complex damping ratios for the lowest five modes for two cases of foundation conditions. The damping ratio coming from the fluid-structure interaction is important and can not be ignored in consideration. The imaginary parts of the generalized complex damping ratios of all major modes are quite small, this is due to the natural frequencies of these modes are well separated. Therefore the mode coupling effect is not important for this example. The imaginary parts of the generalized complex damping ratios might be significant for the minor modes such as the 2nd and 5ht modes, this is due to the coupling effects coming from the major modes.

The integrations of the response spectrum can be obtained easily by applying the formulas developed in this paper. Figures 5.- 12. show the extreme values of the nodal displacement in horizontal direction, member bending moments, member axial forces, and member shear forces, respectively. In order to see the extreme responses from the cross-spectra, two results of the extreme responses are shown in each figure. The solid lines and the dash lines represent the results of the extreme responses including the

Table 1. Member dimensions (outsider diameter × thickness (cm×cm)).

Dimensions	Members
254.00x2.540	25, 26
116.21x3.493	8, 16
114.30x2.540	7, 15, 24
113.03x1.905	4, 5, 6, 12, 13, 14
112.40x1.588	2, 3, 10, 11
111.76x1.270	1, 9, 17, 18, 19, 20, 21, 22, 23
91.44x2.540	51, 52
91.44x1.905	55, 56, 57, 58, 59, 60, 61, 62
76.20x1.905	43, 44, 47, 48
76.20x1.270	39, 40
71.12x1.588	35, 36
60.96x1.270	27, 28, 31, 32
45.72x0.953	63, 64
35.56x0.953	37, 38, 41, 42, 45, 46, 49, 50, 53, 54
32.39x0.953	29, 30, 33, 34

Table 2. Deck mass (kg).

Nodal point	1	2	3
Lumped mass	408520	817040	408520

Table 3. The natural frequencies of offshore platform. (rad/sec).

Foundation condition	Mode				
	1	2	3	4	5
Fixed	3.09	12.11	15.56	25.99	32.24
Pile	2.88	11.64	14.53	25.54	32.15

Table 4. Generalized complex damping ratios $\xi_p = \xi_{1p} + i\xi_{2p}$
(structural damping ratio ξ_s=0.02, $H_{1/3}$=18.3m).

Foundation condition	Damping ratio %	Mode				
		1	2	3	4	5
Fixed	ξ_{1p}	2.670	2.160	3.270	2.116	2.009
	ξ_{2p}	0.003	0.000	-0.035	0.009	0.001
Pile	ξ_{1p}	2.868	2.830	5.285	3.734	2.424
	ξ_{2p}	-0.004	-0.004	-0.146	0.087	0.024

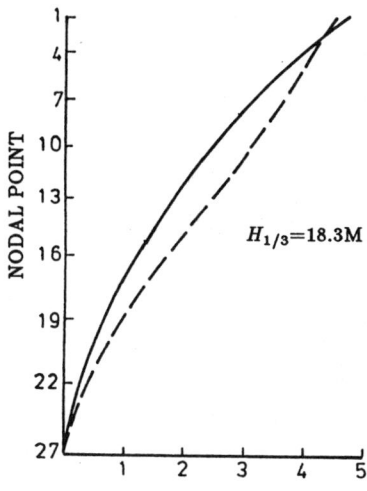

DISPLACEMENT (CM)

Figure 5. Extreme displacement
(fixed foundation).

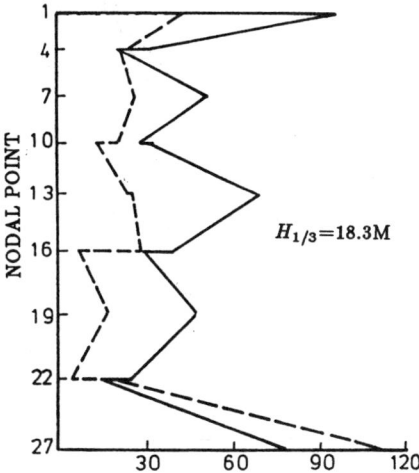

BENDING MOMENT (KN-M)

Figure 6. Extreme member bending moment
(fixed foundation).

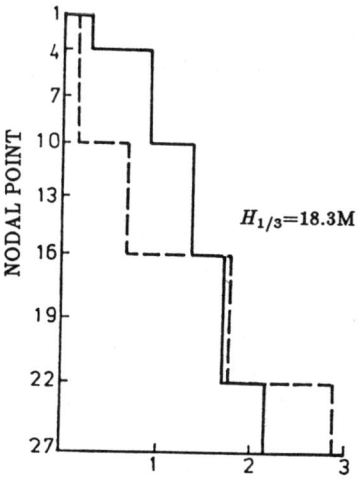

AXIAL FORCE (×10³KN)

Figure 7. Extreme member axial force
(fixed foundation).

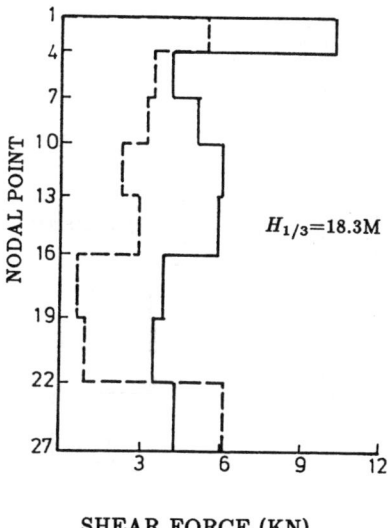

SHEAR FORCE (KN)

Figure 8. Extreme member shear force
(fixed foundation).

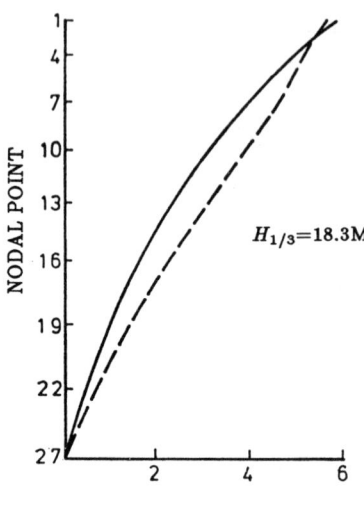

DISPLACEMENT (CM)

Figure 9. Extreme displacement
(pile foundation).

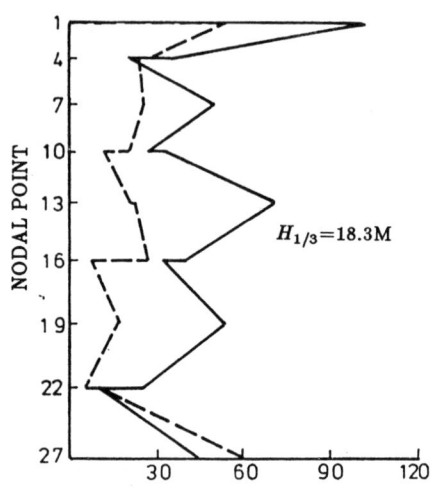

BENDING MOMENT (KN-M)

Figure 10. Extreme member bending moment
(pile foundation).

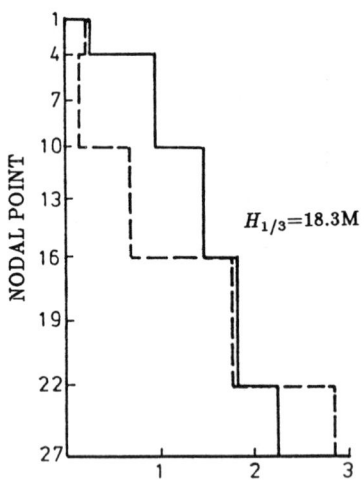

AXIAL FORCE ($\times 10^3$KN)

Figure 11. Extreme member axial force
(pile foundation).

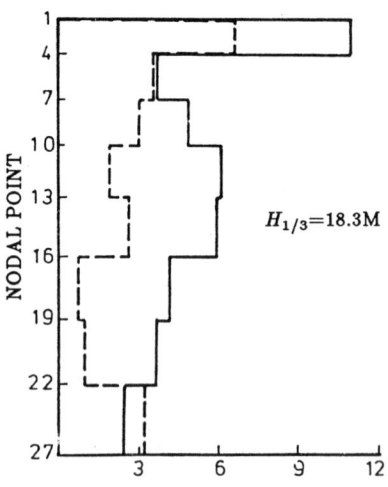

SHEAR FORCE (KN)

Figure 12. Extreme member shear force
(pile foundation).

auto-spectra only and including both the auto-spectra and cross-spectra, respectively. The cross-spectra are significant as shown in each figure. The member axial forces are very large in the lower columns near to the sea bed of this bracing-structure system, conversely the member moment and shear force are large at the upper columns near to the deck due to the high stiffness of the deck structure.

REFERENCES

1. Malhotra, A.K. and Penzien, J. Response of Offshore Structures to Random Wave Forces, J. Struct. Eng., ASCE, Vol.96, No.ST10, pp.2155-2173, 1970.

2. Dao, Van. B. and Penzien, J. Comparison of Treatments of Nonlinear Drag Forces Acting on Fixed Offshore Platform, Applied Ocean Research, Vol.4, No.2, pp.66-72, 1982.

3. Chen, Y.H. Ship Vibration in Random Seas, Journal of Ship Research, SNAME, Vol.24, No.3, pp.156-169, 1980.

4. Chen, Y.H. Dynamic Responses of Three-dimensional Pile-supported Offshore Platforms, pp.13-21, Proceeding of 5th Australian, Conference on Coastal and Ocean Engineering, Perth, W. Australia, November, 1981.

5. Brebbia, C.A. and Walker, S. Dynamic Analysis of Offshore Structures, Newnes-Butterworths, 1979.

6. Sarpkaya, T. and Isaacson, M. Mechanics of Wave Forces on Offshore Structures, Litton Educational Publishing, Inc., 1981.

7. Liou, D.-N. and Penzien, J. Seismic Analysis of an Offshore Structure Supported on Pile Foundations, Report No. UCB/EERC-77/25, U.C. Berkeley, 1977.

8. Novak, M. and Aboul-Ella, F. Impedence Functions of Piles in Layered Media, J. Eng. Mech., ASCE, Vol.104, No.EM4, pp.643-661, 1978.

9. Novak, M., Nogmi, T. and Aboul-Ella, F. Dynamic Soil Reactions for Plane Strain Case, J. Eng. Mech., ASCE, Vol.104, No.EM4, pp.953-959, 1978.

10. Clough, R.W. and Penzien, J. Dynamics of Structures, McGraw-Hill, Inc., 1975.

11. Nigam, N.C. Introduction to Random Vibrations, The Massachusetts Intstitute of Technology Press, 1983.

APPENDIX A: COEFFICIENTS $\hat{C}_1 \sim \hat{C}_2$, $C_1 \sim C_6$, AND $C'_1 \sim C'_4$

The coefficients $\hat{C}_1 \sim \hat{C}_2$, $C_1 \sim C_6$, and $C'_1 \sim C'_4$ appeared in equation (20), (21), and

(22) are given as follows:

$$\hat{C}_1 = r_j$$
$$\hat{C}_2 = \omega_p s_j$$

(28)

$$C_1 = r_{R_j}$$
$$C_2 = a_1 r_{R_j} + \omega_p s_{R_j} - b_1 r_{I_j}$$
$$C_3 = a_2 r_{R_j} + a_1 \omega_p s_{R_j} - b_2 r_{I_j} - b_1 \omega_p s_{I_j}$$
$$C_4 = a_3 r_{R_j} + a_2 \omega_p s_{R_j} - b_3 r_{I_j} - b_2 \omega_p s_{I_j}$$
$$C_5 = a_4 r_{R_j} + a_3 \omega_p s_{R_j} - b_3 \omega_p s_{I_j}$$
$$C_6 = a_4 \omega_p s_{R_j}$$

(29)

and

$$C_1' = -r_{I_j}'$$
$$C_2' = 2\xi_{1p} r_{R_j}' - 2\xi_{2p} r_{I_j}' - \omega_p s_{I_j}'$$
$$C_3' = 2\xi_{1p}\omega_p s_{R_j}' - 2\xi_{2p}\omega_p s_{I_j}' + r_{I_j}'$$
$$C_4' = \omega_p s_{I_j}'$$

(30)

where

$$a_1 = 2(\xi_{2p} + \xi_{2q}\beta)$$
$$a_2 = 4\beta(\xi_{2p}\xi_{2q} + \xi_{1p}\xi_{1q}) - (1 + \beta^2)$$
$$a_3 = -2\beta(\xi_{2q} + \xi_{2p}\beta)$$
$$a_4 = \beta^2$$
$$b_1 = 2(\xi_{1q}\beta - \xi_{1p})$$
$$b_2 = 4\beta(\xi_{2p}\xi_{1q} - \xi_{1p}\xi_{2q})$$
$$b_3 = 2\beta(\xi_{1p}\beta - \xi_{1q})$$
$$\beta = \frac{\omega_p}{\omega_q}$$

(31)

APPENDIX B: INTEGRALS $\hat{I}_0 \sim \hat{I}_3$ AND $I_0 \sim I_7$

If

$$D_p(x) = (1 - x^2 + 2\xi_{2p}x)^2 + 4\xi_{1p}^2 x^2$$
$$D_q(x) = (1 - \beta^2 x^2 + 2\xi_{2q}\beta x)^2 + 4\xi_{1q}^2 \beta^2 x^2$$

(32)

The integrals $\hat{I}_i$ and I_i have the general forms as follows:

$$\hat{I}_i = \int_{x_1}^{x_2} \frac{x^i}{D_p(x)} dx \quad , \quad i = 0 \sim 3$$

$$I_i = \int_{x_1}^{x_2} \frac{x^i}{D_p(x) D_q(x)} dx \quad , \quad i = 0 \sim 7$$

(33)

These integrals have the following results

$$\hat{I}_i = \frac{1}{\bar{G}}\sum_{j=1}^{2}[\bar{R}_{i(2j-1)}L_j + \bar{R}_{i(2j)}T_j] \quad , \quad i = 0 \sim 2$$

$$\hat{I}_3 = \frac{1}{4}ln[\frac{D_p(x_2)}{D_p(x_1)}] + 3\xi_{2p}\bar{I}_2 - a_0\bar{I}_1 + \xi_{2p}\bar{I}_0$$

$$I_i = \sum_{j=1}^{4}\frac{1}{G_j}[R_{i(2j-1)}L_j + R_{i(2j)}T_j] \quad , \quad i = 0 \sim 6 \tag{34}$$

$$I_7 = \frac{1}{2\beta^4}\{\frac{1}{4}ln[\frac{D_p(x_2)D_q(x_2)}{D_p(x_1)D_q(x_1)}] + 7\beta^3(\xi_{2p}\beta + \xi_{2q})I_6$$
$$- 3\beta^2(c_0 + b_0 + a_0\beta^2)I_5 + 5\beta[\xi_{2p}\beta(2b_0 - \beta^2) + \xi_{2q}(2a_0\beta^2 - 1)]I_4$$
$$- [1 + \beta^4 + 4a_0b_0\beta^2 - 2c_0(1 + \beta^2)]I_3 + 3[\xi_{2p}(1 - 2b_0) + \xi_{2p}\beta(\beta^2 - 2a_0)]I_2$$
$$- (a_0 + b_0\beta^2 + c_0)I_1 - (\xi 2p + \xi_{2q}\beta)I_0\}$$

where

$$a_0 = 2\xi_{1p}^2 + 2\xi_{2p}^2 - 1$$
$$a_1 = \xi_{2p} + k_p$$
$$a_2 = \xi_{2p} - k_p$$
$$a_3 = (\xi_{2q} + k_q)/\beta$$
$$a_4 = (\xi_{2q} - k_q)/\beta$$
$$b_0 = 2\xi_{1q}^2 + 2\xi_{2q}^2 - 1$$
$$b_1 = \xi_{1p}(1 + \xi_{2p}/k_p)$$
$$b_2 = \xi_{1p}(1 - \xi_{2p}/k_p)$$
$$b_3 = \xi_{1q}(1 + \xi_{2q}/k_q)/\beta$$
$$b_4 = \xi_{1q}(1 - \xi_{2q}/k_q)/\beta$$
$$c_0 = 8\xi_{2p}\xi_{2q}\beta$$
$$c_1 = a_1 - a_2 = -c_4$$
$$c_2 = a_1 - a_3 = -c_7$$
$$c_3 = a_1 - a_4 = -c_{10}$$
$$c_5 = a_2 - a_3 = -c_8$$
$$c_6 = a_2 - a_4 = -c_{11}$$
$$c_9 = a_3 - a_4 = -c_{12}$$
$$e_1 = c_1^2 - (b_1^2 - b_2^2) = e_4 - 2(b_1^2 - b_2^2) \tag{35}$$
$$e_2 = c_2^2 - (b_1^2 - b_3^2) = e_7 - 2(b_1^2 - b_3^2)$$
$$e_3 = c_3^2 - (b_1^2 - b_4^2) = e_{10} - 2(b_1^2 - b_4^2)$$
$$e_5 = c_5^2 - (b_2^2 - b_3^2) = e_8 - 2(b_2^2 - b_3^2)$$
$$e_6 = c_6^2 - (b_2^2 - b_4^2) = e_{11} - 2(b_2^2 - b_4^2)$$
$$e_9 = c_9^2 - (b_3^2 - b_4^2) = e_{12} - 2(b_3^2 - b_4^2)$$

$$G_j = \beta^4 \Pi_{k=0}^2 [e_{(3j-k)}^2 + 4b_j^2 c_{(3j-k)}^2] \quad , \quad j = 1 \sim 4$$

$$\bar{G} = e_1^2 + 4b_1^2 c_1^2$$

$$k_p = \frac{1}{\sqrt{2}} \{ (\xi_{2p}^2 - \xi_{1p}^2 + 1) + [(\xi_{2p}^2 - \xi_{1p}^2 + 1)^2 + 4\xi_{1p}^2 \xi_{2p}^2]^{\frac{1}{2}} \}^{\frac{1}{2}}$$

$$k_q = \frac{1}{\sqrt{2}} \{ (\xi_{2q}^2 - \xi_{1q}^2 + 1) + [(\xi_{2q}^2 - \xi_{1q}^2 + 1)^2 + 4\xi_{1q}^2 \xi_{2q}^2]^{\frac{1}{2}} \}^{\frac{1}{2}}$$

$$L_j = ln[\frac{(x_1 - a_j)^2 + b_j^2}{(x_2 - a_j)^2 + b_j^2}] \quad , \quad j = 1 \sim 4$$

when $i = 0$ then

$$R_{i(2j-1)} = c_{(3j-2)} e_{(3j-1)} e_{(3j)} + e_{(3j-2)} c_{(3j-1)} e_{(3j)}$$
$$+ e_{(3j-2)} e_{(3j-1)} c_{(3j)} - 4b_j^2 c_{(3j-2)} c_{(3j-1)} c_{(3j)}$$

$$(36)$$

$$R_{i(2j)} = e_{(3j-2)} e_{(3j-1)} e_{(3j)} - 4b_j^2 [e_{(3j-2)} c_{(3j-1)} c_{(3j)}$$
$$+ c_{(3j-2)} e_{(3j-1)} c_{(3j)} + c_{(3j-2)} c_{(3j-1)} e_{(3j)}] \quad , \quad j = 1 \sim 4$$

when $i = 1 \sim 6$ then

$$R_{i(2j-1)} = a_j R_{(i-1)(2j-1)} - 0.5 R_{(i-1)(2j)}$$

$$(37)$$

$$R_{i(2j)} = a_j R_{(i-1)(2j)} + 2b_j^2 R_{(i-1)(2j-1)} \quad , \quad j = 1 \sim 4$$

$$\bar{R}_{01} = c_1$$
$$\bar{R}_{02} = e_1$$
$$\bar{R}_{03} = c_4$$
$$\bar{R}_{04} = e_4$$

$$(38)$$

when $i = 1 \sim 2$ then

$$\bar{R}_{i(2j-1)} = a_j \bar{R}_{(i-1)(2j-1)} - 0.5 \bar{R}_{(i-1)(2j)}$$

$$(39)$$

$$\bar{R}_{i(2j)} = a_j \bar{R}_{(i-1)(2j)} + 2b_j^2 \bar{R}_{(i-1)(2j-1)} \quad , \quad j = 1 \sim 2$$

$$T_j = \frac{1}{b_j} [tan^{-1}(\frac{x_2 - a_j}{b_j}) - tan^{-1}(\frac{x_1 - a_j}{b_j})] \quad , \quad j = 1 \sim 4 \qquad (40)$$

SECTION 8: OFFSHORE OPERATIONS

Computer Controlled Remote Testing of Marine Risers in the Adriatic

A.K. Basu(*) and A. Giuggioli(**)

(*) Brown and Root Vickers Ltd, London, England
(**) Agip S.p.A., Milan, Italy

ABSTRACT

As part of a research project on the dynamic behaviour of rigid risers for tension leg platforms Agip S.p.A. of Italy will conduct extensive at-sea tests on a single and a pair of instrumented risers starting from summer 1991. The risers will be supported from a specially designed test rig installed on one of Agip's production platforms in the Adriatic Sea in 70 metres water depth. As the platform is unmanned the tests will be controlled from an onshore base at Ravenna, which is about 130 km from the platform. To ensure safety of the system at all times and to maximise the amount of collected data and its reliability a highly automated data acquisition system will be employed, which will reduce the need for expert (human) intervention during the tests to a minimum. Apart from the usual software packages needed for such data acquisition tasks, specially written application software will be utilised for identification of the environmental condition and selection of the appropriate test from a predefined Test Plan. Control and monitoring of the test set-up operations and performance of each test, which may involve enforced top motion, for a prescribed duration are also to be done virtually without any manual intervention. The hardware and the software components of the system which will allow these automatic remote operations are described in detail in the paper.

INTRODUCTION

An extensive research project on the dynamic behaviour of rigid risers for tension leg platforms (TLPs) is presently being carried out by Agip with partial funding from the EEC [1] with a Joint Venture (JV) between Tecnomare of Italy and Brown & Root Vickers of London acting as the main contractor. These risers provide the physical connection between the sea bed and the platform for the transport of hydrocarbons. For deep water locations their design presents a challenging problem to the engineers concerned.

Full scale tests on both a single and a pair of instrumented risers form the main task of the project. A specially designed test rig is presently under construction and will be installed next summer on one of Agip's fixed production platforms in the Barbara field of the Adriatic, in 70 metres water depth (Figs. 1 and 2). One of the main reasons for selecting this platform was that its fabrication and installation schedules allowed the necessary modifications to be made to the structure in the fabrication yard, including the incorporation of the additional area required for accommodating the test rig. Since the platform had already been planned for remote operation and was going to be unmanned, the entire strategy for the riser tests and data acquisition had also to be designed on a similar basis.

TEST EQUIPMENT

With a total length of 90m between ball joints, dictated by the platform height and water depth, the outside diameter and thickness of the risers have been chosen as 152mm and 8mm respectively. This is the minimum diameter needed for accommodating all the response sensors inside the risers. At the same time it ensures that the relevant non-dimensional parameters for the test risers governing various fluid dynamic and structural dynamic phenomena are close to those for real TLP risers. This advantage would have been lost with a larger diameter, which would also have necessitated the use of heavier tensioners and support structures.

The test equipment shown in Fig. 3 has been designed to allow a variety of tests involving either of the two riser configurations, under environmental conditions ranging from flat calm to those with up to 2.5m significant wave height and 0.6m/sec current. The following types of tests are planned :

i)   Tests with a range of top tensions
ii) Tests with the riser top(s) stationary, or subjected to harmonic motions in any chosen direction with different combinations of amplitude and frequency
iii) Tests with the top motion simulating the surge response of a model TLP under the coexisting environmental condition
iv) For the riser pair, tests as above but also with a range of spacing and riser plane orientation.

The two main parts of the Test Equipment are (i) the Surface Equipment and (ii) the Bottom Equipment. They are connected to the ends of the test risers by ball joints and allow the required orientation, spacing, tensioning and top motions to be achieved for the test under execution. Most of the operations will be performed hydraulically, and will be controlled and monitored automatically by the Riser Controller (RC) which is a dedicated microprocessor located on the platform. Certain basic actuations may also be carried out by the RC under remote control from the onshore station in Ravenna for diagnostic and fault identification purposes.

Surface Equipment
The Surface Equipment (Fig. 3) consists of (i) a circular railway fixed on the Test Area upper deck, (ii) a monorail along the diameter of the circle supported at its ends by two A-frames resting on the railway and (iii) a wheeled trolley hanging beneath the monorail and supporting the riser(s). The rigid frame consisting of the two A-frames and the monorail is free to rotate about the vertical axis of the railway allowing orientation of the trolley motion direction to any principal wave or current direction occurring during a test. Bearings between the trolley and the riser spacer permit any desired angle to be achieved also between the riser plane and the motion direction.

The tensioning device, one for each riser, is designed to keep the top tension constant during each test but to allow it to vary from test to test if required. Emergency accumulators (not shown) ensure that the risers are left under the prescribed survival tension in case of equipment failure or platform shutdown.

Bottom Equipment

The Bottom Equipment (Fig. 3) consists of three major elements, namely, a main connector, a safety connector and a spacer. The main connector locks the Bottom Equipment to the jacket base under normal conditions, but the latches can be opened (by hydraulic action) when the risers are to be rotated about a vertical axis.

Control Equipment

A number of proximity switches are used for sensing position during equipment set up. Actuations are based on the on-off control of solenoid valves. A position servosystem is used to obtain the motion of the trolley. The position given by a resolver is compared with the demanded value, and the difference is used in a PID control law to drive a hydraulic servovalve controlling the oil flow to the trolley motor.

INSTRUMENTATION

The instrumentation is divided into two main categories of sensors, namely, (i) Environmental Sensors and (ii) Response Sensors (Figs. 2 and 4). The sampling rate for all sensors will be 5Hz.

Environmental sensors

The following environmental sensors have been used :

i)   One wind sensor
ii)   Two wave sensors
iii)   Five pairs of current meters.

Apart from measuring the instantaneous wind speed and direction the wind sensor will be used also to alert the operator of impending storm condition. The current meters will provide the depth-wise distribution of fluid velocity and direction. The wave sensors will measure the sea surface elevation and also provide the tide level values.

Response sensors

The main riser is being comprehensively instrumented to determine its response to hydrodynamic loading and top motion. The instrumentation on the second riser will be limited to that required to compare some characteristic responses of the risers during the two-riser tests with a view to providing a direct assessment of interference effects. The following measurements will be made at sufficient number of locations (Fig. 4), and in two orthogonal planes where appropriate :

i)   Sectional curvatures
ii)   Transverse fluid forces
iii)   Transverse accelerations
iv)   Inclinations to the vertical
v)   Axial tension
vi)   Torque

All the sensors will be housed in specially built riser sections called "pods". The main riser will have eight "main pods" along its length and two "end pods" (Fig. 4). The second riser will be provided with two end pods and four main pods. The pods will in general provide a sealed environment although the tubular sections between the pods will be free flooding. The use of pods will permit rapid replacement of sensors in the event of failure, and will thus minimize experimental downtime.

Due to the size, cost and lack of reliability of watertight electrical connectors, multiplexing units will be used. This will minimize the number of plug/socket connections to be made at each end of the pods. Each pod will have a main and a spare multiplexer. The multiplexers will incorporate a signal conditioning board with operator adjustable gains of 1, 2 and 4 and a digital antialiasing filter to produce data at the required frequency of 5Hz from ones originally sampled at 50 Hz.

OVERVIEW OF DATA ACQUISITION PHILOSOPHY

Tests will be performed according to a Test Plan already worked out in detail. The single riser will be tested first. At the end of a specified period (or on completion of the planned single riser tests if earlier) the second riser will be installed.

During riser installation or removal (which may also be needed when a malfunctioning pod is to be replaced by a spare one) and also for the maintenance and repair of equipment on board the platform, visits to the platform will be necessary. At all other times the test facility on the platform will be unattended. It is the onshore station at Ravenna that will be in charge.

The onshore station is proposed to be manned by a single operator who is not expected to have any special knowledge or expertise on riser behaviour, test equipment operation or the computer system. Most of the tasks will, therefore, be controlled by the Data Acquisition Software (DAS), while the operator would in general have the role of agreeing with or rejecting the computer suggestions but not of imposing one of his own. The exceptions will be the tasks like initiating an analysis of environmental data with a view to performing a test, stopping a test or its preparation for reasons that cannot be known to the computer and handling of acquired data.

However, to allow changes to be made to the rules which govern the operation of the test facility and data acquisition, access to selected areas of the data base will be provided to competent and knowledgeable persons (hereinafter called "engineers") by means of password control. All such changes will be automaticallly recorded in an appropriate database and printed as required.

COMPUTER HARDWARE

In order to meet the above requirements a computer system has been selected, involving processors both offshore and onshore, as shown in Figs. 5a and 5b. The main processor (MP), a DEC Q-bus based MicroVAX 3400, is located on the platform along with a console, a disk drive and a magnetic tape unit. It is locally connected to an Intercole Remote Terminal Unit (RTU) via a ribbon cable, a purpose made Data Concentrator Unit (DCU) via two RS 232 serial links (one for each riser) and to the Riser Controller (RC) via another serial link.

The RTU provides a high speed DMA interface to the MicroVAX for the fourteen environmental sensors and certain other sensors outside of the risers, and incorporates appropriate facilities for signal conditioning, filtering and A/D conversion. The DCU has been provided principally to reduce the loading on the main computer associated with the supervision of the riser pod data polling. The DCU is connected to the pods in each riser by two multidrop RS 485 serial lines. Its function is to poll signals from all the pods of the riser(s) and send them to the MP when it requests for such data, normally every 200 ms, using the command as the synchronisation message. It also resets the gain for all the channels when so commanded by the "engineer" and informs the MP of the current status of the sensor channels and pod muxes when requested.

The RC controls and monitors the trolley motion as well as all the set up operations of the test equipment as mentioned earlier. Once it receives a command from the MP it drives the test equipment to the final set up configuration making use of a preprogrammed sequence of basic actuations, feedback from the field control sensors and periodic data received from the MP (eg. the riser top tension). Data relevant to the operation in progress and system status are sent back to the MP in response to the status request made at 2 sec interval, and are then shown on specially designed screens on the onshore terminal. The safety of the operation is checked at various levels and it can be stopped at any time on operator's command or automatically by the DAS or the RC on the basis of some predetermined rules. On occurrence of critical events, like the loss of relevant communication link, the RC will put the test equipment to the safe standby condition.

The ESD and the SCADA systems of the platform also have direct communication to the MP via serial links. In case of emergency the ESD unit will automatically initiate an emergency shut down procedure cutting the power supply to the container housing all the offshore computer equipment. The emergency condition may be detected by the ESD itself or communicated to it by the SCADA system. The DAS will inform the onshore station of the nature of the emergency and of the progress of the shutdown operation, which involves copying of the system data to disk and shutting down of VMS. An UPS unit will provide back up power for 15 mins for carrying out these operations.

The onshore system consists of a DEC VAXstation 3100 Model 30 incorporating a 19in colour monitor, two hard disks, one Floppy disk drive, one tape streamer and a keyboard, and other equipment shown in Fig. 5b. An Integrex Fast Frame Grabber (not shown) will also be provided to capture screen data from the colour monitor, and then to send it to the printer. The onshore system performs the following main functions :

i)   Provide an operator interface to the offshore system
ii)   Provide colour prints of screen data
iii)   Perform logging tasks of alarm and data information generated by the offshore system with the help of the two LA75 monochrome printers
iv)   Permit sounding of the alarm installed in the onshore control room
v)   Allow direct communication between the DCU and the RC located offshore with their respective test console onshore without having to go through either of the two computers
vi)   Provide a remote development console for the offshore computer via the VT320 monochrome terminal, when the corresponding terminal offshore is switched out
vii)   Transfer of data from offshore disk to onshore tape for archiving.

Communication between the onshore and the offshore system occurs via two radio links connected through 2 Racal ALPHA II asynchronous Modems to an eight channel Racal Omnimux 82 statistical Multiplexer at both ends. The dual link configuration will ensure that uninterrupted transmission will continue even when one of the two links fails, whilst with both the links operational a maximum throughput and a minimum delay can be assured by making use of the equipment's traffic balancing features.

The test consoles for the RC and the DCU are expected to be used by the equipment's maintenance engineers for periodic performance testing of the equipment and for fault diagnostics. This will, hopefully, obviate the need for more frequent visits to the platform and allow better planning of offshore intervention when needed.

SOFTWARE

The software in the two computers runs under the DEC VAX/VMS operating system. The primary data acquisition software resides in the offshore computer and is based around the SETCON Process Control Package produced by SETPOINT Inc.. The main software in the onshore computer is the GCS (Graphics Console System) Display Package, which is also produced by SETPOINT Inc. and works in close relation to the SETCON real-time database. Additionally, there are a number of specially written application programs, some developed by the Joint Venture on behalf of Agip for selecting and conducting tests and others for producing project specific reports and screen displays. These are mounted on one or both the computers as appropriate. Communication between the two computers is achieved using the DECnet software package over a 19.2kbaud serial line connecting each computer to the local stat mux; the higher than usual (9.6kbaud) line speed has been adopted to meet the throughput requirement of data to the GCS screen. Communication between the offshore computer and various instrumentation devices is achieved with specific application software.

SETCON/GCS Package
The main features of the SETCON package utilised in the present project are

i)   Creation and updating of the plant database
ii)   Running of application software with data transferred to and from the database
iii)   Alarm scanning and annunciation, and recording of up to 1000 most recent alarms in a cyclic file in the database
iv)   Creation of status files for equipment and sensors
v)   Event recording in event history files
vi)   Historical recording of processed values

With the help of the GCS package the operator can create and view refreshable graphics based on data in the SETCON database and also view a variety of summary information. The major displays to be made available on the onshore colour monitor are

i)   Fast trend plots of a chosen statistic (eg. RMS * Sign of the Mean) based on successive 3 sec data lengths (ie. 15 data values) for any of the sensors, and similar plots based on longer durations of data
ii)   Status of all equipment and sensors
iii)   Alarm limits and other parameters currently set
iv)   Alarm log and event history files
v)   Continuous updates of parameters related to test equipment (viz. riser top tension, A-frame orientation etc.) on specially designed screens during test set-up operation, and other messages received by the DAS from the RC at 2 second intervals
vi)   Currently set gains for all sensor channels
vii)   Report requests
viii) List of sensors from which twenty are to be selected for special monitoring and trending during actual tests and for generating statistics to be recorded on the relevant "experience" database at the end of the test.

Certain other displays will be available to the "engineer". These include facilities for changing gain of the pod sensor channels through the DCU, carrying out basic actuations of the test equipment through the RC and setting alarm limits for sensors and certain other monitoring parameters.

A few operations will not however involve the GCS, such as initiating a test session; transfer of data from offshore; backups; and printing, archiving or deleting of files generated by the JV written application programs.

JV written Application Software

A number of computer programs have been written by the JV to ensure that tests are conducted according to the agreed Test Plan, to permit real time simulation of riser top motion and to carry out overall check for validity and consistency of acquired data soon after the completion of a test. A short description of the programs follows :

Environmental Analysis Package (EAP) This package consists of 3 programs. Using the most recently acquired time histories of 8192 values, it computes a variety of statistics for all the environmental sensors, the power spectra of the wave elevation and current velocity channels and the directional spectra for waves. Each run of the package is associated with a distinct serial number, which along with the date and time of the analysis and its principal results are written is an ASCII file (one of the "experience" databases) supporting Indexed Sequential Access. The results include the mean speed and direction of wind and current; and the significant height, mean period, mean direction and spreading parameter for waves. The package displays on the operator monitor the main results of the analysis in the form of tables and graphs.

Experimental Riser Operation System (EROS) There are two such packages, one for the single riser tests and the other for the riser pair. Each package is designed to act like an "expert", first determining if the data found from the last EAP run correspond to any of the hundred or so environmental codes listed in the Test Plan. If so, it prepares a list of tests that can be permitted under this condition, arranged in order of priority according to predefined rules and taking into account the tests already carried out.

The sensors and equipment which are essential for carrying out a test vary from test to test and the requirements are again incorporated in the package as a set of rules. The status of all relevant sensors and equipment or system components are read by EROS from the SETCON (or signal) database which is kept continuously updated. It then advises the operator to do the test which has the highest priority and for which all the essential equipment and sensors are working normally. The basis for all such decisions is displayed on the operator screen.

During the at-sea test phase of the project the rules defining the environmental codes and the details of tests for each such code can be changed (by persons with the appropriate privilege) from the onshore station without having to recompile the package.

Motion Simulation Package This package consists of two programs, PREPOLE and INTEPOLE, and computes in real time (ie. within 200ms) the surge movement that a TLP of given response characteristics and located at the top of the riser(s) would experience because of the prevalent wave condition. This is done by using an appropriate Auto Regressive (AR) model for the surge response together with forecasting to allow for noncausality of the response. PREPOLE is run while the test equipment is being set up to generate the AR model parameters corresponding to the most recently acquired time histories for the two wave sensors. INTEPOLE is run during the test. Every 200ms it generates the surge value making use of the most recent measurements from the two wave sensors and sends it to the signal database. The VMS Common Event Flag mechanism is used for achieving synchronisation between INTEPOLE and the rest of the DAS.

<u>Immediate Post-Processing Package (IMPOST)</u> This program analyses selected channels of recently acquired test data to compute certain major statistical parameters of the signals, and presents them to the operator in a summary form. This can then be compared with the previously computed theoretical riser response under similar (if not identical) environmental, top tension and top motion conditions. The computed theoretical results for the test riser, for a range of loading and excitation conditions, will be stored in the onshore VAXstation for ease of comparison.

TEST PLAN & OPERATION OF PLANT

As mentioned earlier the tests will be conducted according to a Test Plan prepared by the JV on behalf of Agip and will be controlled by EROS.   There are altogether 116 environmental conditions in the Plan defined mainly in terms of combinations of mean current speed and wave height. For the single riser a number of tests have been planned for each environmental condition, involving variation in either or both of the two parameters, namely top tension and top motion characteristics. For two-riser tests the additional parameters used in test definition are spacing and riser plane orientation with respect to the main environmental direction. Each set of tests for a given environmental condition is to be performed up to 3 times, maintaining the order of priority specified in the Plan. However, calibration tests, which are to be performed under very calm water conditions, can be repeated any number of times if the environmental condition permits, but the interval between successive tests would not be less than 7 days. The results from these tests as found by IMPOST are to be used to check sensor calibration factors and to estimate zero drifts.

There are principally two modes of operation of the test system, namely, the Monitoring mode and the Test Session mode. There would be periods when in parallel to monitoring the system will be engaged in data transfer from the offshore computer disk to the onshore tape or in running JV programs like IMPOST not incorporated in the automatic operation procedure of the DAS. However, there would be occasions when the DCU will be receiving and executing commands for gain and channel setting or other diagnostics purposes; during these times polling of sensor data from the riser(s) and, hence, monitoring will have to be suspended.

The flow of data during the Monitoring and the Test Session mode is schematically shown in Figs. 6, 7 and 8, and the flow of activities specifically related to the performance of a test is shown in Fig. 9.

The test session starts when the operator initiates a run of EAP, either on schedule or when alerted ('woken up') by the onshore buzzer. This alarm is sounded when the DAS finds from the continuously monitored environmental signals that an interesting condition has developed (Fig. 7). Once the EAP run has been started the DAS takes over control and runs EROS to find if any test is feasible and then to inform the operator. At this point the operator may decide not to proceed. Otherwise, EROS passes on all data necessary for conducting the test to the DAS signal database as PVs (Process Values) and DVs (Discrete Values). The DAS then commands the RC to set up the test equipment. When it knows from the status information received from the RC that the latter is engaged in the operation requiring top tension, it keeps on sending the RC at 1 second interval the most recent value(s) of this parameter acquired via the DCU (Fig. 8)

The DAS is also informed by EROS about the type of test to be done (eg. harmonic top motion or TLP type top motion or no top motion) and the duration of the test (which depends on the type of the test) so that it can take all appropriate actions including the termination of the test, automatically without allowing the operator to depart in any way from the Test Plan. For harmonic top motion tests, the RC is sent the frequency and the amplitude value to generate the motion signals internally. For TLP type top motion tests, PREPOLE and INTEPOLE are run as stated earlier, and the motion computed by INTEPOLE is transferred to the RC via the signal database every 200ms (Fig. 7) during the test. Measured trolley positions during both types of motion tests are sent back to the signal database by the RC every 200ms, to allow comparison of the demanded and the achieved top motion time histories.

During the test the DAS works out at 3 second intervals certain statistical parameters for the twenty sensors selected at the start of the test, and these are made available to the operator for display as trend plots. These are also added to the list of values which are under surveillance throughout the test session (namely, the 3 - second absolute maxima for all sensors), for identifying an emergency condition which may require immediate ending of the test session and putting the system to the survival condition. To eliminate the risk of operator error it is the DAS which will take the necessary action and then report it to the operator.

When a test ends, and if the operator wishes to perform another, the DAS checks whether a specified interval of time has elapsed since the last EAP. If not, it assumes that the environment has not changed and enters EROS directly. Otherwise it starts EAP in order to find the new environmental condition before entering EROS.

If, on the other hand, no further test is to be done, the DAS automatically instructs the RC to put the test equipment to the so called "parking condition" specified by the JV. As for the survival condition, the top tension under this condition is large enough to keep the riser stresses low even when subjected to the worst environmental condition that is likely to occur at the test site. For two-riser configuration, the spacing of the risers is set at its maximum.

At the end of each individual test the acquired time histories for all sensors are converted into engineering units, and copied on disk files. From these data certain overall statistics are computed. These and other details relating to the test are then stored on four "experience" databases (XPSEN.DAT being one such) which can be immediately accessed and printed out for management review. One of these databases gives an updated list of all tests performed, and is used by EROS for giving the "expert" advice.

The main feature of the Monitoring mode is the continuous recording of a fixed length of most recent (and therefore constantly changing) data for all sensor channels, the checking against alarm limits of the successive absolute maximum for every 3 second lengths of data and annunciation of alarms on the operator screen, console printer and buzzer. It also includes continuous checking of status of various components of the test equipment via the RC and of the pod communication links and muxes through the DCU. Another task in this mode could be the automatic but orderly shutdown of the system by the DAS when events so dictate, with the operator being constantly kept informed of the situation.

CONCLUSIONS

The aim of the project described in this paper is to acquire comprehensive and reliable full scale measurements of riser response in the real marine environment. The test equipment and the test programme have been so designed that tests of increasing degrees of complexity can be conducted. Each such test will be performed a number of times under the same or similar environmental conditions, to achieve a high degree of confidence in the data. The acquired data will help in reaching an in-depth understanding of the behaviour of these risers as well as provide, for the first time, an opportunity for a thorough validation of the existing analysis packages.

A major novelty of the project is the use of an unmanned fixed structure to support the test equipment and a computer-controlled procedure for automatic monitoring of test plant and automatic selection and performance of tests and data collection. This solution assures that the tests can be conducted in the real environment at a reasonable cost but with the high degree of control usually possible only in a laboratory. It is hoped that this "full scale laboratory facility" in the sea would be used later for further tests of a similar nature.

ACKNOWLEDGEMENTS

The authors wish to thank Agip S.p.A. and Brown & Root Vickers Ltd for providing facilities for preparation of the paper and for giving permission for its publication.

The Sea-test project is currently in the construction and procurement phase. SD-Scicon UK Ltd are the suppliers of hardware and software for the Data Acquisition System and Tecnomare S.p.A. of Italy are supplying all other parts of the system including the Riser Controller, Data Concentrator Unit and the riser and environmental instrumentation. Brown & Root Vickers Ltd are Agip's Technical Assistant for the procurement of the data acquisition and instrumentation systems. The authors wish to acknowledge the valuable contribution of their colleagues in the above organisations both in the conceptual development of the system and in the execution of the Project. Special thanks are due to P. Campelli of Agip; R.S. Hathaway, S.K. Ray, S. Sarohia and R.W. Robinson of Brown & Root Vickers; M. Bellin of Tecnomare and A.J. Creswell of SD-Scicon.

REFERENCES

1.   Campelli, P., Berta, M. and Basu, A.K. Sea Tests on Risers for Floating Platforms, Proceedings of the 4th Int. Deep Offshore Technology (DOT) Conf., Monte-Carlo, Monaco, October 1987.

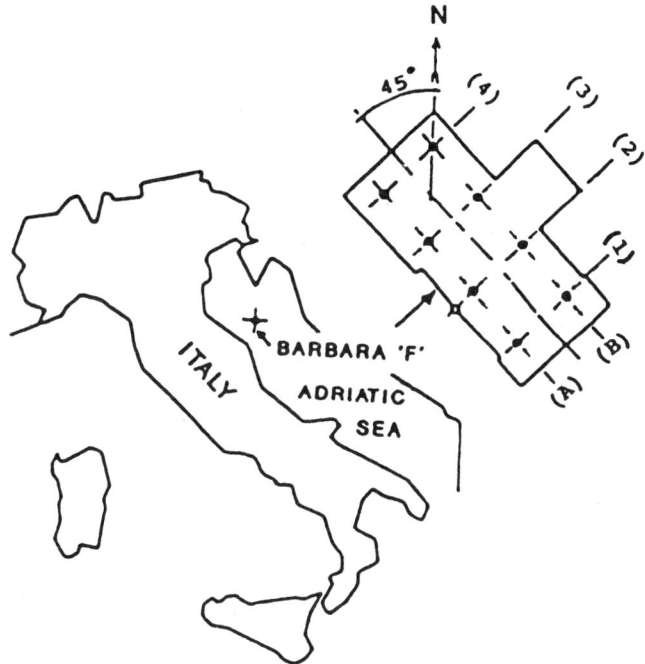

FIG. 1: PLATFORM LOCATION AND ORIENTATION

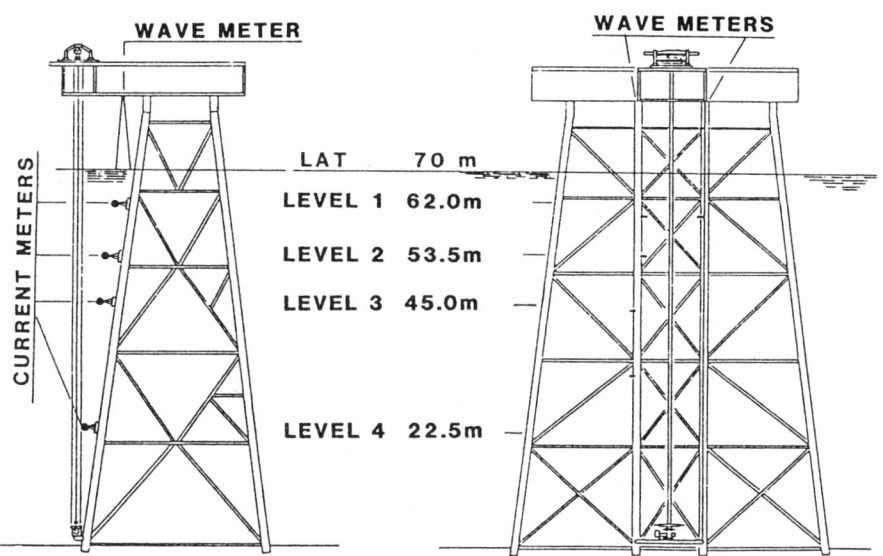

FIG. 2: LOCATION OF CURRENT METERS

AND WAVE ELEVATION SENSORS

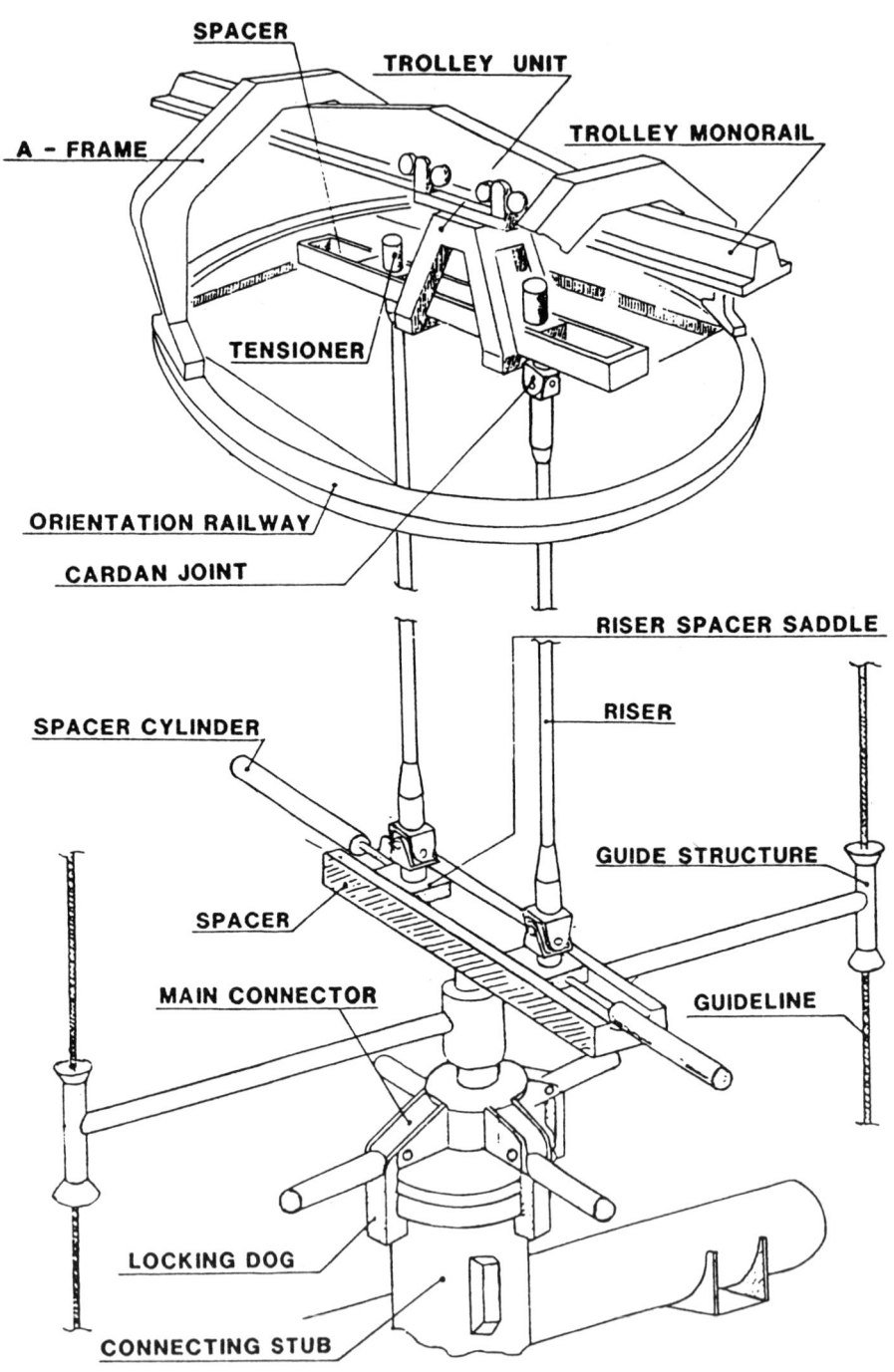

FIG. 3 : SURFACE AND BOTTOM EQUIPMENT

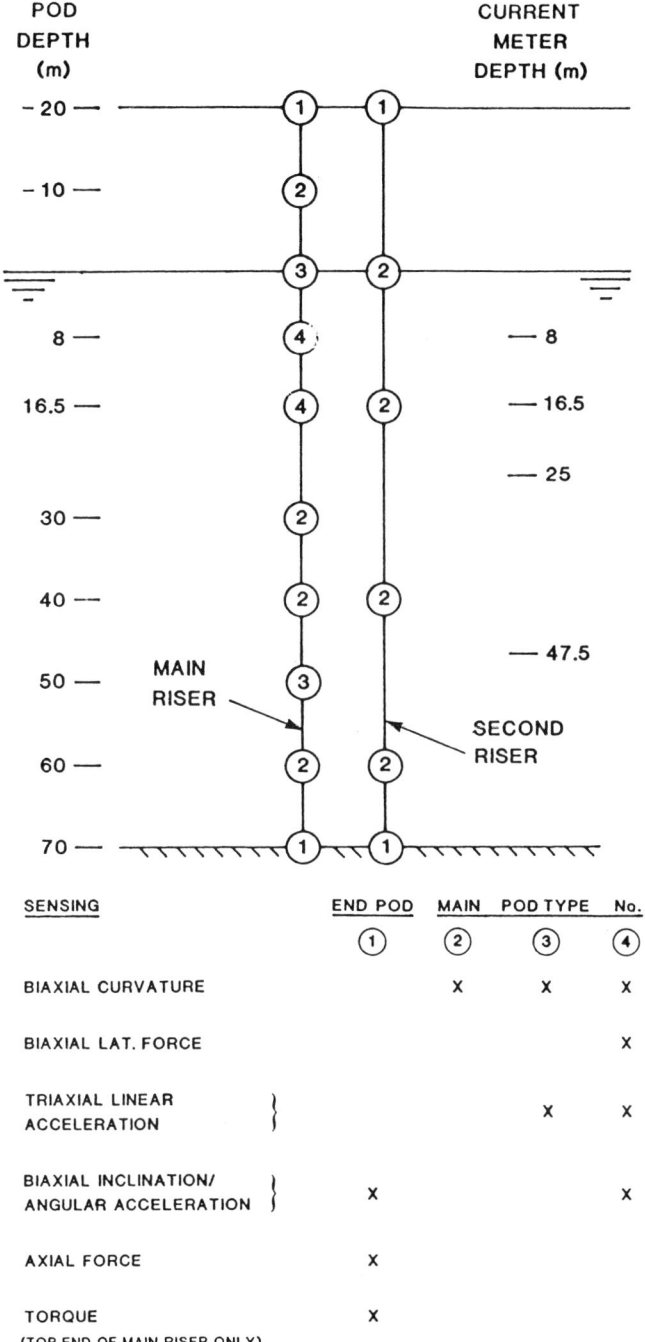

POD DEPTH (m)

CURRENT METER DEPTH (m)

SENSING	END POD ①	MAIN ②	POD TYPE ③	No. ④
BIAXIAL CURVATURE		X	X	X
BIAXIAL LAT. FORCE				X
TRIAXIAL LINEAR ACCELERATION			X	X
BIAXIAL INCLINATION/ ANGULAR ACCELERATION	X			X
AXIAL FORCE	X			
TORQUE (TOP END OF MAIN RISER ONLY)	X			

FIG. 4 : PROPOSED POD CONFIGURATION

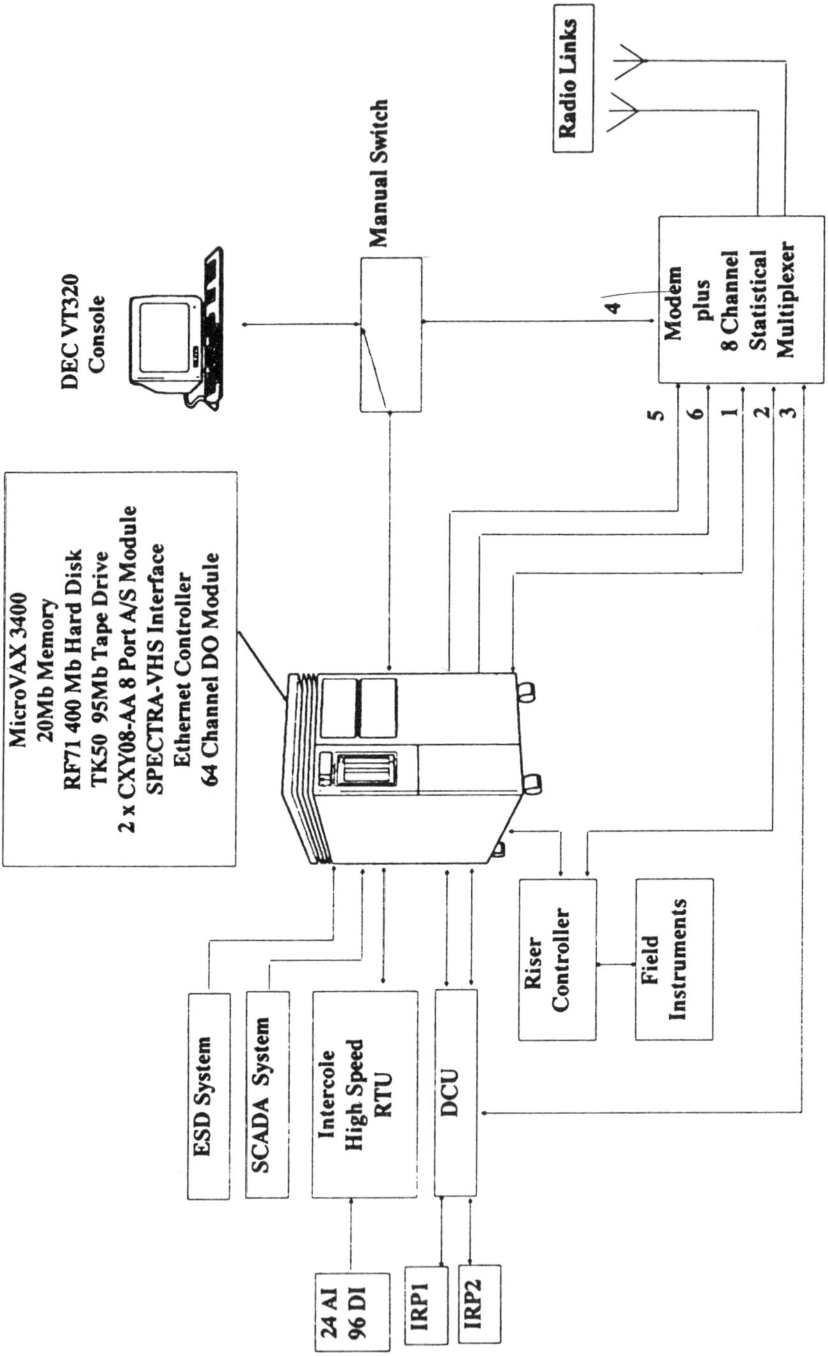

DEC VT320 Console

Manual Switch

Radio Links

MicroVAX 3400
20Mb Memory
RF71 400 Mb Hard Disk
TK50 95Mb Tape Drive
2 x CXY08-AA 8 Port A/S Module
SPECTRA-VHS Interface
Ethernet Controller
64 Channel DO Module

Modem plus 8 Channel Statistical Multiplexer

4

5   6   1   2   3

ESD System

SCADA System

Intercole High Speed RTU

DCU

Riser Controller

Field Instruments

24 AI 96 DI

IRP1

IRP2

FIG. 5a: COMPUTER CONFIGURATION - OFFSHORE

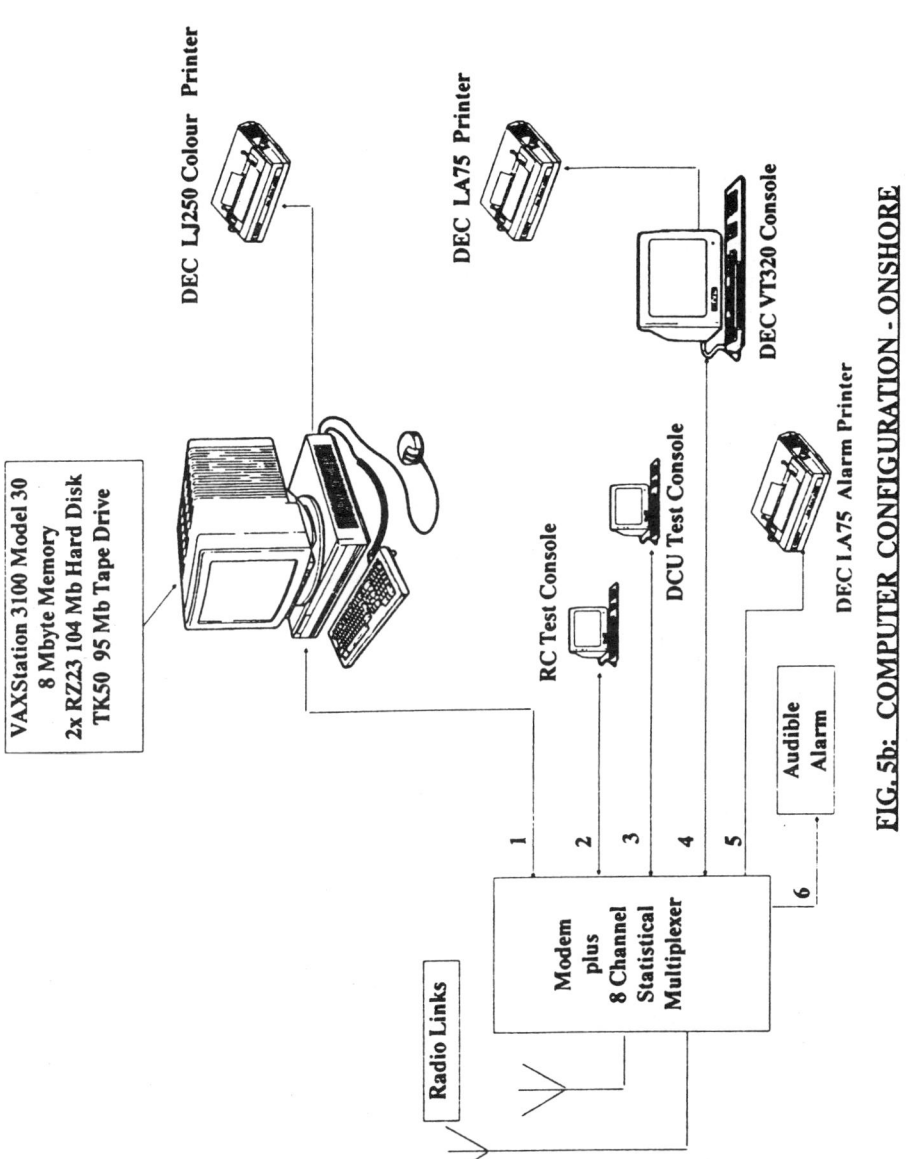

FIG. 5b: COMPUTER CONFIGURATION - ONSHORE

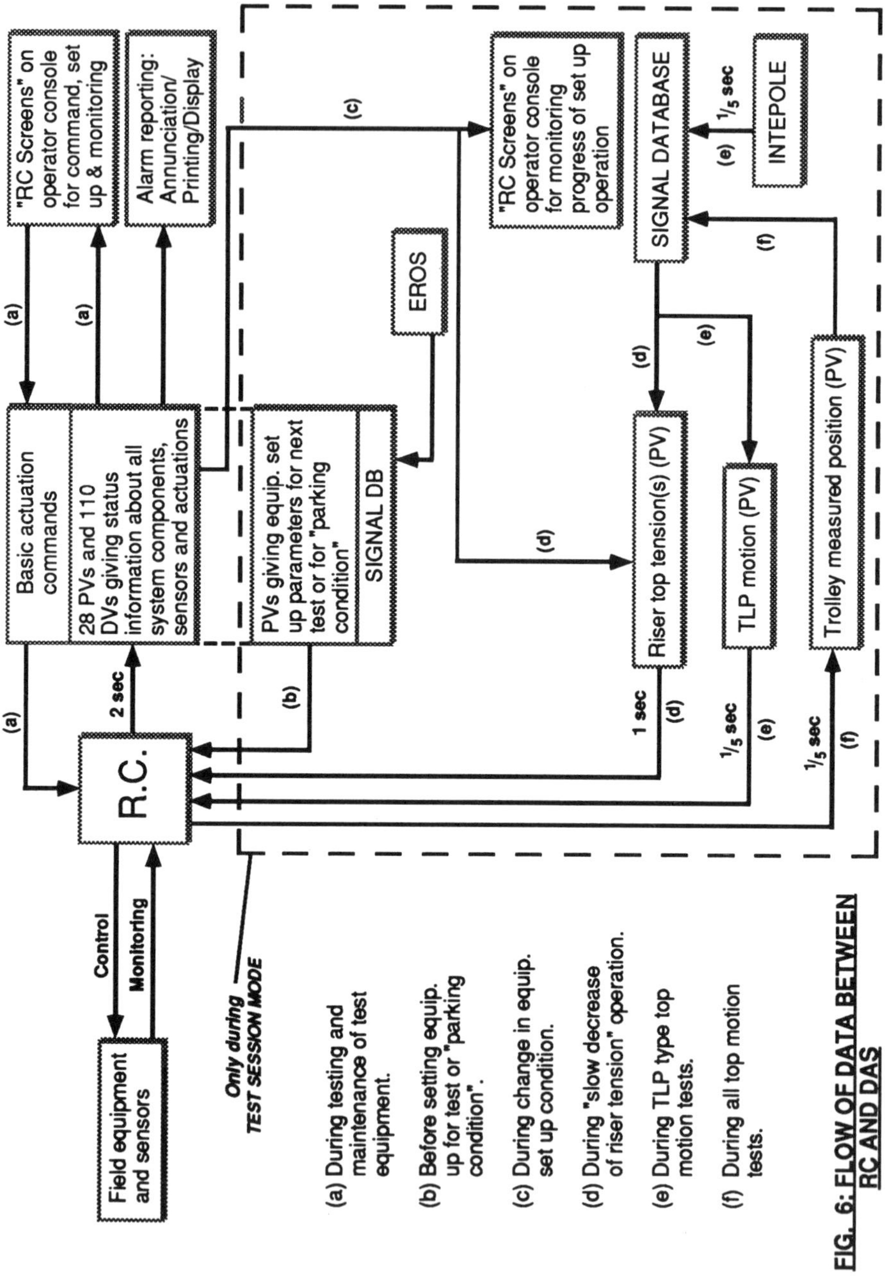

"RC Screens" on operator console for command, set up & monitoring

Alarm reporting: Annunciation/ Printing/Display

"RC Screens" on operator console for monitoring progress of set up operation

SIGNAL DATABASE

INTEPOLE

$^1/_5$ sec

(c)

(e)

Basic actuation commands

28 PVs and 110 DVs giving status information about all system components, sensors and actuations

PVs giving equip. set up parameters for next test or for "parking condition"

SIGNAL DB

EROS

Riser top tension(s) (PV)

TLP motion (PV)

Trolley measured position (PV)

(d)

(a)

(a)

(a)

(b)

(d)

(e)

R.C.

2 sec

1 sec

$^1/_5$ sec

$^1/_5$ sec

(d)

(e)

(f)

(f)

Control

Monitoring

Only during
TEST SESSION MODE

Field equipment and sensors

(a) During testing and maintenance of test equipment.

(b) Before setting equip. up for test or "parking condition".

(c) During change in equip. set up condition.

(d) During "slow decrease of riser tension" operation.

(e) During TLP type top motion tests.

(f) During all top motion tests.

FIG. 6: FLOW OF DATA BETWEEN RC AND DAS

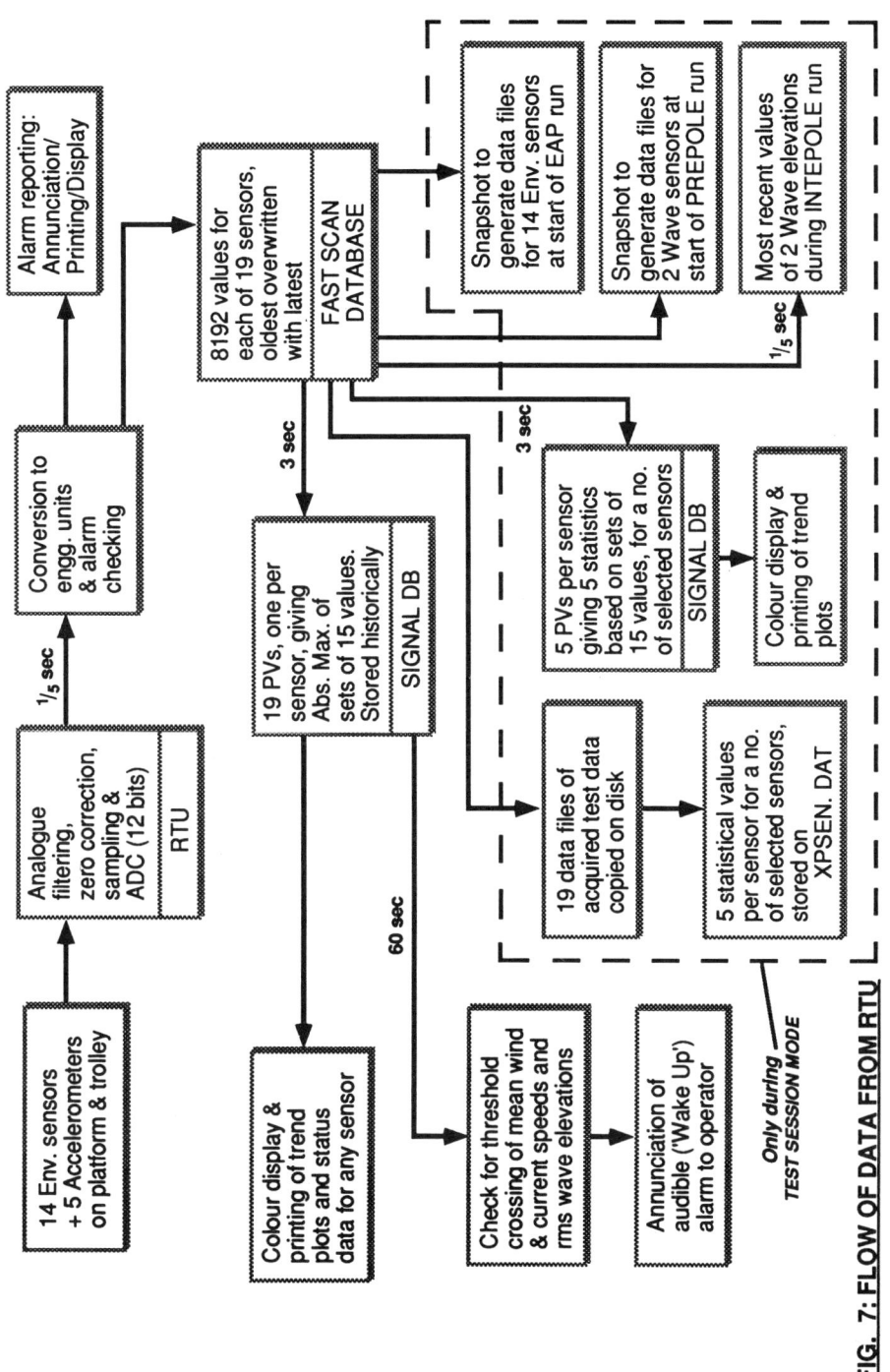

FIG. 7: FLOW OF DATA FROM RTU

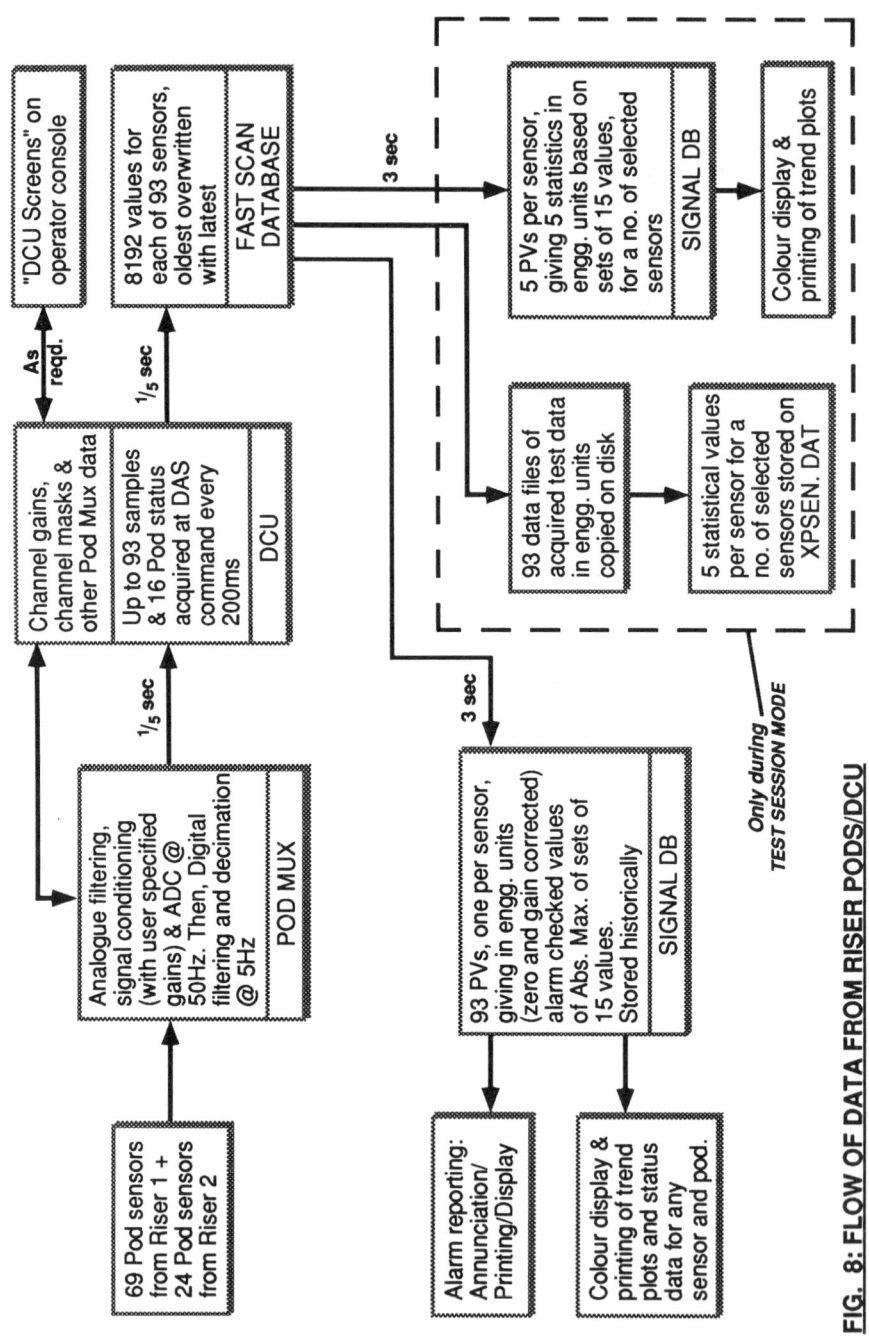

FIG. 8: FLOW OF DATA FROM RISER PODS/DCU

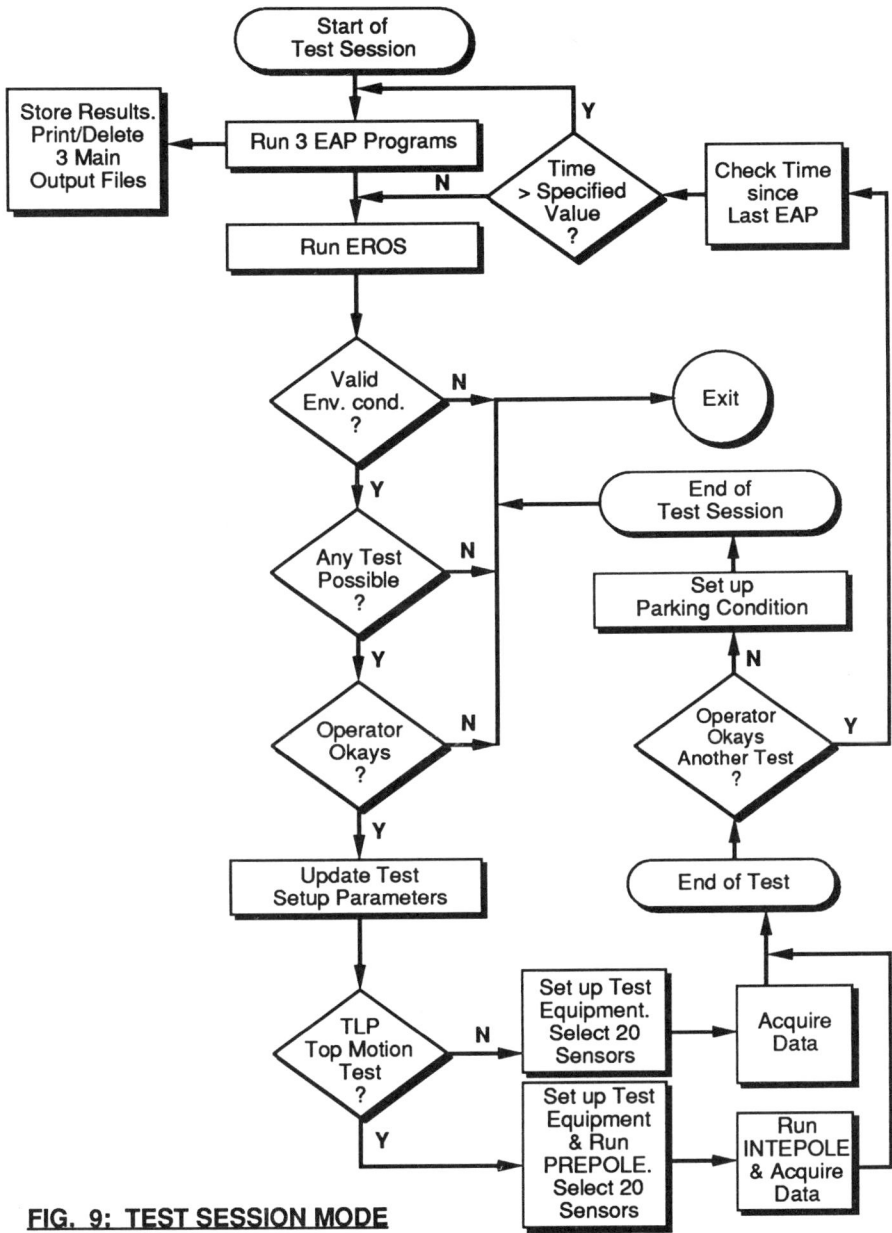

FIG. 9: TEST SESSION MODE

Exploitation of Compound Safety Knowledge for Computer-Based Operational Monitoring and Control

G. Langli(*) and B.A. Bremdal, D. Hirdes(**)

(*) Norwegian Petroleum Consultants, N-1371 Asker, Norway

(**) GeoKnowledge, N-1300 Sandvika, Norway

ABSTRACT

Risk management and control of offshore production installations are achieved through process control, monitoring, and various safety systems. These systems are mainly based on traditional control engineering and computer technology. Intelligent systems are seldom used. Fault diagnosis and decision-making during normal operations and emergencies are normally left to the control room operator. This paper shows that such tasks can be enhanced by integral use of risk and reliability methodologies and knowledge-based systems technology.

INTRODUCTION

General
The main objectives with Process Control and Monitoring Systems (PCMS) include control and monitoring of production systems, and acquisition and storage of data. Today's PCMS are usually open, integrated, modular, and distributed.

The objective of Instrument Safety Systems (ISS) is to ensure that the production systems will not be subjected to excessive loads in case of process upsets and other hazardous situations. ISS actions depend on the location, type and severity of the alarm.

Process safety and process control functions are normally separated and independent (API [1], NPD [11]), but the trend is currently moving towards integrated systems with less built-in redundancy. Examples of such are modern subsea process control systems and wellhead control systems. Both systems also act as a safety system.

Traditional Control, Monitoring and Safety Systems
Both PCMS and ISS are tailored to fit the actual
functional, operational and safety requirements that
are defined prior to design. These systems rely mainly
upon automatic functions. However, fault diagnosis is
mainly part of the operators' scope of work. Neither
ISS nor PCMS are in general able to distinguish between
false and true alarms, even though exceptions exist.
Some of today's PCMS are able to carry out self-
diagnosis upon failure, and the introduction of smart
transmitters will also enhance this aspect, but they
are unable to diagnose failures outside the PCMS.

The operator monitors the process on VDUs (Visual
Display Units) that provide Process Flow Diagrams
(PFDs) and Process and Instrumentation Diagrams
(P&IDs). In order to enhance the readability of the
VDU images, the PFDs and P&IDs are simplified. In
certain situations, this could result in hazardous
decisions.

PCMS and ISS are mainly based upon conventional control
engineering and computer technology. Typically such
systems apply:

* standard control functions
* algorithmic programs based on C, Pascal,
  Fortran, and specially tailored programs
* algorithms are governed by procedural logic
* standard database management systems
* dedicated programming support for icon-based
  graphics development

Database management systems used are either ordinary
shelfware or developed to match special requirements.

Added up, the following characterizes the functionality
of traditional process control and monitoring systems:

* little or no support for decision-making
* improper or no use of available information
  about systems performance (Statoil [15])
* too many false alarms ([15])
* no use of design knowledge in fault detection;
  alarms are announced upon single parameter
  values beyond pre-defined limits
* acquisition of data is for the purpose of
  other issues than risk control and management

Improvements In North Sea development projects there
has been an increasing demand for more simple and reliable solutions in order to reduce both installation and
operational costs. This has also resulted in develop-

ments in PCMS and ISS technology. These improvements mainly comprise utilization of advances in data and information systems technology, new control techniques, and more accurate and reliable instruments. Improvements have hardly been concerned with philosophies and guidelines for design.

Yet the greatest potential for improvements, in our opinion, lies in the formalization and deployment of compound safety knowledge by means of artificial intelligence systems.

Artificial Intelligence in Control and Monitoring

In terms of process control, monitoring and fault diagnosis several knowledge-based systems have demonstrated their applicability (Genesereth [4], Oevre [12], Priha [14], Skatteboe [17], Struss [18]). The basic strength of such systems stems from their ability to contain more than algorithmic expressions of knowledge, to reason independently and to search for and recognize faulty or non-stable patterns of behavior. Especially their ability to handle uncertain information is acknowledged. A general feature of many early systems of this kind was the dominant focus on operational heuristics. Such heuristics is extremely powerful in terms of complicated pattern matching. Further, heuristics provides efficient means of combining various types of domain competence. In later knowledge-based systems knowledge containing deeper causal relationships have been added (Struss [18]). Such causal elements may be derived from process descriptions, functional relationships or structural configurations.

In spite of such developments, system developers are still faced with some problems. The classic acquisition problem addresses a knowledge elicitation issue, where the main bottleneck is how to tap knowledge from humans and to formalize this for later machine application. In terms of deeper knowledge this problem can be extremely hard, yet it is more the grammar of the model described which constitutes the greatest challenge. Another problem concerns issues of response time, time validity of acquired measurements and the need for computing power to treat masses of input data due to the dynamic nature of the processes. Even modern on-line knowledge-based systems have a low level of anticipation. Part of our aim in this paper is to show how it is possible to overcome this.

SOME IMPORTANT SAFETY ISSUES

Accidents - Cause/Consequence Mechanisms

Many offshore accidents would probably have been avoid-

ed, or at least the consequences could have been
reduced, if one had utilized the information available
from the control and monitoring systems (NOU [10]) and
the situation as such. Investigations of accidents in
both sea transportation (Barlow [2], Kristiansen [6])
and offshore industry (Rasmussen [16]) have revealed
that there may be multiple causes to an accident.
Usually the cause mechanism is a combination of independent, random and controlled events which may be
almost harmless when regarded separately.

In order to make correct and safe decisions about an
operational situation, one ought to know the conditions
that may influence the decision in order to evaluate
alternatives and possible consequences. Signs of possible abnormalities in the process, production and
utility systems are such conditions that should be
considered prior to making decisions.

Typical causes of an accident comprise:

* unavailable safety systems (see [9],[13])
* improper design/system failures (see [9])
* human failures and operational/management
  errors (see [9],[10],[13])

Thus, to reduce the probability of occurence for
accidents or the consequences of such, enhanced support
for decision-making is necessary.

Risk and Reliability Analyses in Design
During the design of offshore installations several
risk and reliability analyses are made in order to
provide input to the design, rank alternative solutions, verify systems design, and to identify unwanted
causal effects of every operational aspect. These
analyses provide and generate data and information for
all possible hazards, including empirical knowledge,
mathematical and logical descriptions, and assumptions
on operating conditions.

Deduction of possible causes to and consequences of a
certain scenario is easily described by means of fault
trees and event trees. Whereas fault trees primarily
identify how accidents can occur (i.e. the unwanted top
event), event trees are primarily used for deduction of
possible consequences of a top event. Other safety
analysis techniques are also widely used, such as
Failure Mode and Effect Analysis (FMEA) and Hazard and
Operability Study (HAZOP). Added together, the output
from these and related methods provide a deep understanding about systems behaviour and related
consequences.

A RENEWED APPROACH TO CONTROL AND MONITORING OF PRO-
DUCTION SYSTEMS - INTEGRATED RISK MANAGEMENT

Design documentation is always closely related to the
particular artifact that is depicts. It has much of
the same power as traditional heuristics. In addition,
much documentation supports both structural, functional
descriptions and results from process analyses etc.
All this provides a deep and model-oriented insight
into all aspects of the operational unit.

The engineering phase generates valuable models of
state changes due to some accidental event. Moreover,
safety technology provides a set of methodologies that
incorporate both heuristics and empiricism. Thus, upon
completing the design of an offshore installation, all
the information necessary to enable safe design and
operation is available.

A safety analyst is able to handle both qualitative
and quantitative aspects within the context of a simple
grammar. This is readily adaptable to computing. In
many respects the results of the engineering work
related to safety can be used as skeleton plans or
scenarios. In a knowledge-based system such skeletons
will reduce the need for time-consuming search and
facilitate economical recognition functions. Yet, the
clue is to match a sequence of events to identify a
possible end-state or "purpose". Not the opposite which
is fundamental in traditional planning.

Utilization of risk and reliability documentation/
techniques together with relevant knowledge of systems
design as well as operational data provide the complete
basis for optimum risk management. Due to the nature
of the information and the tasks involved, a computer-
based tool for such purposes seems to be readily
supported by KBS-technology.

An intelligent risk management tool should therefore
be based on design documentation, and furthermore, it
should also be a tool in the design itself. The risk
can then be monitored and controlled by matching an
actual situation with one described in the risk and
reliability documents. When a match is found, causes
and consequences can easily be derived.

A primary effect of an approach like this is the
possibility of making knowledge of processes and plant
design much more available during operations. The
knowledge generated through the engineering phases may
be looked upon as a map benefiting "the navigator"
during operation. As a map this knowledge provides an

abstraction of the problem space. This comprehensive image of the real world is therefore a powerful means of deducing whereabouts and outlooks. As such it is a powerful aid in prognosticating outcomes given the present state of some operational aspect.

It is thus possible to provide a significant reduction in the risk for accidents by inclusion of intelligent systems for fault detection and diagnosis, as well as in supporting the operator in making improved decisions regarding how to handle emergencies and potential hazardous situations.

IRMA - INTELLIGENT RISK MANAGEMENT SYSTEM

Objective and Scope
The objective of the research work presently undertaken is to show that utilization of risk and reliability analysis techniques together with design knowledge and operational data can be used effectively for risk management and control of offshore installations, and furthermore, to show how a renewed KBS-appraoch can support this.

A knowledge-based system for integrated on-line risk management of offshore installations is specified (Langli [8]). This system is called IRMA - Intelligent Risk Management System. In order to test the concepts behind IRMA a prototype of IRMA (IRMA-P) is currently being developed (Langli, Hirdes [9]).

IRMA is not supposed to replace existing PCMS and ISS, but to support these by application of systems know-ledge. IRMA will utilize design and operational data to:
* perform verification of alarms
* provide early detection of faults and potential hazardous situations
* perform fault diagnosis
* carry out consequence evaluations, partly automatic and partly operator-assisted, in order to identify proper corrective actions

In addition, it will enhance the operator's abilities to make more precise and safe decisions during normal operation.

Development of a Prototype of IRMA-P
We have developed a first version of IRMA-P in Smalltalk /V 286. The reason why Smalltalk was chosen was its readiness to support our object-oriented model. More details on this can be found in a separate thesis by Hirdes [5].

IRMA-P consists of two applications, IrmaP and IrmaPES.

IrmaP enhances a model that combines a structural and functional description of a separation process. The implementation utilizes the possibilities that object-oriented programming offers concerning inheritance and specialization. Like a good mental model, it is necessary to define, simplify and organise a consistent description of reality based on the observed and common known conceptual properties of both the separation process and the equipment involved. Much time was spent on identifying the best possible way to relate classes and subclasses for dynamic and static factors and conditions governing the separation process.

IrmaP holds the data representation of the model of our realm. IrmaPES (IRMA-P Expert System Shell) contains the inference engine, the user interface regarding implementation of rules and facts, i.e. the knowledge acquisition part, and the explanation system. In order to ease implementation of rules, if-then statements can be written in "standard" english. A method removes descriptive words before the rule/fact is used in the inference engine.

The formats used on Process Flow Diagrams (PFD) and Process & Instrumentation Diagrams (P&ID) constitute part of the basis for the user interface of IRMA-P.

<u>Interfaces between IRMA and Other Systems</u>
IRMA-P is not intended to be tested against real world systems, but in order to ensure a suitable architecture and data flow etc. it is necessary to keep in mind the purpose of IRMA and its interfaces to other systems.

<u>Verification of Alarms</u> Alarm signals are normally initiated by the ISS, but may also be initiated by the PCMS. A verification of an alarm can be carried out in several ways, but all approaches suggested for IRMA are deductive, and require input from the PCMS. Both real time and historical process data can be needed, depending of the actual scenario. And of course, IRMA must be provided with alarm signals as soon as they appear.

IRMA performs verification based on deductions from consistency, and a result will be returned as FALSE, UNKNOWN, or TRUE. If an alarm is accepted (TRUE) then the safety systems initiate the required actions. In case of a FALSE alarm, the request for safety actions is cancelled. An alarm judged as UNKNOWN will be passed to the operator for final decision, or the signal will be treated as TRUE, depending of the severity of the alarm. In all three cases, the operator is informed, and a log is kept within IRMA.

Fault Detection The main task is to detect (possible)
abnormalities in the process. Fault detection can be
based on

       * results from PCMS or IRMA process simulations
       * results from statistical analysis/trends
       * logged and real-time process data

Access to real-time data can also be necessary in order
to call for alarms in cases where the instrument
systems fail to respond. IRMA will then have to have
access to the various databases connected to or being
part of the PCMS in addition to the connection to the
I/O of the control unit.

Fault Diagnosis Fault diagnosis is or can be initiated
when a fault is detected or an alarm is verified. In
order to carry out diagnosis IRMA must use data
contained in the long-term, or historical, memory of
the PCMS. Besides, IRMA will also utilize data on
maintenance (plans and records), and will thus need a
connection to the management system or another system
that holds this kind of information.

Consequence Evaluation/Identification of Corrective
Actions In design documentation corrective actions/
means are prepared as sets of procedures. The task is
then to retrieve the set of procedures adequate for a
specific situation. The consequence evaluations for
which conclusions will govern the choice of procedures,
are in most cases available from risk and reliability
documentation already contained within IRMA.

Feasibility The feasibilty of IRMA depends on several
factors, including parallel processing and real-time
data technology. Also important is that the communication between IRMA and the other systems is smooth.
A crucial question is how the physical connection
between IRMA and the control system (including signal
transmission/access to databases) should look like.
Informal communications with a PCMS manufacturer
indicate, however, that this is technically feasible.

Search Strategies
A main feature of IRMA-P is that it will be able to
work with standard engineering deliverables such as
event trees and fault trees etc. The reasoning strategy
that is suggested for IRMA-P and currently being implemented in IrmaPES (see Fig 1 below) is based on the
unique inhouse practices related to risk and reliability studies. The first step is always to remove all

facts that do not matter initially. This may be called
heuristic reasoning.
Secondly, the appli-
cation of qualita-
tive reasoning and
physics lead us to
the interesting
aspects in the
study. From there we
continue with model-
based reasoning.
Finally, a con-
clusion or solution

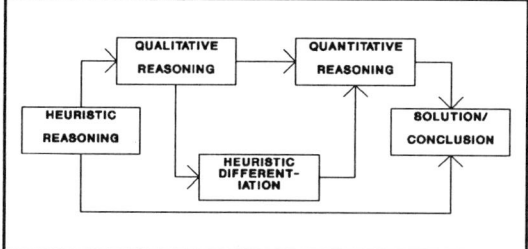

Fig 1 The search strategy

to the problem at hand has been found. "Heuristic
differentiation" is normally not used in risk and
reliability analysis, due to the often coarse level of
detail. This approach is comparable to that of Xiang
and Srihari [20].

IRMA-P: The First Model
The context for development and initial testing of our
propositions concerning improved risk management and
control methodologies has been confined to a simplified

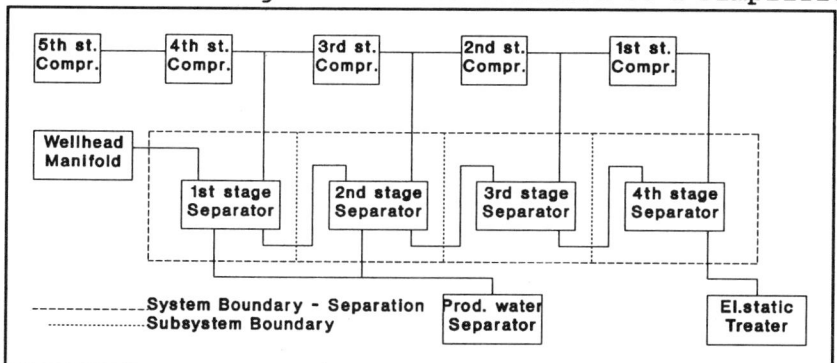

Fig 2 The separator train and interfacing systems

four stage separation process, see Fig 2. Each of the
elements depicted in Fig 2 is equipped with a number
of instruments. Some are used for control, and others
are part of ISS. From the design phase the various
flows and operating conditions are known (see Fig 3).
Deviations from these during operation are normal, but
operating limits are fixed for certain periods and
operations. This constitutes the basic knowledge base
of process behaviour in the system.

Mass conservation yields that anything that enters the
inlet of the 1st stage separator is leaving the system
at different locations. The quality of separation of a
particular well fluid is dependent of the pressure/

temperature relations and the liquid retention time for each of the separators. Hence, the system may be said to be functioning as intended if the oil metering and export (not shown on Fig 2) correspond with the actual production rate. That is, if the parameter conditions of all interfacing systems are good, then the condition of the parameters of the separation system must also balance equally well. (See also Lambert et al [7]).

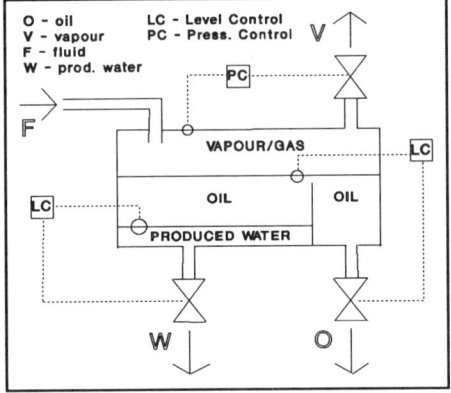

Fig 3 Typical parameters and layout of a multiphase separator

Such reasoning can be performed at many system levels, from the global one as described above and down to a component level. Observed performance can be structured by means of fault trees as shown in Fig 4. And by replacing the AND-gates with OR-gates and vice versa, one can trace backwards from a symptom to reveal possible faults and their locations, or forwards to reveal possible consequences. In other words, if normal, and then consequently also abnormal, process performance can be modelled and augmented with classic risk and reliability analysis methods, the basis of a grammatical vehicle for computers have been found.

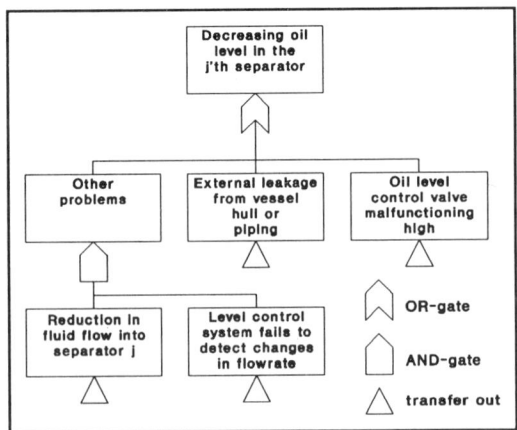

Fig 4 Example on a fault tree model of a possible process upset

Such a representation is further explicit and separates each state as a distinct step in a logic sequence. Associated with each scenario a set of characteristic parameters can be identified. Besides, each scenario can be defined in terms of value ranges that these parameters can span. Instead of cardinal values, process conditions can then be described by

simple descriptions such as LOW, NORMAL, or HIGH. (See
for instance Cohn [3].) A state description applying
this is sufficient to differentiate the various pre-
defined scenarios from each other. Hence, outcomes of
a given state is reduced to a simple match operation.
Hence, IRMA-P accomodates a very powerful mechanism for
recognition.

The development of IRMA-P emphasizes simple models that
easily describe phenomena to operators; a governing
principle is to make IRMA-P work in a way similar to
what operators do when diagnosing faults, but with the
knowledge of experienced operators, process engineers,
safety engineers, and instrument engineers etc.

IRMA-P is concerned with irregularities in systems
performance and relates this kind of information to
safety aspects, such as possible consequences,
reliability data and operational and management data.
Since all accidents are a result of either component
and/or human failures and possible external conditions, it is natural to search these domains for possible embedded inconsistencies. In order to do so one
must answer the following questions:

* which laws govern the system behaviour?
* what are the failure modes of a component and
  what effects do the failure modes have on the
  various subsystems and systems?
* what are the relevant cause-consequence
  relationships?
* what about operational aspects and
  maintenance?

These questions and related ones are answered by the
safety and reliability engineer, the risk analyst and
others, and the answers are thoroughly discussed in the
design documentation.

Having the answers one can start to describe (model)
the system such that the elements and the relations
that provide the answers are included. For instance,
one must know the purpose of each of the elements in
the model. The important thing is to include all that
matters with respect to the intentions of the model.
The models that already are or will be included in
IRMA-P comprise

* structural model (component centered model)
* functional model (process oriented/behaviouristic)
* failure model ("malfunctional" model)

The structural model describes how components are related, and also the characteristics of the components. Characteristics include design data such as capacities, dimensions, and empirical failure data. The functional model describes how the process is affected by the behaviour of each component. The engineering documentation is often organized along the same principles. Hence there are well defined clusters of knowledge organized according to the system description, both structural and functional wise. Knowledge engineering is thus essentially reduced to a transcription task. The object-oriented paradigm in computing lends itself well to this knowledge transcription. Hence, IRMA-P has taken advantage of the feature of Smalltalk /V 286.

CONCLUDING REMARKS

It is our belief that utilization of design knowledge as it appears in risk and reliability documents will prove beneficial in monitoring, control and management of risk of offshore installations. The methodolgy currently being implemented in IRMA-P constitutes a new approach in knowledge engineering. Design knowledge provides compiled concepts about systems performance that would otherwise have to be generared in an on-line process performance monitoring system. IRMA is a project initiated to prove this. The first version IRMA-P using an object-oriented paradigm to deliver this knowledge promises such a proof.

REFERENCES

1.   API - American Petroleum Institute. API RP 14C - Analysis, Design, Installation and Testing of Basic Surface Safety Systems for Offshore Production Installations, 3rd edition, 1984
2.   Barlow, R.E., Lambert, H.E. The Effect of US Coast Guard Rules in Reducing the Probability of LNG Tanker-Ship Collision in Boston Harbour, TERA Corporation, Berkeley, California, May 1979
3.   Cohn, A.G. Qualitative Reasoning, in Lecture Notes in Artificial Intelligence (Ed. Nossum, R.T.), pp. 60-95, ACAI'87, Oslo, Norway, 1987
4.   Genesereth, M.R. The Use of Design Descriptions in Automated Diagnosis, Artificial Intelligence 24, pp.411-436, Elsevier, Amsterdam, 1984
5.   Hirdes, D. Object-Oriented Knowledge Technology in Control of Offshore Processes (in Norwegian), thesis at Dept. of Computer Science at Oestfold Regional College, Halden, Norway, 1990
6.   Kristiansen, S., et al. Risk Assessment of Marine Systems - Lecture Notes, Norwegian Institute of Technology-NTH, Trondheim, Norway, 1985
7.   Lambert, H., Eshelman, L., Iwasaki, Y. Using

Qualitative Physics to Guide the Acquisition of Diagnostic Knowledge, Technical Report, IIRIAM, Marseille, France, 1988

8. Langli, G. IRMA-P System Specification - Draft, unpublished, Nov. 1989

9. Langli, G., Hirdes, D. Application of Safety and Reliability Analysis Techniques in Monitoring and Control of Offshore Production Systems (in Norwegian), presented at NAIS Poster Conference, Oslo, March 1990

10. NOU - Norwegian Public Investigations. The West Vanguard Report/NOU 1986:16 (in Norwegian), Oslo, Norway, 1986

11. NPD - Norwegian Petroleum Directorate. Acts, Rules and Provisions for the Petroleum Activity, Vol. 2, pp. 357-366, Stavanger, Norawy, 1990

12. Oevre, F., Nilsen, S. Integration of a Real Time Expert System and a Modern Graphic Display System in a Swedish Nuclear Power Plant Control Room, to be published in Expert Systems with Applications (Ed. Bremdal, B.A., Fjellheim, R.), Pergamon Press

13. Petroleum Directorate/Government of Newfoundland and Labrador. Technical Investigation of the Ocean Ranger Accident, Vol. 1, April 1983

14. Priha, I. FAKS - An On-Line Expert System Based on Hyperobjects, to be published in Expert Systems with Applications (Ed. Bremdal, B.A., Fjellheim, R.), Pergamon Press

15. Private communications with Statoil personnel, 1989 - 90

16. Rasmussen, J. Human Factors in High Risk Technology, Ch.1.6 in High Risk Safety Technology (Ed. Green, A.E.), John Wiley & Sons Ltd., 1982

17. Skatteboe, R., Lihovd, E., Hystad, R.E. DIAMON - A Knowledge-Based System for Fault Diagnosis and Maintenance Planning for Rotating Machinery, in Proc. of the 6th Int. Workshop on Expert Systems & Their Applications, Avignon, France, 1986

18. Struss, P. Diagnosis from Structure and Behaviour, in Artificial Intelligence in Engineering: Diagnosis and Learning (Ed. Gero, J.S.), Elsevier, 1988

19. Various articles from newspapers and magazines on the "Piper Alpha" disaster 6th July 1988

20. Xiang, Z., Srihari, S.N. A Strategy for Diagnosis Based on Empirical and Model Knowledge, in Proc. of the 6th Int. Workshop on Expert Systems & Their Applications, Avignon, France, 1986